T0212338

# Cybersecurity and Privacy in Cyber-Physical Systems

# Cybersecurity and Privacy in Cyber-Physical Systems

Edited by
## Yassine Maleh
## Mohammad Shojafar
## Ashraf Darwish
## Abdelkrim Haqiq

**CRC Press**
Taylor & Francis Group
Boca Raton London New York

CRC Press is an imprint of the
Taylor & Francis Group, an **informa** business

CRC Press
Taylor & Francis Group
6000 Broken Sound Parkway NW, Suite 300
Boca Raton, FL 33487-2742

First issued in paperback 2022

ISBN 13: 978-1-138-34667-3 (hbk)
ISBN 13: 978-1-03-240151-5 (pbk)

DOI: 10.1201/9780429263897

**Visit the Taylor & Francis Web site at**
**http://www.taylorandfrancis.com**

**and the CRC Press Web site at**
**http://www.crcpress.com**

*For Adam, Lina, and Sabrine*

# Contents

## SECTION III   SECURITY AND PRIVACY IN BIG DATA CYBER-PHYSICAL SYSTEMS

## SECTION IV   CYBERSECURITY IN CYBER-PHYSICAL SYSTEMS

# Preface

Today, everything is digital, and what is not is soon to be digital. As citizens, we all have access to a computer, a tablet or a telephone for our personal and professional use. These means of communication are increasingly connected, gaining accessibility and simplicity for our daily uses. We are living in a data-driven age. Data has been locating or is going to locate every point of our life. Most people think that this influence is a consequence of industry 4.0 that makes our life faster than before, as with all other industrial revolutions. Industry 4.0 enabled the cooperation between the cyber domain and physical systems. This cooperation is called cyber-physical systems (CPSs).

Most critical infrastructures such as the power grid, rail or air traffic control, industrial automation in manufacturing, water/wastewater infrastructure, banking systems and so on are CPSs. Given that the continued availability of their core functions is extremely important to people's normal and economic lives, there is widespread concern that CPSs could be subjected to intense cyberattacks. In fact, a number of these cases have occurred over the past decade. It is therefore extremely important to defend these systems against cyber threats. Due to the cyber-physical nature of most of these systems and the increasing use of networks, embedded computing and supervisory control and data acquisition (SCADA) attack surfaces have increased. In this tutorial, we will examine some of these cyber threats, discuss methodologies, tools and techniques for defending these systems and show how the design of secure cyber-physical systems differs from the previous design.

Cybersecurity attacks are becoming more frequent as cyberattackers exploit system vulnerabilities for financial gain. Nation-state actors employ the most skilled attackers, capable of launching targeted and coordinated attacks. Sony, PumpUp and Saks, Lord & Taylor are recent examples of targeted attacks. The time between a security breach and detection is measured in days. Cyberattackers are aware of existing security controls and are continually improving their attack techniques. To make matters worse, cyberattackers have a wide range of tools to bypass traditional security mechanisms. Malware infection control frameworks, zero-day exploits and rootkits can be easily purchased at an underground market. Attackers can also buy personal information and compromised domains in order to launch additional attacks.

Traditional computer security technologies, previously designed to protect information exchange between cyber components, are deployed to protect physical processes on CPS networks. Due to the size and complexity of many CPS networks, infiltration prevention is far from guaranteed. In addition, exploits have demonstrated that even so-called secure communication protocols can be compromised by zero-day exploits and implementation flaws. As a result, traditional cyber-perimeter defenses are often seen as a response to newly discovered threats and cannot promise protection against unknown exploits.

The objective of this book is to collect and report on recent high-quality research that addresses different problems related to the cybersecurity and privacy in CPSs. High-quality contributions addressing related theoretical and practical aspects are expected. The overall objectives are:

- To improve the awareness of readers about cybersecurity and privacy in CPSs
- To analyze and present the state-of-the-art of CPSs, cybersecurity and related technologies and methodologies
- To highlight and discuss the recent development and emerging trends in cybersecurity and privacy in CPSs
- To propose new models, practical solutions and technological advances related to cybersecurity and privacy in CPSs
- To discuss new cybersecurity and privacy models, prototypes and protocols for CPSs

## TARGET AUDIENCE

The book aims to promote high-quality research by bringing together researchers and experts in CPSs security and privacy from around the world to share their knowledge of the different aspects of CPS security and we have integrated these chapters into a comprehensive book. This book is ideally suited for policymakers, industrial engineers, researchers, academics and professionals seeking a thorough understanding of the principles of cybersecurity and privacy in CPSs. They will learn about promising solutions to these research problems and identify an unresolved and challenging problem for their own research. Readers of this book will have an overview of CPS cybersecurity and privacy design.

# Editorial Advisory Board and Reviewers

# Editors

**Yassine Maleh** received a PhD in computer science from the University Hassan 1st, Morocco, in 2013. Since December 2012, he has been working as an IT senior analyst at the National Port Agency in Morocco. He is a senior researcher in the Faculty of Sciences and Techniques, Settat. He is a senior member of the Institute of Electrical and Electronics Engineers (IEEE), member of the International Association of Engineers (IAENG) and Machine Intelligence Research Labs. Dr. Maleh has made contributions in the fields of information security and privacy, Internet of Things security and wireless and constrained networks security. His research interests include information security and privacy, the Internet of Things, networks security, information systems, and IT governance. He has published more than 40 papers (book chapters, international journals, and conferences), two edited books and one authored book. He has served as an associate editor for the *International Journal of Digital Crime and Forensics (IJDCF)* and the *International Journal of Information Security and Privacy (IJISP)*. He was also a guest editor of a special issue on recent advances in cybersecurity and privacy for cloud of things in *International Journal of Digital Crime and Forensics (IJDCF)*, July–September 2019, Vol. 11, Issue. 3. He has served and continues to serve on executive and technical program committees and as a reviewer of numerous international conference and journals such as *Elsevier Ad Hoc Networks, IEEE Sensor Journal, ICT Express, Springer Cluster Computing, International Journal of Computers and Applications, IEEE Transactions on Network Science and Engineering, Journal of Cases on Information Technology (JCIT)* and the *International Journal of Cyber Warfare and Terrorism (IJCWT)*. He is the General Chair of the MLBDACP 19 symposium. He received the Publon Top 1% Reviewer of the Year 2018 award.

**Mohammad Shojafar** received a PhD in information communication and telecommunications (advisor, Professor Enzo Baccarelli) from Sapienza University of Rome, Italy, the second-rank university in QS Ranking in Italy and top 100 in the world with an excellent degree in May 2016. He was a senior researcher participating in the European Horizon 2020 SUPERFLUIDITY project working with the Consorzio Nazionale Interuniversitario per le Telecomunicazioni (CNIT) partner under the supervision of Professors Nicola Blefari-Melazzi and Luca Chiaraviglio in the Department of Electronic Engineering at the University of Tor Vergata, Rome, Italy, for 14 months. From December 2015 to December 2016, he was a postdoctoral researcher (supervised by Dr. Riccardo Lancellotti and Dr. Claudia Canali) working on the SAMMClouds project at the University of Modena and Reggio Emilia, Italy. In January 2018, he joined the University of Padua, Italy (among the best Italian universities) as a senior researcher (Researcher Grant B). He has contributed to H2020 European projects such as SUPERFLUIDITY and TagSmartIt, and has been sponsored by the Italian Public Education Ministry (i.e., the Ministry of Education, Universities, and Research (MIUR)). In addition, he has worked as a researcher in scheduling/resource allocation in grid/cloud computing under the supervision of Professors Rajkumar Buyya, Jemal Abawajy, Ajith Abraham, and Mukesh Singhal since 2011. His research interests are mainly in the areas of security and privacy. In these fields, he has published more than 90 papers in the topmost international peer-reviewed journals and conferences, including IEEE TCC, IEEE TNSM, IEEE TSUSC, IEEE TGCN, IEEE ICC, IEEE GLOBECOM, IEEE ISCC, IEEE SMC, IEEE PIMRC, and IEEE Network. He is an associate editor for Springer—*Cluster Computing* and KSII—*Transactions on Internet and Information Systems* and *International Journal of Computers and Applications (IJCA)*, and he served as a program committee member of several conferences, including IEEE ICCE, IEEE UCC, IEEE SC2, and IEEE SMC. He was the general chair for INCoS in 2018.

**Ashraf Darwish** received a PhD in computer science in 2006 from the computer science department at Saint Petersburg State University (specialization in artificial intelligence) and joined as lecturer (assistant professor) in the computer science department, Faculty of Science, Helwan University, on June 25, 2006. He is a member of such notable computing associations as the IEEE, ACM, EMS (Egyptian Mathematical Society), QAAP (Quality Assurance and Accreditation Project, Egyptian Supreme Council), Quality Assurance and Accreditation Authority (Egypt), and the board of the Russian-Egyptian Association for graduates. Dr. Darwish is the author of many scientific publications, including papers, abstracts, and book chapters concerning computational intelligence. He keeps in touch with his mathematical background through his research. His consulting, research, and teaching mainly focuses on artificial intelligence, information security, data and web mining, intelligent computing, image processing (in particular image retrieval, medical imaging), modeling and simulation, intelligent environments, and body sensor networking. This has resulted in his articles appearing in journals by such publishers as IEEE and Springer, including *IJCSN* and *Advances in Computer Science and Engineering*. Dr. Darwish is one of the founders of Cyber Security Research Lab (CSRL), organized by Professor Aboul Ella Hassanein, Faculty of Computers and Information, Cairo University, Egypt, and he is a member of research projects in Egypt in the areas of bio-inspired HCV analysis, breast cancer, and e-learning. Dr. Darwish has been solicited as a reviewer and a speaker, organiser, and session chair for many international journals and conferences. He has worked in a wide variety of academic organisations and supervises many master and PhD theses in different areas of computer science. Dr. Darwish practices four languages on a daily basis.

**Abdelkrim Haqiq** has received a High Study Degree and a PhD, both in the field of modeling and performance evaluation of computer communication networks, from the University of Mohammed V, Agdal, Faculty of Sciences, Rabat, Morocco. Since September 1995, he has been working as a professor in the Department of Mathematics and Computer Science at the Faculty of Sciences and Techniques, Settat, Morocco. He is the director of the Computer, Networks, Mobility and Modeling Laboratory. He is also the general secretary of the electronic Next Generation Networks (e-NGN) Research Group, Moroccan section. He is an IEEE senior member and an IEEE Communications Society member. He was a co-director of a NATO multi-year project entitled "Cyber Security Analysis and Assurance Using Cloud-Based Security Measurement System", having the code SPS984425. Dr. Abdelkrim Haqiq's interests are in the areas of modeling and performance evaluation of communication networks, mobile communications networks, cloud computing and security, queueing theory and game theory. He is the author and co-author of more than 150 papers for international journals and for conferences and workshops. He is an associate editor of *International Journal of Computer International Systems and Industrial Management Applications* (*IJCISM*) and an editorial board member of the *International Journal of Intelligent Engineering Informatics* (*IJIEI*). He was also a guest editor of a special issue on next-generation networks and services of the *International Journal of Mobile Computing and Multimedia Communications* (*IJMCMC*), July–September 2012, Vol. 4, No. 3, and a special issue of the *Journal of Mobile Multimedia* (*JMM*), Vol. 9, No.3(4), 2014.

# Contributors

**Amir Ahmad**
United Arab Emirates University
Al Ain, UAE

**Chuadhry Mujeeb Ahmed**
Singapore University of Technology
and Design
Singapore

**Mohammad Saad Alam**
Department of Computer Engineering
Aligarh Muslim University
Aligarh, Uttar Pradesh, India

**Rashid Ali**
Department of Computer Engineering
Aligarh Muslim University
Aligarh, Uttar Pradesh, India

**M. M. Sufyan Beg**
Department of Computer Engineering
Aligarh Muslim University
Aligarh, Uttar Pradesh, India

**M. Roopa Chandrika**
Malla Reddy Engineering College
Hyderabad, Telangana, India

**Pushpita Chatterjee**
Old Dominion University
Suffolk, Virginia

**Precilla M. Dimpe**
Tshwane University of Technology
Pretoria, South Africa

**Guillermo A. Francia, III**
Center for Cybersecurity
University of West Florida
Pensacola, Florida

**Uttam Ghosh**
Vanderbilt University
Nashville, Tennessee

**Banu Günel**
Department of Information Systems
Middle East Technical University
Ankara, Turkey

**Hisham Haddad**
Kennesaw State University
Marietta, Georgia

**Abdelkrim Haqiq**
e-NGN Research Group, Africa and
Middle East
Hassan 1st University
Settat, Morocco

**Md. Muzakkir Hussain**
Department of Computer Engineering
Aligarh Muslim University
Aligarh, Uttar Pradesh, India

**Georgi Iliev**
Department of Communication
    Networks
Faculty of Telecommunications
Technical University of Sofia
Sofia, Bulgaria

**A. Kamaraj**
Department of Electronics and
    Communication Engineering
Mepco Schlenk Engineering
    College
Sivakasi, Tamil Nadu, India

**Charles Kamhoua**
Army Research Lab
Adelphi, Maryland

**Okuthe P. Kogeda**
Tshwane University of Technology
Pretoria, South Africa

**Aswani Kumar Cherukuri**
VIT University
Vellore, Tamil Nadu, India

**J. Senthil Kumar**
Department of Electronics and
    Communication Engineering
Mepco Schlenk Engineering
    College
Sivakasi, Tamil Nadu, India

**Safaa Mahrach**
Hassan 1st University
Settat, Morocco

**Aditya P. Mathur**
Singapore University of Technology
    and Design
Singapore

**Dimitriya Mihaylova**
Department of Communication
    Networks
Faculty of Telecommunications
Technical University of Sofia
Sofia, Bulgaria

**J. Jesu Vedha Nayahi**
Department of Computer Science and
    Engineering
Anna University Regional Campus-
    Tirunelveli
Tirunelveli, Tamil Nadu, India

**S. Selva Nidhyananthan**
Department of Electronics and
    Communication Engineering
Mepco Schlenk Engineering College
Sivakasi, Tamil Nadu, India

**Laurent Njilla**
Air Force Research Lab
Rome, New York

**Gretchen Richards**
School of Education
Jacksonville State University
Jacksonville, Alabama

**Hossain Shahriar**
Kennesaw State University
Marietta, Georgia

**Sachin S. Shetty**
Old Dominion University
Suffolk, Virginia

**Jay Snellen**
MCIS Department
Jacksonville State University
Jacksonville, Alabama

**Ferda Özdemir Sönmez**
Department of Information
  Systems
Middle East Technical University
Ankara, Turkey

**D. Sumathi**
Malla Reddy Engineering
  College
Hyderabad, Telangana, India

**Md. Arabin Islam Talukder**
Kennesaw State University
Marietta, Georgia

**Sumaiya Thaseen**
VIT University
Vellore, Tamil Nadu, India

**L. Josephine Usha**
Department of Information Technology,
St. Xavier's Catholic College of
  Engineering
Chunkankadai, Nagercoil,
Tamil Nadu, India

**Zlatka Valkova-Jarvis**
Department of Communication
  Networks
Faculty of Telecommunications
Technical University of Sofia
Sofia, Bulgaria

**Jianying Zhou**
Singapore University of Technology
  and Design
Singapore

# CYBER-PHYSICAL SYSTEMS: VULNERABILITIES, ATTACKS AND THREATS

**I**

Chapter 1, "Improving Security and Privacy for Cyber-Physical Systems," analyzes the threats and attacks in various cyber-physical system domains and discusses the defensive approaches for avoiding such attacks. Different security studies have been compared in this chapter with a focus on issues and solutions by considering the impact of cyber-attacks, attack modeling and security architecture deployment.

Chapter 2, "Vulnerability Analysis for Cyber-Physical Systems," provides a brief overview of the vulnerability analysis of cyber-physical systems. The chapter introduces cyber-physical systems (CPS) and discusses the different attacks and threats that prevail in CPS and countermeasures to prevent or mitigate risks. A detailed assessment of vulnerabilities in various CPSs such as industrial control systems (ICSs), distributed control systems (DCSs), and smart grids is analyzed.

Chapter 3, "State Estimation-Based Attack Detection in Cyber-Physical Systems: Limitations and Solutions," presents a detailed case study regarding model-based attack detection procedures for CPSs. In particular, data from a real-world water treatment plant is collected and analyzed.

# CYBER-PHYSICAL SYSTEM VULNERABILITIES ATTACKS AND THREATS

# Chapter 1

# Improving Security and Privacy in Cyber-Physical Systems

Sumaiya Thaseen, Aswani Kumar Cherukuri
and Amir Ahmad

## Contents

## 1.1  Introduction

The integration of different cyber and physical components utilizing modern computing technologies is termed as cyber-physical systems (CPSs). The Internet of Things
(IoT) paradigm ensures secure and energy-efficient transfer of information between
the physical and the cyber world. Many applications such as smart medicine, smart
vehicles, smart city, mobile systems and defense systems use CPS as the basis for the
development. The development of CPS was originally considered as an innovation to
improve the quality of human life and to get rid of routine work. The CPS has been
a key target in some of the highly publicized security breaches over the last decade.
Cyber- and physical-security concepts cannot protect CPSs from unexpected vulnerabilities arising due to the crossover effects and complex interdependencies; physical
attacks may cause compromise or damage to the information system; cyber attacks
can cause physical malfunctions. Cyber threats seen in nature which originate in
cyberspace are scalable but create an impact on the physical space of the system.

Barriers of CPS include security issues, various protocols and standards, and
power supply devices. The risk of intrusions and attacks in CPS is also due to the
self-sufficiency and remote location of CPS devices. A critical challenge is encountered when there is human interaction with CPSs. Interpreting human–machine
behavior and developing appropriate models considering the situational and environmental changes are huge and challenging tasks. Such changes are essential in
military and air traffic systems (Sztipanovits et al., 2012).

The primary problems identified for CPS security (Xinlan et al., 2010) are (1)
awareness of threats and possible consequences of attacks for modeling security
threats, (2) understanding the unique characteristics of CPSs and their variations from

traditional information technology security, and (3) discussion of the security mechanisms applicable to CPSs to design reliable and fault-tolerant architectures for detection and prevention of cyber and physical threats. Another issue with security is there are many stakeholders. They all distinguish various security threats and risks and focus towards different goals. Thus, the consideration is to ensure multiple levels of security for the stakeholders involved in their respective contexts and environments. CPS requires an improved infrastructure with sensor-enabled automatic systems including computational devices and physical processes. There are different kinds of data transported to and from different devices, in the format that is required by the devices using specific communication mechanism which the device can handle. As a result, security and privacy are the key concerns for CPS design, development and operation. The aim of this chapter is to discuss the potential attacks in CPSs and help designers of emerging CPSs to build more secure, privacy-enhanced products in the future.

The sections of the chapter are organized as follows: Section 1.2 highlights the security and privacy in CPS with regard to security objectives. Major threats of CPS and defense mechanisms for mitigating the attacks are described in Section 1.3. Section 1.4 analyzes the security perspective of CPSs among various domains and highlights the literature pertaining to intrusion detection systems, cyberattack consequences, attack modeling, attack detection and security architecture deployment. Section 1.5 presents the architecture of CPS and each subsection of 1.5 describes the attacks in each layer; Section 1.6 describes the attack mitigation strategy for CPS; Section 1.6.1 briefs the risk assessment for CPS; Sections 1.6.2 and 1.6.3 discuss the single-layer and multi-layer solutions for CPS; Section 1.6.4 presents the security framework for CPS. The conclusion and the future research areas mainly to assist designers in the development of secure, privacy-enhanced CPS are covered in the final section.

## 1.2 Security Objectives for Cyber-Physical Systems

### 1.2.1 Security

The security witnesses adverse effects from integrity loss of information or devices in the system. The effect of the attacks could be direct on the physical part of the process or the cyber elements. The two attacks considered in our chapter are as follows:

#### 1.2.1.1 Physical

Physical elements in the CPS are directly tampered by the attack: for example, the batteries of an implantable medical device are changed.

#### 1.2.1.2 Cyber

The type of attacks that are deployed through malware, software, or through access to elements of the communication network: for example, faking sensor information.

## 1.2.2 Privacy

The privacy of the users may be compromised as CPS relies on granular and diverse sensors. Privacy attacks are mostly passive and may require access to private data, or make inferences about specific information from public data.

The different security objectives of CPS are as follows:

### 1.2.2.1 Confidentiality

The capability to prevent disclosure of information to unauthorized individuals or systems is termed confidentiality. For example, a healthcare CPS on the Internet requires personal health records to be transferred from the personal health record system to the doctor or medical devices. Confidentiality can be ensured by encrypting the personal health record during transmission and by restricting the access to the places where it is stored (databases, log files and backups). A confidentiality breach occurs when any unauthorized party tries to access the personal healthcare records. Confidentiality is necessary for maintaining the user's privacy in cyber-physical systems (Pham et al., 2010). It is achieved by preventing the adversary from modifying the state of the physical system by eavesdropping on the communication channels between the sensors and the controller and also between the controller and the actuator.

### 1.2.2.2 Integrity

*Integrity* refers to the preservation of data without modification unless it is done by an authorized user. Integrity is violated when an adversary accidentally or with the intent of causing harm modifies or deletes important data. Thus, receivers receive false data and assume that to be true. Integrity is achieved by preventing, detecting or blocking deception attacks on the information sent and received by the sensors and actuators or controllers (Madden et al., 2010).

### 1.2.2.3 Availability

For any system to serve its purpose, the service must be available when it is needed. High availability of CPSs aims to provide service by preventing any computation, control and communication corruptions (Work et al., 2008). The system should have the ability to provide service even in the event of a hardware failure, system upgrade, power outage or denial-of-service attacks.

### 1.2.2.4 Authenticity

It is necessary to ensure that data, transactions and communications are genuine during the computation and communication process. It is important to validate both parties involved during authentication (Stallings William, 2017). In CPSs, the authentication is performed during sensing, communication and actuation.

## 1.3 CPS Attacks and Threats

Threats may be deliberate, accidental or environmental according to the ISO/IEC 270001:2013 standard. Examples of typical threats include natural events, compromise of data (software tampering, eavesdropping, etc.), technical failures, compromise of functions (abuse of rights), and unauthorized actions. The following are the different types of attack on CPS targeted to disrupt confidentiality, integrity, authenticity and availability of data.

Figure 1.1 illustrates the tree diagram of various attacks and threats on cyber-physical systems. The different branches of the tree include the following types of

**Figure 1.1 Tree diagram of attacks and threats on cyber-physical systems technologies.**

attack: (a) attacks on sensor devices, (b) attacks on actuators, (c) attacks on computing components, (d) attacks on communications, and (e) attacks on feedback.

a. The researchers have identified the various threats and vulnerabilities that affect CPS sensors. They are GPS spoofing, injecting false radar signals and dazzling cameras with light. The accuracy of data acquisition process has to be ensured as CPSs are closely related to the physical process. Physical authentication has to be provided by the sensors to ensure trustworthiness for data received from a physical entity (Krotofil et al., 2015).

b. The impact of cyber attacks on actuators was analyzed by Djouadi et al. (2015), who considered the various potential attacks, namely, the finite energy attack which includes the loss and modification of personal packets, the finite time attack, impulse attack and the bounded attack, which suppresses the control signal.

c. Computing resources face attacks from viruses, worms, Trojans and DoS attacks (Singhal, 2007) such that the CPS is secretly damaged. As there are violations and measurement errors in control systems, the detection techniques must ensure that these regular errors do not lead to false alarms. This can indirectly provide the attacker a space to hide.

d. Different kinds of communication attacks like selective forwarding, packet spoofing, Sybil, and packet replaying can disrupt resource allocation between nodes through the introduction of malware. Thus, the routing of system packages is violated (Krotofil et al., 2015). Data intervention may lead to future errors while processing requirements.

e. A system comprising a three-layered logical model of CPSs and a meta-model of cyber attacks was developed (Hahn et al., 2015). The system is targeted by Feedback Integrity Attack. The protection of control systems in CPSs which provide feedback for actuation is termed *feedback security*.

Thus, the literature shows the probability of the major attacks having a huge impact on CPS is described as follows:

## 1.3.1 Eavesdropping

Eavesdropping refers to an attack such that the adversary can intercept any information by the system. It is seen as a passive attack because the attacker does not interfere with the working of the system and only observes the operation. CPS is vulnerable to eavesdropping through traffic analysis such as intercepting the monitoring data transferred in sensor networks. User privacy is violated by eavesdropping.

## 1.3.2 Compromised-Key Attack

The key is compromised when an attacker gets hold of a secret key. An attacker can gain access to a secured communication without the knowledge of sender or receiver by utilizing the compromised key. He can decrypt or modify data and also compute additional keys thereby accessing other secured communications or resources. For example, the attacker could monitor the sensors to execute the task of reverse engineering in order to compute the keys. He can also pretend to be a valid sensor node to cheat to agree on keys with other sensors.

### 1.3.3 Man-in-the-Middle Attack

False messages sent to the operator can take the form of false positive or false negative in this attack. The operator assumes that everything is fine and does not take action when required. When the operator follows normal procedures and attempts to perform a system change, his action could cause an undesirable event. There are many variations to the modification and replay of control data, which could affect the system operations.

Figure 1.2 is a schematic representation of the man-in-the-middle attack in CPSs which disrupts data integrity. A communication is established between the physical plant to the feedback controller through the network. The original data is modified by the adversary and the false data is injected into the network. The feedback controller receives false data and processes it accordingly.

### 1.3.4 Denial-of-Service (DoS) Attack

Denial-of-service attack is one of the major attacks that prevent legitimate traffic or requests for network resources from being processed by the system. In this type of attack, a huge volume of data is transmitted to the network to make the server busy, thereby causing disruption in normal services. After gaining access to the network of cyber-physical systems, the attacker can perform any one of the following:

- Flood a controller or the entire sensor network with traffic until a shutdown occurs due to overload.
- Invalid data is sent to the controller or system networks causing abnormal termination or behavior of the services.
- Traffic is blocked resulting in loss of access to network resources by authorized elements of the system.

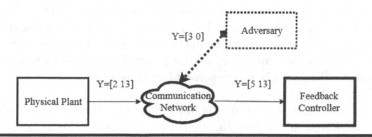

**Figure 1.2  Man-in-the-middle attack in CPS.**

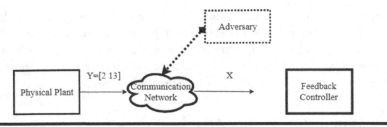

**Figure 1.3   Denial-of-service attack in CPS.**

Figure 1.3 is a schematic representation of a denial-of-service attack in CPS. An established communication between the physical plant and the feedback controller through the communication network is disrupted by the adversary. As a result, no information from the physical plant is available to the feedback controller.

## 1.3.5 CPS Defense Mechanisms

After identifying the vulnerabilities of a CPS, it is necessary to develop defenses that can prevent access to adversaries. A few defense mechanisms that are implemented for CPSs are discussed in the remainder of this section. Researchers have implemented cryptographic techniques, secure routing and anonymous routing for avoiding eavesdropping attack and also analyzed important issues in designing cryptosystems such as key management, authentication and encryption/decryption algorithms (Kao et al., 2006).

Literature shows proposals from many investigators regarding the key establishment protocols for preventing the compromised-key attack. Chalkias et al. (2007) point at protocols of two kinds. The first relates to the key transport protocols in which a session key is created by an entity and is transmitted in a secure manner by the other. The second category is the key agreement protocols in which information from both entities is utilized for the derivation of a shared key for preventing the compromised-key attack.

Yuan et al. (2013) have designed resilient controllers for cyber-physical control systems targeted by a DoS attack. A coupled design framework is incorporated into the cyber configuration policy of intrusion detection systems (IDS) and also the proposed design algorithms based on linear matrix inequalities to enable computation of an optimal cybersecurity policy and control laws for preventing DoS attacks.

Table 1.1 shows the security methods for preventing the major attacks in CPS. The defense mechanisms can be categorized into three groups:

■ *Prevention*: The security mechanisms that prevent attacks by providing authentication, access controls, security policies, and network segmentation.

**Table 1.1 Security Techniques for Preventing Attacks in CPS**

| Type of Attack | Security Methods |
|---|---|
| Eavesdropping (Kao et al., 2006) | Cryptosystem (symmetric and asymmetric), secure routing and anonymous routing. |
| Compromised-key attack (Chalkis et al., 2007) | Cryptography, key transport protocol, key agreement protocols and two-party key establishment protocols. |
| Man-in-the-middle attack (Adhikari, 2015; Lyn, 2015; Saltzman and Adi, 2009) | Message digest, digital signature, MAC, biometrics and trusted platform module (TPM). |
| Denial-of-service attack (Yuan et al., 2013) | Access control lists (ACLs), Role-based access control (RBAC), discretionary access control (DA) and mandatory access control (MAC). |

■ *Detection*: Despite the imperative need for preventive techniques, adversaries with enough resources, time and commitment can bypass them and launch successful attacks against the CPS system. Detection strategies are designed to identify anomalous behavior and attacks in the system.

■ *Response*: A security mechanism may need to act automatically (without waiting for human intervention) to mitigate detected attacks as most of the CPS have real-time constraints. Reactive response refers to the actions executed after an attack is detected. It aims to mitigate the impact of the attack and, if possible, restore the system. In particular, the focus is on those that are launched online (in response to the detection of the attack) and are mostly automatic, as, for example, switching between real sensors to simulated sensors, or switching to redundant systems.

# 1.4 CPS Security Perspectives

The major domains of CPS considered for the study are as follows:

■ Smart grids
■ Medical devices
■ Industrial control systems (ICSs)
■ Intelligent transport systems (ITSs)
■ Miscellaneous

The authors of this chapter have perused and done an analysis of a few papers from each domain to understand and identify the types of attack in each domain and their prevention strategy.

The different cyber attacks that can occur during various stages of the grid are summarized (He et al., 2016). For instance, attacks on generation, such as the Aurora attack, can desynchronize power generators and damage them; attacks on transmission can affect substations and relays; and also one of the most studied types of attacks, state estimation attacks, where a modification of some measurements gets the probability of injecting stealthy false information to mislead the estimation of phase angles. Finally, attacks on distribution include electricity theft and information/privacy leakages by compromising smart meters. They also summarize defensive strategies as follows:

- Protection, by enhancing communication and introducing encrypted devices optimally located
- Detecting attacks by deploying signature-based and behavior-based IDS
- Mitigation, which minimizes the potential disruptions and damages caused by an attack

One of the most comprehensive summaries of problems associated with privacy in smart grid has been analyzed (Mo et al., 2009). The authors have focused on adversaries that eavesdrop information to make inferences about users. Legal frameworks try to impose laws and regulations to enhance privacy for users. The major privacy risks are nonintrusive load monitoring (attacker infers which device is being used) and utilize mode detection (attacker infers the activity performed inside the device such as identifying television channels). The authors have highlighted several techniques to mitigate the privacy risks:

- Anonymization
- Trusted computing (attestation)
- Cryptographic approaches
- Perturbation (differential privacy [DP])
- Verifiable computation

Rushanan et al. (2014) and Camara et al. (2015) have described the types of adversaries that medical devices will be subject to, including the ability to eavesdrop all communication channels (passive) or read, modify, and inject data (active). The threats are mainly focused on the telemetry interface. However, Rushanan et al. (2014) have also analyzed account software, hardware, and sensor interfaces. They have proposed authentication (e.g., biometric, distance bounding, and out-of-band channels), and the presence or absence of the extra wearable device that allows or denies access to the medical device for mitigating the possible attacks on telemetry interface. In addition to prevention, they also discuss attack-detection by observing patterns to distinguish between safe and unsafe behavior.

McLaughlin et al. (2016) have suggested an eight-step process for an exhaustive vulnerability assessment, from document analysis to final testing, to implement security in ICS. The mitigation strategies focus largely on control architectures that customize their mechanism according to the domain: checking of control code, reference monitor architecture at run-time, an architecture that provides an estimate of time for reaching unsafe states, and architecture with a trusted computing base. Urbina et al. (2016) focused on leveraging the physical characteristics of the system itself to detect attacks (i.e., physics-based intrusion detection). The authors discuss the need to have a clearly defined:

- Model of the physical system
- Trust assumptions
- Statistical test used for anomaly detection
- Method for the evaluation of effectiveness of the anomaly detector (metrics)

These four characteristics are then identified in a wide variety of publications across several CPS domains, including power systems, ICSs, control theory, automated vehicles, video cameras, electricity theft, and medical devices. The authors have reached conclusions after discussing the common assumptions and shortcomings of this class of research and suggested several improvements, including where to place the security monitor. They have also proposed a new metric for evaluating the effectiveness of physics-based intrusion detection models. Security and privacy concerns of the industrial Internet of Things (IoT) have been addressed (Sadeghi et al., 2015). As mitigation, the authors propose the use of integrity checking techniques through software and hardware.

Van der Heijden et al. (2016) compared mechanisms to identify malicious nodes and malicious data in ITSs. The authors analyzed misbehavior detection by defining a domain-specific taxonomy of misbehavior depending on whether the anomaly arose from inconsistent nodes or data. Their paper surveys attack-detection schemes on three levels: local, global and cooperative. The authors state that the link between the onboard units (OBU) and messages is crucial for identifying misbehavior and have developed schemes based on the degree of link ability available. A survey of the mechanisms proposed for defense in Vehicular Adhoc Networks (VANETs) is analyzed (Sakiz and Sevil, 2017). The survey has a high inclination toward model-based security mechanisms, such as mechanical modeling, trust-based modeling, and Markov-chain modeling apart from watchdog-based defenses. The survey also provides a comparative study of the infrastructure of the systems and the attacks that the particular technique can thwart. The study also mentions that most of the defenses aim at discovering misbehavior at a particular layer and urges researchers to work on the reliability of links. One of the major challenges facing modeling a transportation system relates to is its agility. The problem becomes magnified while developing security features for them. Additional problems of security are related to scalability, lack of a clear line of defense, and real-time operations (Sakiz and Sevil, 2017). Various possible attacks that may affect privacy are discussed and also analyzed is how most of the attacks like falsifying GPS information or deploying control commands can cause a malfunction in drones (Altawy et al., 2017). The authors also point out the possible solutions to mitigate the impact of these attacks, such as encryption and IDS.

Nguyen et al. (2017) have studied security issues in CPSs and shown how software models can help in the design and verification of CPSs. Model-based strategies bring several benefits as the level of abstraction is higher than the code level. A systematic mapping study has been developed by the authors that discovered various trends on model-based security analysis. Thus, there has been an increasing interest on model-based security to analyze vulnerabilities and threats in the past 2 years; however, there is a lack of research focusing on mitigating the vulnerabilities.

## 1.4.1 Intrusion Detection Systems

Han et al. (2014) and Mitchell et al. (2014) have focused on the topic of intrusion detection systems for CPSs. The first paper refers to the ability of CPSs to self-maintain, self-repair, and self-upgrade themselves and that self-detection of intrusion is now a forefront research topic. First, the authors explain the background of external and internal vulnerabilities, which the attackers tap, and the detection techniques are explained. The available detection techniques could be broadly classified into signature based, anomaly-based, and stateful protocol analysis-based mechanisms. The authors propose five necessary characteristics of intrusion detection in CPS:

- They should be able to function in a distributed topology,
- Should provide runtime data,
- Should have the ability to thwart both unknown and known attacks,
- Should be system fault tolerant, and
- Should not hamper privacy.

Mitchell et al. (2014) have proposed a different classification of IDS, with two main groups: detection technique and audit material. The first involves knowledge-based detection, which includes techniques that have prior knowledge of bad behavior and have the ability to determine whether the system (or data) is misbehaving. Another detection class is behavior-based, which is useful for zero-day attacks since it do not look for something specific. On the other hand, audit material can be classified into host-based that focuses on analyzing logs, and network-based audit, which monitors network activity to determine if a node is compromised (e.g., via deep packet inspection).

**Table 1.2  Advantages of IDS Techniques for CPSs**

| Dimension | Type | Advantages |
|---|---|---|
| Detection technique | Behavior | Detect unknown attacks. |
| | Behavior specification | Detect unknown attacks, low false-positive rate. |
| | Knowledge | Low processor demand, low false-positive rate. |
| Audit material | Host | Distributed control and ease of specifying/detecting host-level misbehavior. |
| | Network | Reduced load on resource-constrained devices. |

Table 1.2 summarizes the advantages of various IDS techniques as applying to CPS. The major advantages are:

■ The advantage of behavior-based detection techniques is they detect zero-day attacks. The importance of detecting unknown attacks cannot be overstated. The most sophisticated adversaries will target the most critical systems, and these attackers will not rely on previously disclosed vulnerabilities.

■ Behavior-specification-based detection techniques can detect zero-day attacks and they yield a low false positive rate.

■ Knowledge-based detection techniques yield a low false positive rate and they make minimal demands on the host microprocessor.

■ Host-based auditing is distributed control and ease of specifying/detecting host-level misbehavior.

■ Network-based auditing reduces the demand for processor and memory on resource-constrained nodes.

Table 1.3 summarizes the drawbacks of various detection techniques/CPS materials. The major drawbacks are as follows:

■ The drawback of behavior-based detection techniques is their high false positive rate. For unattended CPSs operating in hostile or inaccessible locations, unnecessary evictions will reduce lifetime and increase operating cost.

■ The drawback of behavior-specification-based detection techniques is that a human must implement the state machine or grammar that represents safe system behavior. This activity is expensive, slow and prone to error.

**Table 1.3  Weaknesses of IDS Techniques for CPSs**

| Dimension | Type | Drawbacks |
|---|---|---|
| Detection technique | Behavior<br>Behavior specification<br><br>Knowledge | High false-positive rate.<br>Human must implement the model.<br>Attack dictionary must be stored and updated, misses unknown attack. |
| Audit material | Host<br><br><br><br>Network | Load increase on resource-constrained devices, audit material is vulnerable and generality is limited.<br>Visibility limits the effectiveness. |

- Knowledge-based detection techniques are helpless against zero-day attacks and they rely on an attack dictionary, which must be stored and updated. The most sensitive CPSs operate on isolated networks which are a hindrance to attack dictionary maintenance.
- The drawbacks of host-based auditing are increased processor and memory demand on resource-constrained nodes, vulnerability of audit material and limited generality based on OS or application.
- The drawback of network-based auditing is the visibility of nodes collecting audit material limits the effectiveness.

In addition to the study on various CPS domains for identification and prevention of attacks, literature given below summarizes the consequence of cyber attacks, CPS attack modeling and detection, and deployment of security architecture among various domains.

## 1.4.2 Estimation of Consequences of Cyber Attacks

Complex and sophisticated attacks are designed to cause significant damage to the cyber and physical characteristics of CPSs. For instance, Stuxnet (Collins et al., 2012; Karnouskos, 2011), which was specifically developed to cause physical damage on infrastructures, was the first malicious software. Therefore, it is very critical to evaluate the impact of cyber attacks and their consequences. Table 1.4 shows an extensive analysis of different cyberattack consequences.

**Table 1.4  Cyberattack Consequences**

| Proposed Approach | Contribution | Future Direction |
|---|---|---|
| Identifying few appropriate issues in the cyber-physical security of WAMPAC (Ashok et al., 2014) | • Dynamic cyberattack situations are modeled based on attacker/defender model.<br>• Traditional risk assessment techniques cannot be modeled however application of game theory to model cyber attacks. | The framework can be extended to complicated scenarios. |
| Evaluation of the impact of network and installation (Genge et al., 2012) | The resilience of physical process is increased to confront the cyber attacks by two key parameters, namely control valve speed and control code task scheduling. | This solution should be considered while the process is designed as it will result in a robust physical process. |
| Assessing the impacts of cyberattacks on infrastructures (Genge et al., 2015) | Performs better in comparison to graph theory methods and electrical centrality measures. | Evaluation and integration of CAIA result in the control network design methodologies. |
| Analysis of the attack effect on the physical environment and threat models for control systems (Huang et al., 2009) | The influence of various cyber attacks such as DoS and integrity attacks on CPS was analyzed. The results show the attacks on control signals as more serious than sensor signals. | Evaluation of the impact caused by a combination of attacks on CPS. |
| Evaluating the consequences of security attacks on the physical process (Krotofil et al., 2014) | Compares the behavior of CPS at different time instants and varied disturbances on various control parameters. The approach can be applied to a group of attacks. | The proposed method can be analyzed in various CPS. |

*(Continued)*

**Table 1.4 (*Continued*) Cyberattack Consequences**

| Proposed Approach | Contribution | Future Direction |
|---|---|---|
| Data source assigned with robustness level with regard to confidentiality, integrity, and privacy (Sicari et al., 2016) | The reliability of registered and non-registered IoT data sources is evaluated through an algorithm. | Key management in the platform can be introduced. |
| CPS security assessment by aspect-oriented modeling (AOM) (Wasicek et al., 2014) | System models and associated attacks are assessed within the same environment. | Executable attack models and attack pattern development. |
| Risk Assessment technique (Wu et al., 2015) | Systemic risk in real time and response to the risk is analyzed better due to the proposed risk change curve. To predict risk in the future. | Risk assessment by automatic identification and quantitative analysis methods, thereby dealing with a large number of real-time assets, threats, and vulnerabilities of CPS. |
| Cyber-physical attack captured by cyber-physical attack description language (Yampolskiy et al., 2015) | Qualitative and quantitative attacks on CPS are analyzed. | The knowledge base of known attacks on CPS can be developed. |

## 1.4.3 Modeling of CPS Attacks

Attack and vulnerability models are used for identification of weaknesses in CPSs to support their search strategy and the understanding of the attacks. It is necessary to develop attack models for their assessment and to take appropriate countermeasures to ensure CPS security, as shown in Table 1.5. The attacker needs to understand the failure conditions of the equipment, control principles, process behavior, and so on.

**Table 1.5  Modeling the CPS Attacks**

| Proposed Approach | Main Contribution | Future Direction |
|---|---|---|
| Dynamic probabilistic model for simulating physical security attacks on CPSs (Khalil, 2016) | Visual flowchart applied as a programming language. | Defender countermeasures are incorporated. The prediction probability of the proposed model has to be validated. |
| Identifying the potential threats at the design phase of Martins et al. (2015) | Real-world railway temperature monitoring system utilized as the case study. Threat modeling for CPS performed by a tool. | Various threat modeling techniques are merged to enable the extension of threat identification and vulnerabilities. |
| An IPV6 spoofing attack is described which corrupts the border router's routing table of the 6LoWPAN network (Mavani and Krishna, 2017) | The attack success probability is affected by the path loss exponent. Performed a mathematical analysis using an attack tree model. | The impact of multiple attackers on the network is assessed and countermeasure to be proposed. |
| The analytical model developed based on SPN techniques for analyzing and modeling attacks and countermeasures for CPSs (Mitchell and Chen, 2016) | Allows optimal design parameter settings for maximizing the mean time to failure (MTTF) of the CPS. | Analysis of the countermeasures for improving CPS survivability. |
| Modeling the attacks using the vulnerability of data, communication and network Cyber vulnerability index based on discovery, feasibility, threats, access and speed (Srivastava et al., 2013) | Cyber and physical vulnerability models are integrated based on the basis of incomplete information. | Mitigation techniques to be developed to avoid cyber-physical attacks that are coordinated on the smart grid. |

## *1.4.4 CPS Attacks Detection*

It is important to develop detection algorithms and countermeasures for all well-known attacks in advance to reduce the impact of attacks for a limited time and minimize system damage. Table 1.6 summarizes the papers on CPS attacks detection, the main contributions, and future research directions.

**Table 1.6  CPS Attacks Detection**

| Proposed Approach | Main Contribution | Future Direction |
|---|---|---|
| Intrusion detection and attack classification in sensor networks (Finogeev and Finogeev, 2017) | Reduced energy consumption due to a simultaneous exchange of key information while data is transmitted. | Hidden transfer of open or encrypted key details utilizing steganography techniques. |
| Anomaly detection technique that utilizes log-lines produced by various modules and systems in ICT networks (Friedberg et al., 2015) | Real-world evaluation by the anomaly detection model. | Intelligent approach for event class generation. |
| Efficient algorithm to determine unobservable attacks (Giani et al., 2013) | Complex attacks that involve compromise of two-power injection detected. | Comprehensive and realistic analysis of cybersecurity threats under normal and contingency conditions. |
| Distributed host-based collaborative detection (DHCD) technique to identify and prevent data injection attacks in smart grid CPS (Li et al., 2016) | Real-time measurement data is analyzed and distributed mode reduces the central computation burden. | The proposed method can be extended to capture power system faults. |
| Attack detection using a cyber-physical fusion in smart grid utilizing ATSE (Liu et al., 2015) | Easy-to-implement and low-cost technique. Integrated to heterogeneous data in smart grids. | Interaction and correlation between the cyber domain and power system to be investigated. Abnormal detection techniques and IDS tools to be integrated into ITSE. |

*(Continued)*

**Table 1.6 (*Continued*)   CPS Attacks Detection**

| Proposed Approach | Main Contribution | Future Direction |
|---|---|---|
| Model-based technique for identifying the integrity attacks on the sensors (Mo et al., 2014) | Improve the detection probability by compromising the control performance. | Enhancement of the proposed technique to sophisticated attacks and distributed systems. |
| Automatic recognition of the integrity attack class targeting a CPS (Ntalampiras, 2016) | Novel feature set and customized pattern recognition algorithms. Identifying integrity attacks by integrating the characteristics of two varied signal representations. | Online clustering technique for identifying novel data. |
| The holistic approach of existing literature on intrusion detection in VANETs (Sakiz and Sevil, 2017) | Review of different detection mechanisms along with advantages and limitations. | Detection of attack in VANETs. |
| A novel design approach to identify real-time attack to enhance the limitations of quality control systems (Vincent et al., 2015) | Identification of compromised manufactured parts without interrupting the process flow. | New manufacturing approaches for identifying cyberattacks that incorporate the physical nature of manufacturing systems. |
| Analysis of data injection attacks against Kalman filtering during dynamic estimation of power systems Countermeasures to defend against the attacks (Yang et al., 2016) | Best performance is achieved using enhanced unscented Kalman filter (UKF) technique in comparison to other Kalman filtering techniques. This technique reduced the impact of attacks. | Analyzing the impact of data injection attacks against the assessment of power grid systems. |

## 1.4.5 Development of Security Architecture

The development of CPSs is constrained by security factors. The main task of designing complex CPS architectures is to test and validate "secure design" to ensure the security and reliability of physical and cyber components. It is necessary to develop new reliable control and evaluation algorithms that consider findings from realistic attack models. Table 1.7 shows a summary of the literature relating to the development of security architecture.

**Table 1.7  Security Architecture Development**

| Proposed Approach | Main Contribution | Future Direction |
|---|---|---|
| Framework for network robust analysis and extending to large-scale networks (IoT, CPS and M2M) (Chen et al., 2014) | Integrated defense mechanism to minimize the damage caused by deliberate attacks. | Enhanced to multistage hierarchical network consisting of several autonomous fusion centers. |
| Study on the principle of building a robust CPS (Friedberg et al., 2015) | Review of the entire design process. Qualitative and quantitative description of CPS robustness. Analysis of sensor-actuator interaction and CPS security issues. | Discussion on critical research issues for the development of robust CPS. |
| Secure authentication and authorization architecture for healthcare using IoT. (Moosavi et al., 2015) | Blocking the malicious activity before entering into the secure health domain. Distributed smart e-health (SEA) gateways utilized for the IOT-based healthcare architecture. | A scalable and reliable end-to-end security for IoT-based healthcare models. |
| Analytical framework for information security (Venkitasubramaniam et al., 2015) | Addresses the challenges and describes the recent advancement utilizing the framework. | Deeper investigation of cyber-secure control and cyber-physical security. |

*(Continued)*

**Table 1.7 (*Continued*)   Security Architecture Development**

| Proposed Approach | Main Contribution | Future Direction |
|---|---|---|
| Security architectures in the heterogeneous CPS environment (Yoo and Taeshik, 2016) | Security issues that can occur in the environment are classified and security countermeasures are suggested. | A method to ascertain the establishment of a protocol conversion in a normal fashion. |
| Security design for CPS that considers that considers specific characteristics such as the CPS environment, real-time requirements and geographic distribution (Eisenbarth et al., 2007) | Developing new security tools that are designed according to specific system requirements. | Security issues to be identified in the design phase and considered as the main part of CPS developing process. |

CPSs have a high potential for creating new solutions to social risks but impose high demands on quality, safety, security and privacy (Barnum et al., 2010; Baheti and Helen, 2011; Derler and Alberto, 2012; Kim and Panganamala, 2012). A predictable level of verification and quality is achieved using fundamental scientific research for effective combating of external and internal threats. An extensive study of detection and isolation mechanisms has been made for different types of attacks, and numerous findings relating to strategies for CPS are seen in the literature. After identifying an attack or vulnerabilities in a CPS, it is necessary to respond to those attacks so that their impact over CPS is attenuated. Combita et al. (2015) have analyzed different trends on automatic attack detection and responses, namely, preventive and reactive responses. Preventive response results when vulnerabilities in CPS are identified. Thus, the system structure can be modified in order to improve the resiliency of the system to attacks. Reactive response performs an action only when an alarm is raised. The attack is counteracted by modifying the control mechanism online. The interaction between attacker and defender is typically modeled using game theory, as the adversary is intelligent and can also respond to the defense action. Thus, it is necessary to build a secure architecture for CPS which is discussed in the section below.

## 1.5  CPS Architecture and Attacks

Figure 1.4 is the schematic representation of the three layers of CPS, namely, the perception layer, transmission layer and the application layer. The first layer, the perception layer, is also called the recognition layer or sensors layer

**Figure 1.4   Typical three layers of CPS.**

(Mahmoud et al., 2015). In this layer, multiple terminal types of equipment such as sensors, actuators, cameras, global positioning systems (GPSs), laser scanners, intelligent devices, radio frequency identification (RFID) tags with 2-D bar code labels and readers are present (Lu et al., 2015; Zhang et al., 2011). Real-time data is collected by the devices for different purposes such as monitoring and tracking. The information collected from the physical world is interpreted and commands from the application layer are performed. The collected information relates to sound, light, mechanics, chemistry, heat, electricity, biology and location (Peng et al., 2013; Zhao and Lina, 2013). Real-time data with node cooperation generated by sensors in wide and local network domains (Mahmoud et al., 2015) is aggregated and analyzed in the application layer. Aggregation of the information depends on the type of sensors which sense temperature, acceleration, humidity, vibration, location and air chemical changes.

The interchange and processing of data between the perception and application are the responsibility of the second layer which is the transmission layer. It is also known as the transport layer (Lu et al., 2015) or network layer. Data interaction and transmission in this layer are achieved using local area networks, communication networks, the Internet or other existing networks. The technologies used are Bluetooth, 4G and 5G, universal mobile telecommunications service (UMTS), Wi-Fi, Infrared and ZigBee, depending on the sensor devices. However, most of the interconnections are achieved via the Internet due to availability and cost-effectiveness. Hence real-time operations should be supported by the networks. The transmission layer can initially process and handle a vast volume of data and realize real-time transmission (Lu et al., 2015). This is achieved as a result of reliable communication support (Zhang et al., 2011). The layer is also responsible for data routing and transmission through various devices and hubs over the used networks (Mahmoud et al., 2015). Cloud computing platforms, routing devices, switching and Internet Gateways also work at this layer using technologies such as Wi-Fi, LTE, Bluetooth, 4G/5G or ZigBee. The network gateways serve as the connector point for different nodes that collect, filter, transmit and receives data among the nodes (Mahmoud et al., 2015) and other layers of the CPS. Traffic and storage are the other two issues which affect security within the CPS because of the increased number of connected devices. Although this traffic can be managed by protocols including firewalls, the security of devices with limited capabilities cannot be guaranteed since their computational capabilities and storage are very limited (Mahmoud et al., 2015).

The third and most interactive layer is the application layer. This layer is responsible for processing the received information from the data transmission level and issues instructions to be implemented by the physical units, sensors and actuators (Peng et al., 2013). This layer works by deploying complex decision-making algorithms on the collected data to provide correct decisions (Saqib et al., 2015). Control commands are also used for corrective actions. The application layer also receives and processes information from the perception layer and then determines the necessary automated actions to be invoked. The implementation of connected devices at the physical layer (Zhang et al., 2011) can be achieved by utilizing cloud computing, middleware and data mining algorithms. This layer also saves past actions to enable the provision of the feedback from any previous action for ensuring future operational improvements. A smart environment (Mahmoud et al., 2015) was created and CPS was integrated with industry professional applications. This resulted in development of extensive and intelligent applications in domains that may include private and secure data, such as smart cities and homes, smart power grid; smart health, intelligent transportation (Peng et al., 2013), smart farming, smart auto, environmental monitoring and industry control (Lu et al., 2015). The applications must provide mechanisms to protect the data as they collect users' private data, such as health information and habits. On the other side, application systems are distinct and require appropriate security policies. Therefore, the implementation of a security policy for each application system on an individual basis is a challenging job. With increase in the usage of CPS, there is also an increase in security issues, which needs consideration.

## 1.5.1 Attacks at the Perception Layer

The perception layer consists of end devices, namely tags in RFID and sensors, which are constrained by computing resources and memory capabilities. In addition, these devices are prone to physical attacks such as tampering with the devices' components or replacing the devices. This is due to the location of the devices mostly in external and outdoor environments. Common attacks at the perception layer include equipment failure, line failure, witch, electromagnetic interference, information disclosure, information tracking, perceptual data corruption (Peng et al., 2013), tampering, sensing information leakage (Gou et al., 2013), differential power analysis (Zhao and Lina, 2013), physical destruction and energy-exhaustion attacks (Bhattacharya, 2013). Common forms of these attacks are as follows:

- *Node Capture*: Leaks information by taking control over the node that could possess encryption keys, which is then used to threaten the security of the entire system. This attack targets confidentiality, integrity, availability and authenticity (Bhattacharya, 2013; Mahmoud et al., 2015; Zhao and Lina, 2013).
- *False Node*: Data integrity is compromised through the addition of another node to the network which sends malicious data. This results in a DoS attack through the consumption of the energy of the nodes in the system (Mahmoud et al., 2015; Zhao and Lina, 2013).
- *Node Outage*: Node services are stopped and hence reading and gathering of information from these nodes is a difficult job. As a result, a variety of other attacks are launched which affect availability and integrity (Bhattacharya, 2013).
- *Path-Based DOS*: A large number of flooding packets are sent along the routing path to the base station causing network disruption and battery exhaustion of the node. This results in a reduction in the availability of the nodes (Bhattacharya, 2013).
- *Integrity*: Injection of external control inputs and false sensor measurements causing system disruption (Mo et al., 2014; Saqib et al., 2015).

## 1.5.2 Attacks at the Transmission Layer

Attacks on this layer occur in the form of data leakage during the transmission of information. This occurs as a result of the openness of the transmission media, especially in wireless communication. Such attacks utilize a radio interface to capture a transmitted message, modify and retransmit it, or exchange information between heterogeneous networks. Thus, the legitimate user is impersonated. The other factors that could increase the chance of being attacked (Mahmoud et al., 2015; Peng et al., 2013; Shafi, 2012) are remote access mechanisms causing traffic congestion among network nodes of a huge number. Common attacks at this layer include response and Sybil, flooding, trap doors, tampering, black hole, wormhole, exhaustion, traffic analysis, sink node, direction misleading sinkhole, collision, wrong path selection, tunneling (Peng et al., 2013; Raza, 2013) and illegal access (Gou et al., 2013; Mitchell and Ing-Ray, 2014). The following are examples of common forms of attacks at the transmission layer:

- *Routing*: Loops are created while routing resulting in resistant network transmission, increased transmitting delay or extended source path (Raza, 2013; Zhao and Lina, 2013).

- *Wormhole*: False paths through which all the packets are routed are announced. This develops information holes in the network (Gaddam et al., 2008).
- *Jamming*: The wireless channel jammed between sensor nodes and the remote base station introduces noise or a signal with the same frequency. Intentional network interference created could lead to a DoS attack (Li et al., 2013; Maheshwari, 2016; Raza, 2013).
- *Selective Forwarding*: A compromised node is created to drop and discard packets and forward selected packets. In some situations, the compromised node forwards chosen messages only or stops forwarding packets to the intended destination and discards all other packets. However, this node is considered legitimate (Raza, 2013).
- *Sinkhole*: The best routing path to be used is announced to route the traffic to other nodes. Other attacks such as selective forwarding and spoofing are launched by this attack (Raza, 2013).

### 1.5.3 Attacks at the Application Layer

The attacks in this layer result in data damage, unauthorized access to devices and privacy loss such as user habits and health conditions. This is due to a large amount of users' information gathered at this layer (Peng et al., 2013). Common attacks at the application layer include malicious code, unauthorized access, user privacy leakage, database and control command forgery attacks (Lu et al., 2015; Peng et al., 2013; Suo et al., 2012). Common examples of attacks at this layer include the following:

- *Buffer Overflow*: The vulnerable features in the software lead to buffer overflow vulnerabilities and exploit it to launch attacks (Zhao and Lina, 2013).
- *Malicious Code*: The user application is attacked by launching various malicious codes, such as viruses and worms which causes the network to slow down or damage (Suo et al., 2012).

Risk assessment, single-layer and multilayer solutions for mitigating the attacks on CPS are discussed in the section that follows.

## 1.6 Attack Mitigation Strategy

### 1.6.1 Risk Assessment for CPS

A well-defined risk assessment has to be designed for CPS to provide an overall view of CPS security status and protect the resources. The four elements that require consideration while designing risk assessment for CPS are asset, threat, vulnerability and damage. The final risk value has a positive relation to all the four elements. Assets have a direct value for the organization and their presence can be in tangible or intangible form. Quantization of assets can be considered from three aspects: direct economic losses, indirect economic losses and casualties. The threat is an event or factor that can damage the assets of an organization from outside. Threat quantification can be conducted through the threat matrix, which includes intense, stealth, time, technical information, physical knowledge and access implemented by US Sandia Lab (Xie et al., 2013). The vulnerability is a condition or environment whose existence causes loss to corporate or institutional assets through mitigation of the underlying threats. Expert evaluation method or best practices in the industry can be compared for vulnerability quantification. The real components, data stream and entity stream of CPS can be simulated by methods that anticipate the effects to the whole system and analyze the possible damage (Peng et al., 2013).

## 1.6.2 Single-Layer Solutions for CPS

An improved identity-based key distribution scheme for wireless sensor network (WSNs) using elliptic curve cryptography (ECC) key management has been proposed by Yang et al. (2016). Asymmetric encryption for WSNs has been presented by Premnath and Zygmunt (2015) through analysis of small cryptographic keys. In this study, the computational process is decreased using smaller key sizes among nodes. Key-breaking cost estimation is also provided where only limited resources are available (costs in dollars and time in number of days). There exists a trade-off between the processing load for a node and the required time of privacy protection. The results show that using a small 1024-bit public key modulus requires a node to perform only 3.1% of the computations relative to a typical 3248-bit modulus.

A lightweight authentication protocol has been proposed for securing RFID tags to prevent attackers from gaining access to the network (Trappe et al., 2015). This is accomplished by sniffing the Electronic Product Key of the victim tag and programming it to another tag. This attack can be prevented by ensuring mutual authentication among RFID readers and tagged items with low overhead on devices. Wang et al. (2011) have proposed a cyber-physical enhanced secured WSN that integrates cloud computing for healthcare application and also u-life care architecture. The system monitors and performs decision making too. The three fundamental parts of this architecture are communication, computation and resource scheduling and management. The protection against attacks is implemented by a combination of a source sensor node with an encrypted random number. The focus of the security core is on enhancing WSNs and integrating them into cloud computing.

Three critical security requirements, namely, authentication, message integrity and authorization, are required for enhancing security for a smart grid application. A lightweight two-step mutual authentication scheme at different hierarchical networks for distributed smart meters has been proposed by Fouda et al. (2011). Diffie–Hellman exchange protocol has been used for shared key session exchange. Messages communicated among smart meters have been authenticated using hash-based authentication code and shared a session key.

There is need for a multifactor authentication for CPS devices, a new hardware-based security technique for CPS which is specifically for devices with limited computing power. Device access restriction is achieved using the Physical Random Function also known as the Physically Unclonable Function (PUF) with assigned keys. PUF provides a unique value depending on the physical properties of the hardware of the used device. This mechanism creates a zero-knowledge identity proof which is used as a unique identifier to certain devices. PUF is implemented in the hardware, such as the use of static random access memory (SRAM), for explicit identification of the devices. This technique can also be utilized for location base access control and encryption. The advantage of using PUF is that it produces a unique value which can be used for confirming the unique identity of CPS devices, thereby ensuring the authenticity and integrity of the associated device. Unique cryptographic keys can also be created by PUF. The storage of cryptographic keys is secured using this technology. Therefore, it would be difficult for an adversary to obtain these keys as they will be bound to the hardware.

IDS techniques can also be included as another approach for enhancing security. These techniques are developed for discovering adversaries in the transmission layer which can timely monitor node behavior to identify any suspicious behaviors (Zhao and Lina, 2013).

## 1.6.3 Multilayer Solutions for CPS

The solutions listed in the previous subsection deal with a single measure which might not be enough in solving security issues, as the solution has to be from a multi-measure perspective. Implementing a robust security solution at the sensor level of a system with a weak application layer can fulfill security needs in one layer whereas will not satisfy the required security objectives. Therefore, it is necessary to build cross-domain security solutions with cooperation between the three layers of the system. The focus of some researchers has been on developing security solutions as a framework for all CPS layers, together because the security for CPS will not be entirely accomplished through independent implementation of a single solution in each layer. However, the complexity of any solution developed increases as a result of variations in the requirements between the layers. Thus, the focus must be on analyzing the limitations of used devices and developing an alternative mechanism.

Combined public-key (CPK) utilized for the development of an off-line authentication mechanism has been presented by Zhang et al. (2011). The main objective of this mechanism is to solve the security issues that are associated with cross-domain authentication. The proposed security architecture provides security preservation for tag privacy, sensor data and data transmission. This approach includes distinct identification and the authentication validity of the integrating nodes. The three layers (application, transmission and perception) are taken into consideration for building the trusted system to enhance cybersecurity. At the application layer, trusted access control is used for ensuring the process of non-repudiation, improvement to legal access and distinctive authentication of the connected devices. A trusted database is used for providing mutual authentication data access. Third-party certification is eliminated due to the presence of the proposed authentication mechanism. A CPK special communication chip is integrated with wireless-oriented or wired communication equipment in the transmission layer; thus avoiding the need for certification from a third party. Tags are embedded with ECC algorithms to provide authorized access in the perception layer and used with the CPK, which is identity-based authentication, to provide fast authentication.

Security solutions for CPS are designed considering some of the characteristics, such as feedback, real-time requirements, geographic distribution, distributed management and control has been presented by Neuman (2009). The focus of this study is on the physical control of the CPS, with techniques for protecting real-time requirements, communication channels and applications. A security framework for CPS providing a comprehensive analysis of three aspects of security objectives, security methodologies and security in specific CPS applications has been proposed by Lu et al. (2014).

## 1.6.4 Security Framework for CPS

A secure cyber-physical system model designed and proposed by Vegh and Liviu (2014) involves a combination of both cryptography and steganography. The security level is increased through hierarchical access to information by modeling the security framework. This method encompasses encrypting and hiding data while hiding the secret key in a different cover file. Data protection in CPS is enhanced by the pattern of combining security algorithms in the same system. A multi-agent idea is utilized for building the system and each agent has incomplete decentralized information to solve the tasks. This implies that each agent (user) has no chance to view the entire information of that system due to the presence of only a local view of the system. For example, tree root has full access to the data inside the system, while the rest have only a limited access. Access to information is restricted by hierarchical access. The ElGamal algorithm is used in the proposed system for securing CPS which consists of three main stages: key generation, encryption, and decryption.

The probability of performing the required actions by objects of trust has not been considered by many researchers. Saqib et al. (2015) have proposed a trust-based approach to the creation of a reliable and secure CPS with two-tier blankets, which consist of internal and external trust layers. The authors have taken into consideration the following points for ensuring secure and reliable communications in CPS: users authentication prior to joining the network, a trust relationship between different nodes of the CPS, malicious nodes that may attack the key nodes of the CPS (sensors or actuators) and reconfiguring the CPS system in an aggression situation. The idea of the proposed approach is to comprise security as an integral part of CPS architecture rather than applying it as a complementary solution. The author's proposal is of great help in enhancing the security task apart from CPS security objectives, considering a trust-oriented approach as the fifth objective.

Xie and Dong (2014) have discussed the importance of trust among users with proper permissions and access control by considering the ability of the devices to do a physical movement from one location to another. An item-level access-control framework is implemented to present a mutual trust idea for intersystem security. A key and a token are created by the owner or the manufacturer of the RFID and assigned to the device for the creation of the trust. Permission by the owner or the device itself can be changed while assigning the device to a new user. Thus, the permission of the device can be replaced without any additional overhead between the previous user and the new user. A privilege system is used for reducing the overhead of assigning keys, generated by the manufacturer of the RFID device. A CPS security framework has been presented by Lu et al. (2013) based on three-layer architecture: interpretation (perception), transmission (network) and cyber (application). The security levels are increased by utilizing multiple security mechanisms which are set in the information (cyber) field using a hierarchical network structure, whereas control domain security is treated using tolerant control or distributed estimation. This method primarily provides a security framework for CPS, depending on the analysis of a potential threat. A risk assessment operation is taken into consideration from the perspective of threats, assets (values), vulnerabilities and damage.

Wang et al. (2010) have proposed a context-aware security framework for general CPS. It is a set of environmental situations and settings for the regulation of the behavior of a user or an application's event. The proposed security framework comprises three essential security parts: sensing, cyber and control. There are parameters for the determination of the behavior of the system, situational information and environmental situations for computing the level of security of the system and improving the decision on information security. Relevant context information is integrated into multi-security measures, for example, encryption, key agreement and access control, to develop an adapted CPS security for the physical environment. Confidentiality, integrity, availability and authenticity are the major objectives of the proposed context-aware security framework. The function of CPS is divided into the following four stages: monitoring physical processes and the environment; networking, which includes data accumulation and distribution; computing for the collection and analysis of data during the monitoring phase; and an actuating stage for the implementation of a determined action in the computation stage. The aim of this technique is to make a security mechanism for CPS that can dynamically adapt to the physical environment.

Information security of CPS faces the problems of information gathering, processing and distributing nondestructively in large-scale, high-mix, collaborative autonomous network environment. Security solves the problems in the networking systems with open-interconnection and loosely coupled architecture. The aim of this architecture is to overcome the influence of attacks on system estimation and control algorithms. In the domain of cybersecurity, multiple security mechanisms are realized in depth for the same problem. Security threats are analyzed through use of traditional delay, intrusion and fault model. The utilization of tolerant control, distributed estimation and robust estimation techniques help to achieve the required level of security.

- Security architecture in the perception layer: In the perception layer, the major security issue lies with the sensor network. It is vulnerable to external cyber attacks. Hence building an intrusion detection and recovery mechanism to improve the robustness of the system is a matter for concern in this layer. A behavioral assessment of suspicious nodes has to be performed to establish a creditability model. A mutual trust mechanism between sensor nodes and external networks guarantees the availability of secure transmission of sensory information.
- Security architecture in the transmission layer: In this layer, both controlling commands and sensory data are sensitive to time. A large number of heterogeneous networks with distinct defense and performance capabilities are targeted for cyberattacks. Thus, special security protocols for improving the network specificity are in great demand. There are two sublayers, namely, point-to-point and end-to-end in the security layer. The data is securely transmitted during the hop transmission by the point-to-point security layer. The security mechanisms at this sublayer include mutual authentication between nodes, hop encryption and across-network certification. End-to-end confidentiality and the network are protected by the end-to-end security sublayer. Other security mechanisms include end-to-end authentication and key agreement, key management and cryptographic algorithm selection, identification and prevention of DoS and DDoS attacks. Special security mechanism has been assigned on the basis of the communication among nodes for sending unicast, broadcast and multicast messages.
- Security architecture in the application layer: The main objective of this layer is to provide targeted security services according to the requirement of the user. CPS applications which are known for many varieties have different security requirements. Even for the same security service, the definition is different from the user perspectives. The challenges in this layer include hierarchical access to the sensory data and ensuring privacy during the authentication process.

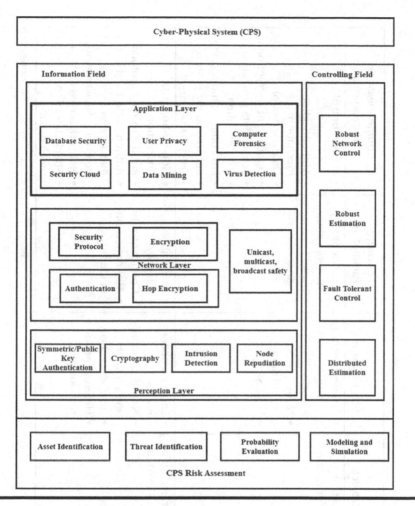

**Figure 1.5   Secure framework for CPS.**

Figure 1.5 shows the illustration of implementing a secure framework in CPS by deploying security mechanisms in each layer of CPS.

Table 1.8 shows the security requirements for each CPS layer as well as the security techniques that require utilization while designing security solutions (Lu et al., 2015; Suo et al., 2012; Zhao and Lina, 2013). Thus, building a multi-layered approach for CPS is an efficient approach for ensuring the security of CPS. This is due to the fact that the security of the system is considered at the initial stage of design for each layer.

**Table 1.8  Summary of CPS Security**

| CPS Layer | Components | Objective | Security Issues | Security Parameters | Countermeasures Mechanisms |
|---|---|---|---|---|---|
| Perception layer | • RFID tag and readers<br>• WSN<br>• Smart card<br>• GPS | Collecting information | • Terminal security<br>• Sensor network security<br>• Node repudiation<br>• Privacy | • Confidentiality<br>• Authentication<br>• Trust management | • Access control<br>• Certification<br>• Authentication mechanism<br>• Data encryption<br>• Lightweight encryption<br>• Sensor data protection<br>• Key agreement<br>• Environment monitoring<br>• Secure routing protocol<br>• Trust management |
| Transmission layer | • Wireless networks<br>• Wired Networks<br>• Computers<br>• Components | • Transmitting information | • Large number of nodes<br>• Network routing and security<br>• Internet security<br>• Heterogeneous technology | • Availability<br>• Integrity<br>• Confidentiality<br>• Authentication of identity | • Hop-by-hop data encryption<br>• Robust routing protocol<br>• Authentication and key management across heterogeneous network<br>• Network access control<br>• Attack detection mechanism |

*(Continued)*

**Table 1.8 (Continued)  Summary of CPS Security**

| CPS Layer | Components | Objective | Security Issues | Security Parameters | Countermeasures Mechanisms |
|---|---|---|---|---|---|
| Application layer | • Intelligent devices | • Analyzing information<br>• Control decision making | • Processing information<br>• Access control problem<br>• Interception of information<br>• Safety<br>• Privacy | • Privacy<br>• Authentication and key agreement<br>• Cloud security | • End-to-end encryption<br>• P2P<br>• Intrusion detection<br>• Trust management<br>• Authorization and authentication of user |

## 1.7 Conclusion and Future Research Direction

CPSs are expected to have a major impact on the real world for the development of current and future engineering systems. Therefore, security and privacy in CPS must be taken seriously considering the continuous exponential growth of the population of the interconnected devices. This chapter provides the background of CPS with regard to the security characteristics, threats and defensive strategies. The security studies in various domains of CPS are discussed along with their preventive techniques. A summary of cyberattack consequences estimation, CPS attacks modeling, detection of CPS attacks and development of security architecture is tabulated. The future research direction of each article is also outlined in this manner. CPS layered architecture is discussed along with attacks and solutions for each layer. A secure framework for CPS is illustrated for ensuring the security and privacy in CPS. The authors of the chapter conclude with the following directions for future research:

- Methods to be developed for authentication of CPS components. The security of CPS can be increased by the inclusion of authenticated components, thereby creating a secure channel between the sensors and the controllers (Nourian and Stuart, 2015).
- Metrics to be developed for the evaluation of the level of trust in CPS components. The integrity of CPS against diverse attacks requires performance on the basis of a certain level of trust. Data from various sensors is utilized by the CPS for full information. The user may receive false information when there is a conflict between reliable and faulty sensors (Tang et al., 2013).
- Data privacy may be compromised due to unauthorized access to confidential data. Thus, the development of methods ensures security to private data. Machine learning algorithms utilize intelligent data analysis for the identification of malicious users accessing the private data. One of the solutions is security through transparency (Ouaddah et al., 2017).
- Analyzing the major CPS problems that arise due to cyber and physical threats and development of a fault-tolerant architecture to ensure a high level of security and cost-effectiveness.
- Increasing the survivability of CPS by the development of countermeasures. Development of a defensive mechanism to improve the reliability and flexibility of CPS.
- Development of smart security protocols which incorporate the self-controlling CPS architecture and integratation into the state-of-the-art devices are the high-priority activities. Special attention is required while providing built-in security and privacy from components to the CPS.

# References

Adhikari, Uttam. *Event and Intrusion Detection Systems for Cyber-Physical Power Systems.* Mississippi State University, 2015.

Altawy, Riham, and Amr M. Youssef. "Security, privacy, and safety aspects of civilian drones: A survey." *ACM Transactions on Cyber-Physical Systems* 1, no. 2 (2017): 7.

Ashok, Aditya, Adam Hahn, and Manimaran Govindarasu. "Cyber-physical security of wide-area monitoring, protection and control in a smart grid environment." *Journal of Advanced Research* 5, no. 4 (2014): 481–489.

Avizienis, Algirdas, Jean-Claude Laprie, Brian Randell, and Carl Landwehr. "Basic concepts and taxonomy of dependable and secure computing." *IEEE Transactions on Dependable and Secure Computing* 1, no. 1 (2004): 11–33.

Baheti, Radhakisan, and Helen Gill. "Cyber-physical systems." *The Impact of Control Technology* 12, no. 1 (2011): 161–166.

Barnum, Sean, Shankar Sastry, and John A. Stankovic. "Roundtable: Reliability of embedded and cyber-physical systems." *IEEE Security & Privacy* 8, no. 5 (2010): 27–32.

Bhattacharya, Rina. "A comparative study of physical attacks on wireless sensor networks." *International Journal of Research in Engineering and Technology* 2, no. 1 (2013): 72–74.

Camara, Carmen, Pedro Peris-Lopez, and Juan E. Tapiador. "Security and privacy issues in implantable medical devices: A comprehensive survey." *Journal of Biomedical Informatics* 55 (2015): 272–289.

Cardenas, Alvaro, Saurabh Amin, Bruno Sinopoli, Annarita Giani, Adrian Perrig, and Shankar Sastry. "Challenges for securing cyber physical systems." In *Workshop on Future Directions in Cyber-Physical Systems Security*, vol. 5 no. 1 (2009).

Chalkias, Konstantinos, Foteini Baldimtsi, Dimitrios Hristu-Varsakelis, and George Stephanides. "Two types of key-compromise impersonation attacks against one-pass key establishment protocols." In *International Conference on E-Business and Telecommunications*, pp. 227–238. Springer, Berlin, Germany, 2007.

Chen, Pin-Yu, Shin-Ming Cheng, and Kwang-Cheng Chen. "Information fusion to defend intentional attack in Internet of Things." *IEEE Internet of Things Journal* 1, no. 4 (2014): 337–348.

Collins, Sean, and Stephen McCombie. "Stuxnet: The emergence of a new cyber weapon and its implications." *Journal of Policing, Intelligence and Counter Terrorism* 7, no. 1 (2012): 80–91.

Cómbita, Luis F., Jairo Giraldo, Alvaro A. Cárdenas, and Nicanor Quijano. "Response and reconfiguration of cyber-physical control systems: A survey." In *Automatic Control (CCAC), 2015 IEEE 2nd Colombian Conference on*, pp. 1–6. IEEE, Piscataway, NJ, 2015.

Derler, Patricia, Edward A. Lee, and Alberto Sangiovanni Vincentelli. "Modeling cyber–physical systems." *Proceedings of the IEEE* 100, no. 1 (2012): 13–28.

Djouadi, Seddik M., Alexander M. Melin, Erik M. Ferragut, Jason A. Laska, Jin Dong, and Anis Drira. "Finite energy and bounded actuator attacks on cyber-physical systems." In *Control Conference (ECC), 2015 European*, pp. 3659–3664. IEEE, 2015.

Eisenbarth, Thomas, Sandeep Kumar, Christof Paar, Axel Poschmann, and Leif Uhsadel. "A survey of lightweight-cryptography implementations." *IEEE Design & Test of Computers* 6 (2007): 522–533.

Finogeev, Alexey G., and Anton A. Finogeev. "Information attacks and security in wireless sensor networks of industrial SCADA systems." *Journal of Industrial Information Integration* 5, (2017): 6–16.

Fouda, Mostafa M., Zubair Md Fadlullah, Nei Kato, Rongxing Lu, and Xuemin Sherman Shen. "A lightweight message authentication scheme for smart grid communications." *IEEE Transactions on Smart Grid* 2, no. 4 (2011): 675–685.

Friedberg, Ivo, Florian Skopik, Giuseppe Settanni, and Roman Fiedler. "Combating advanced persistent threats: From network event correlation to incident detection." *Computers & Security* 48 (2015): 35–57.

Gaddam, Nishanth, G. Sudha Anil Kumar, and Arun K. Somani. "Securing physical processes against cyber attacks in cyber-physical systems." In *Proceeding National Workshop Research High-Confidence Transportation Cyber-Physical Systems, Automotive Aviation Rail*, pp. 1–3. 2008.

Genge, Béla, Christos Siaterlis, and Marc Hohenadel. "Impact of network infrastructure parameters to the effectiveness of cyber attacks against industrial control systems." *International Journal of Computers Communications & Control* 7, no. 4 (2012): 674–687.

Genge, Béla, István Kiss, and Piroska Haller. "A system dynamics approach for assessing the impact of cyber attacks on critical infrastructures." *International Journal of Critical Infrastructure Protection* 10 (2015): 3–17.

Giani, Annarita, Eilyan Bitar, Manuel Garcia, Miles McQueen, Pramod Khargonekar, and Kameshwar Poolla. "Smart grid data integrity attacks." *IEEE Transactions on Smart Grid* 4, no. 3 (2013): 1244–1253.

Gou, Quandeng, Lianshan Yan, Yihe Liu, and Yao Li. "Construction and strategies in IoT security system." In *Green Computing and Communications (GreenCom), 2013 IEEE and Internet of Things (iThings/CPSCom), IEEE International Conference on and IEEE Cyber, Physical and Social Computing*, pp. 1129–1132. IEEE, Piscataway, NJ, 2013.

Hahn, Adam, Roshan K. Thomas, Ivan Lozano, and Alvaro Cardenas. "A multi-layered and kill-chain based security analysis framework for cyber-physical systems." *International Journal of Critical Infrastructure Protection* 11 (2015): 39–50.

Håkansson, Anne, Ronald Hartung, and Esmiralda Moradian. "Reasoning strategies in smart cyber-physical systems." *Procedia Computer Science* 60 (2015): 1575–1584.

Han, Song, Xie, Miao, Chen, Hsiao-Hwa and Ling, Yun, 2014. Intrusion detection in cyber-physical systems: Techniques and challenges. *IEEE Systems Journal* 8 (4), pp. 1052–1062.

He, Haibo, and Jun Yan. "Cyber-physical attacks and defences in the smart grid: A survey." *IET Cyber-Physical Systems: Theory & Applications* 1, no. 1 (2016): 13–27.

Huang, Yu-Lun, Alvaro A. Cárdenas, Saurabh Amin, Zong-Syun Lin, Hsin-Yi Tsai, and Shankar Sastry. "Understanding the physical and economic consequences of attacks on control systems." *International Journal of Critical Infrastructure Protection* 2, no. 3 (2009): 73–83.

Kao, Jung-Chun, and Radu Marculescu. "Eavesdropping minimization via transmission power control in ad-hoc wireless networks." In *Sensor and Ad Hoc Communications and Networks, 2006. SECON'06. 2006 3rd Annual IEEE Communications Society on*, vol. 2, pp. 707–714. IEEE, 2006.

Karnouskos, Stamatis. "Stuxnet worm impact on industrial cyber-physical system security." In *IECON 2011-37th Annual Conference on IEEE Industrial Electronics Society*, pp. 4490–4494. IEEE, 2011.

Khalil, Yehia. F. "A novel probabilistically timed dynamic model for physical security attack scenarios on critical infrastructures." *Process Safety and Environmental Protection* 102 (2016): 473–484.

Kim, Kyoung-Dae, and Panganamala R. Kumar. "Cyber–physical systems: A perspective at the centennial." *Proceedings of the IEEE* 100, no. Special Centennial Issue (2012): 1287–1308.

Krotofil, Marina, Alvaro Cardenas, Jason Larsen, and Dieter Gollmann. "Vulnerabilities of cyber-physical systems to stale data—Determining the optimal time to launch attacks." *International Journal of Critical Infrastructure Protection* 7, no. 4 (2014): 213–232.

Krotofil, Marina, Jason Larsen, and Dieter Gollmann. "The process matters: Ensuring data veracity in cyber-physical systems." In *Proceedings of the 10th ACM Symposium on Information, Computer and Communications Security*, pp. 133–144. ACM, New York, 2015.

Li, Weize, Lun Xie, Zulan Deng, and Zhiliang Wang. "False sequential logic attack on SCADA system and its physical impact analysis." *Computers & Security* 58 (2016): 149–159.

Li, Yuzhe, Ling Shi, Peng Cheng, Jiming Chen, and Daniel E. Quevedo. "Jamming attack on cyber-physical systems: A game-theoretic approach." In *Cyber Technology in Automation, Control and Intelligent Systems (CYBER), 2013 IEEE 3rd Annual International Conference on*, pp. 252–257. IEEE, 2013.

Liao, Weixian. "*Security and Privacy of Cyber-Physical Systems*." PhD dissertation, Case Western Reserve University, 2018.

Liu, Ting, Yanan Sun, Yang Liu, Yuhong Gui, Yucheng Zhao, Dai Wang, and Chao Shen. "Abnormal traffic-indexed state estimation: A cyber–physical fusion approach for smart grid attack detection." *Future Generation Computer Systems* 49 (2015): 94–103.

Lu, Tianbo, Bing Xu, Xiaobo Guo, Lingling Zhao, and Feng Xie. "A new multilevel framework for cyber-physical system security." In *First International Workshop on the Swarm at the Edge of the Cloud*. 2013.

Lu, Tianbo, Jiaxi Lin, Lingling Zhao, Yang Li, and Yong Peng. "A Security Architecture in Cyber-Physical Systems: Security Theories, Analysis, Simulation and Application Fields." *International Journal of Security and Its Applications* 9, no. 7 (2015): 1–16.

Lu, Tianbo, Jiaxi Lin, Lingling Zhao, Yang Li, and Yong Peng. "An Analysis of Cyber Physical System Security Theories." In *Security Technology (SecTech), 2014 7th International Conference on*, pp. 19–21. IEEE, 2014.

Lyn, Kevin G. "Classification of and resilience to cyber-attacks on cyber-physical systems." PhD diss., Georgia Institute of Technology, 2015.

Madden, Jason, Bruce McMillin, and Anik Sinha. "Environmental obfuscation of a cyber physical system-vehicle example." In *Computer Software and Applications Conference Workshops (COMPSACW), 2010 IEEE 34th Annual*, pp. 176–181. IEEE, 2010.

Maheshwari, Piyush. "Security issues of cyber physical system: A review." *International Journal Computer Application* (2016): 7–11.

Mahmoud, Rwan, Tasneem Yousuf, Fadi Aloul, and Imran Zualkernan. "Internet of things (IoT) security: Current status, challenges and prospective measures." In *Internet Technology and Secured Transactions (ICITST), 2015 10th International Conference for*, pp. 336–341. IEEE, 2015.

Martins, Goncalo, Sajal Bhatia, Xenofon Koutsoukos, Keith Stouffer, CheeYee Tang, and Richard Candell. "Towards a systematic threat modeling approach for cyber-physical systems." In *Resilience Week (RWS)* (2015) pp. 1–6. IEEE, 2015.

Mavani, Monali, and Krishna Asawa. "Modeling and analyses of IP spoofing attack in 6LoWPAN network." *Computers & Security* 70 (2017): 95–110.

McLaughlin, Stephen, Charalambos Konstantinou, Xueyang Wang, Lucas Davi, Ahmad-Reza Sadeghi, Michail Maniatakos, and Ramesh Karri. "The cybersecurity landscape in industrial control systems." *Proceedings of the IEEE* 104, no. 5 (2016): 1039–1057.

Mitchell, Robert, and Ing-Ray Chen. "A survey of intrusion detection techniques for cyber-physical systems." *ACM Computing Surveys (CSUR)* 46, no. 4 (2014): 55.

Mitchell, Robert, and Ray Chen. "Modeling and analysis of attacks and counter defense mechanisms for cyber physical systems." *IEEE Transactions on Reliability* 65, no. 1 (2016): 350–358.

Mo, Yilin, Rohan Chabukswar, and Bruno Sinopoli. "Detecting integrity attacks on SCADA systems." *IEEE Transactions on Control Systems Technology* 22, no. 4 (2014): 1396–1407.

Mo, Yilin, and Bruno Sinopoli. "Secure control against replay attacks. In *2009 47th Annual Allerton Conference on Communication, Control, and Computing (Allerton)*, pp. 911–918. IEEE, 2009.

Moosavi, Sanaz Rahimi, Tuan Nguyen Gia, Amir-Mohammad Rahmani, Ethiopia Nigussie, Seppo Virtanen, Jouni Isoaho, and Hannu Tenhunen. "SEA: A secure and efficient authentication and authorization architecture for IoT-based healthcare using smart gateways." *Procedia Computer Science* 52 (2015): 452–459.

Neuman, Clifford. "Challenges in security for cyber-physical systems." In *DHS Workshop on Future Directions in Cyber-Physical Systems Security*, pp. 22–24. 2009.

Nguyen, Phu H., Shaukat Ali, and Tao Yue. "Model-based security engineering for cyber-physical systems: A systematic mapping study." *Information and Software Technology* 83 (2017): 116–135.

Nourian, Arash, and Stuart Madnick. "A systems theoretic approach to the security threats in cyber physical systems applied to stuxnet." *IEEE Transactions on Dependable and Secure Computing* (2015).

Ntalampiras, Stavros. "Automatic identification of integrity attacks in cyber-physical systems." *Expert Systems with Applications* 58 (2016): 164–173.

Ouaddah, Aafaf, Hajar Mousannif, Anas Abou Elkalam, and Abdellah Ait Ouahman. "Access control in the internet of things: Big challenges and new opportunities." *Computer Networks* 112 (2017): 237–262.

Peng, Yong, Tianbo Lu, Jingli Liu, Yang Gao, Xiaobo Guo, and Feng Xie. "Cyber-physical system risk assessment." In *Intelligent Information Hiding and Multimedia Signal Processing, 2013 Ninth International Conference on*, pp. 442–447. IEEE, 2013.

Pham, Nam, Tarek Abdelzaher, and Suman Nath. "On bounding data stream privacy in distributed cyber-physical systems." In *Sensor Networks, Ubiquitous, and Trustworthy Computing (SUTC), 2010 IEEE International Conference on*, pp. 221–228. IEEE, 2010.

Premnath, Sriram N., and Zygmunt J. Haas. "Security and privacy in the internet-of-things under time-and-budget-limited adversary model." *IEEE Wireless Communications Letters* 4, no. 3 (2015): 277–280.

Raza, Shahid. "Lightweight security solutions for the internet of things." PhD diss., Mälardalen University, Västerås, Sweden, 2013.

Rushanan, Michael, Aviel D. Rubin, Denis Foo Kune, and Colleen M. Swanson. "Sok: Security and privacy in implantable medical devices and body area networks." In *2014 IEEE Symposium on Security and Privacy (SP)*, pp. 524–539. IEEE, 2014.

Sadeghi, Ahmad-Reza, Christian Wachsmann, and Michael Waidner. "Security and privacy challenges in industrial internet of things." In *Design Automation Conference (DAC), 2015 52nd ACM/EDAC/IEEE*, pp. 1–6. IEEE, 2015.

Sakiz, Fatih, and Sevil Sen. "A survey of attacks and detection mechanisms on intelligent transportation systems: VANETs and IoV." *Ad Hoc Networks* 61 (2017): 33–50.

Saltzer, Jerome H., and Michael D. Schroeder. "The protection of information in computer systems." *Proceedings of the IEEE* 63, no. 9 (1975): 1278–1308.

Saltzman, Roi, and Adi Sharabani. "Active man in the middle attacks." *OWASP AU* (2009).

Saqib, A., Raja Waseem Anwar, Omar Khadeer Hussain, Mudassar Ahmad, Md Asri Ngadi, Mohd Murtadha Mohamad, Zohair Malki et al., "Cyber security for cyber physical systems: A trust-based approach." *Journal of Theoretical Applied Information Technology* 71, no. 2 (2015): 144–152.

Shafi, Qaisar. "Cyber physical systems security: A brief survey." In *Computational Science and Its Applications (ICCSA), 2012 12th International Conference on*, pp. 146–150. IEEE, 2012.

Sicari, Sabrina, Alessandra Rizzardi, Daniele Miorandi, Cinzia Cappiello, and Alberto Coen-Porisini. "A secure and quality-aware prototypical architecture for the Internet of Things." *Information Systems* 58 (2016): 43–55.

Singhal, Anoop. *Data Warehousing and Data Mining Techniques for Cyber Security*. Vol. 31. Springer Science & Business Media, London, UK, 2007.

Srivastava, Anurag, Thomas Morris, Timothy Ernster, Ceeman Vellaithurai, Shengyi Pan, and Uttam Adhikari. "Modeling cyber-physical vulnerability of the smart grid with incomplete information." *IEEE Transactions on Smart Grid* 4, no. 1 (2013): 235–244.

Stallings, William. *Cryptography and Network Security: Principles and Practice*. Pearson Education, Upper Saddle River, NJ, 2017.

Suo, Hui, Jiafu Wan, Caifeng Zou, and Jianqi Liu. "Security in the internet of things: A review." In *Computer Science and Electronics Engineering (ICCSEE), 2012 International Conference on*, vol. 3, pp. 648–651. IEEE, 2012.

Sztipanovits, Janos, Susan Ying, I. Cohen, D. Corman, J. Davis, H. Khurana, P. J. Mosterman, V. Prasad, and L. Stormo. "Strategic R&D opportunities for 21st century cyber-physical systems." Technical report, *Technical Report for Steering Committee for Foundation in Innovation for Cyber-Physical Systems* (2012).

Tang, Lu-An, Xiao Yu, Sangkyum Kim, Quanquan Gu, Jiawei Han, Alice Leung, and Thomas La Porta. "Trustworthiness analysis of sensor data in cyber-physical systems." *Journal of Computer and System Sciences* 79, no. 3 (2013): 383–401.

Trappe, Wade, Richard Howard, and Robert S. Moore. "Low-energy security: Limits and opportunities in the internet of things." *IEEE Security & Privacy* 13, no. 1 (2015): 14–21.

Urbina, David I., David I. Urbina, Jairo Giraldo, Alvaro A. Cardenas, Junia Valente, Mustafa Faisal, Nils Ole Tippenhauer, Justin Ruths, Richard Candell, and Henrik Sandberg. *Survey and New Directions for Physics-Based Attack Detection in Control Systems*. US Department of Commerce, National Institute of Standards and Technology, 2016.

Van der Heijden, Rens W., Stefan Dietzel, Tim Leinmüller, and Frank Kargl. "Survey on misbehavior detection in cooperative intelligent transportation systems." *Arxiv preprint arXiv*:1610.06810 (2016).

Vegh, Laura, and Liviu Miclea. "Enhancing security in cyber-physical systems through cryptographic and steganographic techniques." In *2014 IEEE International Conference on Automation, Quality and Testing, Robotics (AQTR)*, pp. 1–6. IEEE, 2014.

Venkitasubramaniam, Parv, Jiyun Yao, and Parth Pradhan. "Information-theoretic security in stochastic control systems." *Proceedings of the IEEE* 103, no. 10 (2015): 1914–1931.

Vincent, Hannah, Lee Wells, Pablo Tarazaga, and Jaime Camelio. "Trojan detection and side-channel analyses for cyber-security in cyber-physical manufacturing systems." *Proceeding Manufacturing* 1 (2015): 77–85.

Wang, Eric Ke, Yunming Ye, Xiaofei Xu, Siu-Ming Yiu, Lucas Chi Kwong Hui, and Kam-Pui Chow. "Security issues and challenges for cyber physical system." In *Green Computing and Communications (GreenCom), 2010 IEEE/ACM Int'l Conference on & Int'l Conference on Cyber, Physical and Social Computing (CPSCom)*, pp. 733–738. IEEE, 2010.

Wang, Jin, Hassan Abid, Sungyoung Lee, Lei Shu, and Feng Xia. "A secured health care application architecture for cyber-physical systems." *arXiv preprint arXiv:1201.0213* (2011).

Wasicek, Armin, Patricia Derler, and Edward A. Lee. "Aspect-oriented modeling of attacks in automotive cyber-physical systems." In *Design Automation Conference (DAC), 2014 51st ACM/EDAC/IEEE*, pp. 1–6. IEEE, 2014.

Work, Daniel, Alexandre Bayen, and Quinn Jacobson. "Automotive cyber physical systems in the context of human mobility." In *National Workshop on High-Confidence Automotive Cyber-Physical Systems*, pp. 3–4. 2008.

Wu, Wenbo, Rui Kang, and Zi Li. "Risk assessment method for cyber security of cyber physical systems." In *Reliability Systems Engineering (ICRSE), 2015 First International Conference on*, pp. 1–5. IEEE, 2015.

Xie, Feng, Tianbo Lu, Xiaobo Guo, Jingli Liu, Yong Peng, and Yang Gao. "Security analysis on cyber-physical system using attack tree." In *Intelligent Information Hiding and Multimedia Signal Processing, 2013 Ninth International Conference on*, pp. 429–432. IEEE, 2013.

Xie, Yi, and Dong Wang. "An item-level access control framework for inter-system security in the internet of things." In *Applied Mechanics and Materials*, vol. 548, pp. 1430–1432. Trans Tech Publications, 2014.

Xinlan, Zhang, Huang Zhifang, Wei Guangfu, and Zhang Xin. "Information security risk assessment methodology research: Group decision making and analytic hierarchy process." In *Software Engineering (WCSE), 2010 Second World Congress on*, vol. 2, pp. 157–160. IEEE, 2010.

Yampolskiy, Mark, Péter Horváth, Xenofon D. Koutsoukos, Yuan Xue, and Janos Sztipanovits. "A language for describing attacks on cyber-physical systems." *International Journal of Critical Infrastructure Protection* 8 (2015): 40–52.

Yang, Qingyu, Liguo Chang, and Wei Yu. "On false data injection attacks against Kalman filtering in power system dynamic state estimation." *Security and Communication Networks* 9, no. 9 (2016): 833–849.

Yoo, Hyunguk, and Taeshik Eisenbarth. "Challenges and research directions for heterogeneous cyber–physical system based on IEC 61850: Vulnerabilities, security requirements, and security architecture." *Future Generation Computer Systems* 61 (2016): 128–136.

Yuan, Yuan, Quanyan Zhu, Fuchun Sun, Qinyi Wang, and Tamer Başar. "Resilient control of cyber-physical systems against denial-of-service attacks." In *Resilient Control Systems (ISRCS), 2013 6th International Symposium on*, pp. 54–59. IEEE, 2013.

Zhang, Bing, Xin-Xin Ma, and Zhi-Guang Qin. "Security architecture on the trusting internet of things." *Journal of Electronic Science and Technology* 9, no. 4 (2011): 364–367.

Zhao, Kai, and Lina Ge. "A survey on the internet of things security." *Computational Intelligence and Security (CIS), 2013 9th International Conference on*. IEEE, 2013.

## Chapter 2

# Vulnerability Analysis for Cyber-Physical Systems

D. Sumathi and M. Roopa Chandrika

## Contents

## 2.1 Introduction to Cyber-Physical Systems

An amalgamation of components of cyber and physical systems that integrates various computational resources is termed cyber-physical systems (CPSs). The main intention of CPS is to provide a good support in real life processes and in addition, the Internet of Things (IoT) objects are controlled by the physical devices that are used to sense the environment and modify it. It could also be stated as interdisciplinary systems in which feedback control is obtained from the communication, computation and control technologies, which are combined together. The real-time capability and accuracy of the data that has been collected from various distributed devices which are in CPS system are maintained with the help of data acquisition modules. Data that has been collected this way is transmitted to the information processing layer as per the service requests made. Certain tasks such as statistical signal processing, data security processing, and feedback controlling are performed with the data collected. The collaboration of physical and computational components, especially the use of CPSs, is directed toward the IoT implementations. Sensors deployed in CSPs are connected to all distributed intelligence in the environment so that the knowledge about the environment has been captured and therefore this facilitates more precise actuation (Ledwaba & Venter, 2017). The structure of a CPS system is shown in Figure 2.1 below and the way it operates in Figure 2.2.

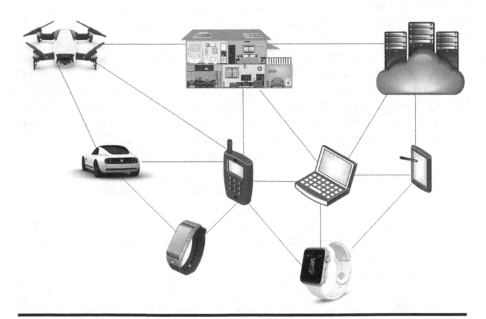

**Figure 2.1  Cyber-physical Systems adopted by NIST.**

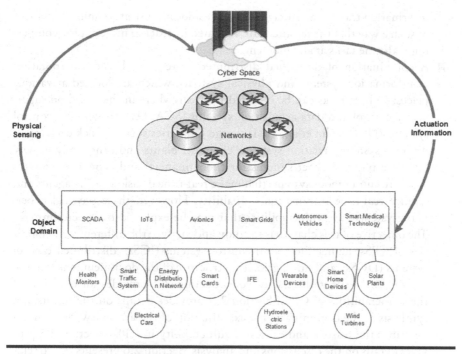

**Figure 2.2  Operational structure of a CPS.**

The important features of CPS:

- The information system is the heart of the CPS: The information that has been collected from various physical devices is transformed into rules of software systems.
- The most important component in the CPS is physical systems: A deep knowledge about the sensors and their workings is required as the measurement data that has been collected from various physical devices is used for further analysis and process. Physical systems that have been deployed in this CPS must meet the requirements such as time, temporal and spatial, in order to carry out several tasks to obtain a high degree of automation.
- The information that has been collected is subjected to various properties such as security and privacy, authentication, envisaging resources and capability. Certain issues such as invasion, tampering, delay and other attacks might occur as the network and physical system are kept open. Hence, CPS must be built in such a way that all these issues must be resolved. Information that has been collected from various devices must be stored securely. For that efficient encryption and decryption techniques could be applied during the

information transit and storage. Resource allocation strategy must be framed in such a way that the resources are allocated in order to meet out the competing real-time tasks at any moment.

■ A combination of diversified distributed systems with the incorporation of information systems and physical systems, which are located at various places: Cyberattacks can be initiated from anywhere in the world on supervisory control and data acquisition systems (SCADA) and network control systems (NCS) (Zhu et al., 2011). Large industries such as stock exchanges, military systems, aviation systems, chemical plants and others rely heavily on these physical cyber systems for their financial and economic growth. They are autonomous systems that make real-time decisions using agents and require real-time information availability. However, these physical cybernetic systems have vulnerabilities that may be exploited by cybercriminals. The objective of this chapter is to study and analyze the different vulnerabilities of CPSs, namely industrial control systems (ICSs), distributed control systems (DCSs), and smart grid. The structure of the document is as follows. Section 2.1 deals with related work. Section 2.2 provides an overview of the vulnerabilities of CPSs. Section 2.3 provides details on the results and synthesis at the beginning of the examination of the art review. Section 2.4 describes the reasons and causes of vulnerability in CPSs. Section 2.5 provides details on the discussions and analysis. Section 2.6 presents the conclusions and Section 2.7 presents the research challenges.

## 2.2 Related Works

Humayed et al. (2017) systemized a framework which consists of three features that have been studied from various works:

■ A well-known taxonomy of threats, vulnerabilities, attacks and controls must be formed from the security perspective.
■ Cyber, physical and cyber-physical components have to be viewed.
■ Discovery of general CPS features and representative systems such as smart cars, medical devices and smart grids etc.

A model has been recommended which might be suitable for various heterogeneous applications.

Giraldo et al. (2017) classified surveys on CPS security and privacy based on the following features:

■ Attacks
■ Network security
■ Research trends

- CPS domains
- Defenses
- Security-level implementation and computational strategies

In addition to this, the authors have presented a survey on various domains like medical devices, smart grids, manufacturing, ICS and ITS.

Wan, Canedo and Faruque (2015) proposed a security-aware functional modeling method, cybersecurity functions, their related attacks, countermeasure models, system-level modeling and simulation of attacks which are beneficial for the concept design stage of CPS. This work is mainly focused on the analysis of the security problem during the early design stage. The developed model has been demonstrated for several attacks such as fuzzy attack model, replay attack model, man-in-the-middle attack model, interruption attack model, downsampling attack model and overflow attack model.

Lu et al. (2015) proposed a security framework that ensures the security of CPS and analyzes the relations of CPS security in three levels as CPS security approaches, CPS security objectives and security in specific CPS applications.

Yeboah-ofori et al. (2018) presented a study on the risks that are related to the components such as human elements, digital systems and physical systems. This is mainly to solve risk mitigation goals that are raised in business value, organizational requirements, threat agent and impact based on the review results. The author has implemented the analytical hierarchical process (AHP) to establish the relative importance of these goals that contribute to developing cybercrime and richness from CPSs.

## 2.3 Vulnerabilities in CPSs

Vulnerabilities in CPSs could be categorized as vulnerabilities in management and policy, vulnerabilities in network and vulnerabilities in the platform (Gao et al., 2013).

### 2.3.1 Vulnerabilities of Platform

When the operating system, the hardware, is not configured properly, it leads to vulnerability (Zhu et al., 2011). This could be resolved by applying various security controls, application patching and security software. Table 2.1 shows the vulnerabilities of the platform and its countermeasures.

### 2.3.2 Vulnerabilities Encountered in the Network Layer

Vulnerabilities in the network layer could be mitigated by applying security controls such as encryption, network design defense, defense in depth, physical access control on network components. A list of vulnerabilities and its countermeasures are shown in Table 2.2.

**Table 2.1    Platform Vulnerabilities and Countermeasures**

| S. No | Classification | Vulnerability | Countermeasure |
|---|---|---|---|
| 1 | Vulnerabilities due to malware | Malware protection is not installed. | Need to install protection software and it must be updated regularly. Data drive must be scanned and analyzed. Basic internet security applications must be installed. |
| | | Malware protection software is not updated. | |
| 2 | Vulnerabilities due to platform software | Buffer overflow | OS hardening could be used to mitigate the overflow. Mitigating OS permissions of software. |
| | | Authentication and access control of Inadequate configuration and programming software. | Configuration and authentication policies must be framed. |
| | | Intrusion detection and prevention software has not been installed. | It must be installed. |
| | | Using plain text | Digest values created by various security algorithms could be used. |
| | | Logs are not maintained and the event is not excavated. | Need to maintain log files for further clarification and events must be classified for excavation. |
| 3 | Platform configuration vulnerability | No backup of the important configuration. | Certain important configurations must be stored. |
| | | Password policy is not sufficient. | Proper policies have to be framed while framing passwords. |
| | | Data protection is not proper in portable devices. | Protection policies have to be framed. |
| | | No proper maintenance of security for data in portable devices. | Proper security measures must be provided for data storage. |

**Table 2.2   Vulnerabilities in the Network Layer and Its Countermeasures**

| S. No | Classification | Vulnerability | Countermeasures |
|-------|----------------|---------------|-----------------|
| 1 | Communication vulnerability | Critical monitoring and control path might not be identified during the communication. | Maintain complete map and tracking of up-to-date status of all access points must be done. |
| | | The device might be defective or authentication details do not match. | Device must be checked for its defectiveness before its connection. Authentication details must be stored properly in a secured manner. |
| 2 | Network perimeter vulnerabilities | The security perimeter is not defined. | Proper procedural or technical controls at the access points must be defined in order to establish a security perimeter. |
| | | No proper configuration of firewall or it might not exist | Firewall establishment must be in accordance with security policy and procedures. |
| 3 | Network monitoring and logging vulnerabilities | Lack of firewalls and routers log | Exporting of log files would be done by default and it would be available for editing. Once if firewalls and routers are reconfigured, log files would be updated. |
| | | ICS network with no security monitoring | Regular monitoring is required in order to detect intrusions, detections and network anomalies. |
| 4 | ICS network with no security monitoring | Network security architecture is fragile. | Wrap all security policies as a single policy. Impose protection on the host through the distributed firewalls. |

*(Continued)*

**Table 2.2 (*Continued*)  Vulnerabilities in the Network Layer and Its Countermeasures**

| S. No | Classification | Vulnerability | Countermeasures |
|---|---|---|---|
| | | Security devices are not configured properly. | The configuration of security devices is not a onetime event and regular reevaluation is needed and approval is also required. |
| | | Insufficient access control application | Settings to applications and privileges must be managed. Integration of mechanisms that organize privilege separation functionality and construction of architecture that rely on the least privilege principle should be done. |
| 5 | Network hardware vulnerability | Physical protection to the physical device is not adequate. | Minimization of risk of resource theft and destruction could be done by safeguarding the building sites. |
| | | Nonessential people can access devices and network connections. | Authentication and authorization must be given to the required people who need to access. |

## 2.3.3  Vulnerabilities in Management and Policy

Vulnerabilities could be mitigated by framing suitable security policies and this is shown in Table 2.3.

Some applications of CPS have been stated as follows:

- Smart building environments: The comfort of people could be improved; safety and security also could be improved by implanting smart devices and CPSs in an interactive manner so that energy consumption is reduced. For example, a monitoring system could facilitate the CSP in obtaining an Zero-Energy building or the extent of the damage that a building might suffer after unanticipated events could be identified and thus, structural failures could be prevented further.
- Agriculture systems: Information about the climate, ground and other data are collected by CPSs in order to maintain precise and modern agriculture. Various

**Table 2.3 Vulnerabilities in Management and Policy and Countermeasures**

| Vulnerabilities | Countermeasures |
| --- | --- |
| Inadequate ICS security policies | Establishment of security policies in a proper manner. |
| ICS equipment installation guide does not exist or is absent. | Support from the senior experts and browsing through the Internet would help. |
| Lack of safety measures to implement management mechanisms | Safety measures must be framed and awareness regarding the safety measures must be given. |
| There is no disaster recovery plan (DRP) or operational continuity. | Disaster recovery plan is required to get back the services on track rapidly and protect the infrastructure in the event of a disaster. |
| ICS security review might be lacking. | Security review must be conducted in order to assess the current state vs. security best practices. |

different resources such as soil moisture, humidity, plant health, humidity are monitored constantly so that the environment is maintained in an ideal manner.

■ Industrial Control Systems: CSPs are used to control and monitor the entire production processes and customize the production process according to the preferences of the customer. Through this, a high degree of visibility and control on supply chains could be accomplished so that the security of goods and traceability also could be improved.

■ Healthcare environment: To monitor the physical conditions of patients remotely and in real time, CSPs are implemented. In addition, through this CPS, treatments for disabled and elderly patients could be done at the proper time. Moreover, many research works have been carried out in the neuroscience field so that the therapeutic robotics and brain-machine interface were used to study human functions in detail.

■ Transportation systems: Vehicles communicate with each other in order to share real-time information about the location such as traffic, so that accidents could be prevented and therefore safety could be improved. In addition, time could be saved ultimately.

■ Smart grid: This is defined as electric networks that utilize control and communication technologies so that reliable and secure energy, frequent monitoring, extension in operation efficiency for distributors and generators are supplied, and proposing suitable choices for the consumers to select based on their preferences (Mo et al., 2012; Yu & Xue, 2016).

## 2.3.4 Security

The predominant issue in CPS is security. It mainly lies in the heterogeneity of the building blocks. CSP comprises different hardware components like actuators, sensors, embedded systems and software products that are used for monitoring and controlling the systems. The components used in the systems could be considered as a factor of a CSP attack. Due to the complex nature of CPSs and the combination of various CSP components, various difficulties exist for the security and privacy protection of CPSs. Due to the complex nature of cyber-physical interactions, it becomes multifaceted to analyze and this leads to the new security issue. Moreover, it also becomes complicated to determine, track the attacks that initiate from, move between and target various CPS components.

# 2.4 Reasons/Causes for Vulnerability in CPS

## 2.4.1 Increased Connectivity

Due to the increase in a number of devices that are deployed in various fields, connectivity gets increased which are prone to attacks from the external sources. Devices that are connected in the network could be easily attacked.

Heterogeneity: CPS comprises heterogeneous components that are deployed and integrated to build a CPS application. This kind of integration invites vulnerabilities from the external world. The internal implementation details of the heterogeneous components are not visible and hence they might generate unexpected performance when they are deployed in the network.

## 2.4.2 Vulnerabilities in Medical Cyber-Physical Systems

Due to the increase in the interconnection of medical devices with other clinical systems, medical devices are vulnerable to security infringe. It is observed that the medical devices that are connected in the network might create a direct impact on patients' data. From the report generated by the SANS Institute, it is identified that the major intention of cyberattackers is to hack the data recorded from the healthcare, and 94% of healthcare organizations have been attacked by the cyberattackers.

## 2.4.3 Vulnerabilities in Industrial Control Systems (ICSs)

The term *ICS* denotes an extensive group of control systems that comprises:

- Automation System (AS)
- Energy Management System (EMS)

- Process Control System (PCS)
- Distributed Control System (DCS)
- Safety Instrumented System (SIS)
- Supervisory Control and Data Acquisition system (SCADA).

As the automation gets increased nowadays, the cyberattacks on critical infrastructure also attract the attackers to disturb and distract the industrial activities for political, social and monetary gain. Several vulnerabilities that exist in the ICS are as follows:

The security policy for the ICS might not be framed in a proper manner and there is no particular procedure in maintaining the documents.

There must be an awareness program and safety training measures for the company staff.

- Security guidelines that are used to implement the ICS are not sufficient or adequate (Stouffer et al., 2011).
- ICS technology audit must be conducted regularly.
- Contingency planning during the disaster might be lacking.
- Due to the high interconnection and inter-reliance among the systems, chances of vulnerability are more. Failure of one device might lead to the failure of another device which results in the malfunction of other devices. This, in turn, disrupts the network.
- Services that are not needed might run default.
- Processors and network services are made to run with the help of OS platforms. Due to the ports that are kept open during this process, chances for vulnerability exist and therefore there is a possibility of attack such as buffer overflow.
- Security updates and frequent testing of software is necessary for certain attributes such as availability and critical response time. These types of actions must be given priority and shall be allowed only in occasional scenarios.
- Critical timings might not allow the ICS components to endure with the security software.
- Delay in response and threats to system stability might occur due to the security software that has been deployed in control system components of ICS.
- In spite of regular updates of antivirus software, new malware threats might exist due to the current innovations in technological developments.
- Installation of standard security tools in the control system might lead to prolonged response time. Hence, operational errors might exist and there is a chance for system failure. Therefore, installation of software requires prior approval from the vendor.
- In addition, any installation of new techniques into the control system must be in such a way that the performance of the ICS must not be affected.

■ Due to the occurrence of misconfiguration, flaws and poor maintenance of platforms, which includes operating systems, hardware vulnerabilities occur. It could be reduced by ensuring security controls such as access control policy adherence, security software, OS and application patching.

Several potential platform vulnerabilities that exist are described as follows in Table 2.4.

**Table 2.4  ICS Systems and Vulnerabilities**

| S. No | Platform Vulnerabilities | Countermeasures |
|---|---|---|
| 1 | OS and application patches might be developed earlier and they might not be maintained properly. | Regular testing methodologies must be followed in such a way to monitor it and frequent updates are required. |
| 2 | Configurations that are default are used. | Based on the necessity of the process, configurations must be modified and used. |
| 3 | Critical configurations are not backed up. | It is mandatory to maintain a documentation regarding the configurations that are crucial and restoring process of configurations must be documented. |
| 4 | Portable device that carries the sensitive data is prone for attacking. | Protection mechanisms must be framed in order to access the device since it stores sensitive data. |
| 5 | Issues such as password guessing, password disclosure and inadequate policy for framing passwords | A proper and strong mechanism for framing the passwords has to be framed. While disclosing the password, certain measures could be followed: 1. Strong authentication mechanisms 2. Authorization mechanisms must be properly framed. Password must be created in such a way that it must not have common words that might lead to an easy guess. Users must be educated towards the good and bad passwords. |

*(Continued)*

**Table 2.4 (*Continued*)   ICS Systems and Vulnerabilities**

| S. No | Platform Vulnerabilities | Countermeasures |
|-------|--------------------------|-----------------|
| 6 | Buffer overflow: A few examples of buffer overflow that has been encountered in ICS products are:<br>• Stack-based buffer overflow in ICS Web service<br>• Stack-based buffer overflows allowed remote code execution on ICS hosts<br>• Heap-based buffer overflows allowed remote code execution on ICS hosts<br>• Multiple buffer overflows identified in network packet parsing application<br>• Buffer overflows in application that accepts command line and process control arguments over the network | Input data has to be validated. Programmers must be given training in secure coding. Periodic review for coding must be done and testing should be performed by senior professionals in order to check for the input functions that are vulnerable to buffer overflow attacks. Validation on the inputs must be done. Restrictions on network data and integrity checking must be performed by the user. A code review of all ICS applications that are responsible for handling the network traffic must be done regularly. |

## 2.4.4 SCADA Systems and Vulnerabilities

A process or a group of processes are controlled by a system and it is termed a SCADA system. Various processes that are controlled by SCADA systems are:

■ Train track switching
■ Power relay control
■ Opening and closing of the water valve

SCADA systems consist of four components, namely sensors, a human interface, communication network and a system that could handle the data that is collected and control commands on the local area network (LAN). A traditional SCADA architecture is shown in Figure 2.3.

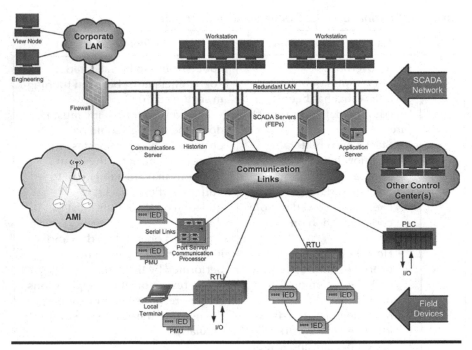

**Figure 2.3  Traditional Architecture of SCADA.**

Data collected from various sensors is transmitted through the communication networks to the administration systems. Several analog and digital sensors are connected. In addition, the communication networks could vary from serial lines and radio links to Ethernet LANs or wide area networks (WANs). Interoperation is made successful by deploying a dedicated line for backup and a separate corporate LAN that joins the remaining part of the systems is implemented. The vulnerability arises from various components such as emergency shutdown systems, industrial video surveillance systems, protection relays and environment monitoring systems (Robles & Choi, 2009).

Communication protocols used in SCADA systems are IEC60870-5-101 and IEC60870-5-104, which are found to be lacking in security mechanisms that provide security at the application and data link layers. The communication vulnerabilities (Ayaburi Emmanuel, 2015) at data transit level are:

■ Media used for communication purpose is found to be unreliable.
■ Data transfer in terms of frame length is limited since there are limitations in the bandwidth.
■ Checksum size used for verifying data integrity is not sufficient in IEC 60870-5-101 protocol.
■ Relying on checksum alone for data integrity verification is not advisable since it is prone to modifications.

Vulnerabilities must be identified and should be mitigated prior to the exploitation (Ayaburi Emmanuel, 2015).

There are two ways of vulnerability assessment and they are categorized as active assessment and passive assessment (David et al., 2011; Samtani et al., 2016):

*Active Assessment*: Active probing of devices is done to identify vulnerabilities such as port scanning, SQL injections, HTML injections and monitoring network traffic etc.

*Passive Assessment*: OS version, specific software versions and so on are identified with databases of known vulnerabilities.

Existing active and passive scanning methods with reference to the tools and technologies used in SCADA environments are collated in Table 2.5.

**Table 2.5 Active and Passive Assessments in SCADA**

| S. No. | Author | Paper | Findings |
|---|---|---|---|
| 1 | Xu et al. (2010) | CYunjing X: Context-Aware Network Vulnerability Scanning | Interaction with the devices or continuous monitoring of the network traffic is performed. Extraction of important information and identifiers from each packet has been taken. |
| 2 | Gonzalez and Papa (2008) | Passive scanning in Modbus networks | Information is extracted from Modbus traffic and thus, information about master and slave device is gained. In addition, the status of Modbus transactions is monitored. |
| 3 | Coffey et al. (2018) | Vulnerability Analysis of Network Scanning on SCADA Systems | Nessus has the capability of probing each service so that potential vulnerabilities that cause accidental DoS on SCADA systems have to be reported. |
| 4 | Coffey et al. (2018) | Nmap and Zmap | Version detection or full TCP connection could cause legacy systems to misbehave. |

## 2.4.5 Vulnerabilities in Distributed Control Systems (DCSs)

Distributed control systems that are deployed in real-time environments comprise agents that are networked. Agents such as actuators, control processing units, sensors and communication devices are present in DCSs.

### 2.4.5.1 Communications Vulnerabilities

Vulnerabilities occur due to the insufficient implementation of the design or improper design. Communicating systems are potentially vulnerable to intrusion. This vulnerability can be due to improper design or due to an inadequate implementation of the design.

A centralized supervisory control loop is used to arbitrate a set of controllers that have been deployed so that the overall tasks are shared in order to smoothen the entire production process. Production systems that are situated within the same geographical location for several industries such as wastewater treatment, chemical manufacturing plants, pharmaceutical processing facilities, oil refineries and electric power generation plants utilize DCSs. Nowadays, the DCS is incorporated with the network to enable a view of production that could be given. Certain vulnerabilities exist during the planning, design, installation and operation phases in DCS. These vulnerabilities have to be resolved to reduce impact. Vulnerabilities and their resolving ways are tabulated as shown in Table 2.6.

**Table 2.6 Vulnerabilities in DCS**

| S. No. | Issues | Resolving Process |
|---|---|---|
| 1 | Design of networks with inadequate defense in depth | 1. Coarse-grained security could be applied.<br>2. Firewall technology could be implemented for fine-grained security.<br>3. Privacy could be ensured by deploying encryption throughout the network.<br>4. Threats to the integrity of the network are detected and remediated.<br>5. End-point security in policy enforcement must be included.<br>6. Authenticate and authorize all network users. |
| 2 | Separate auditable administration mechanisms | Monitoring the control system ensures that the software is kept up-to-date. |

*(Continued)*

**Table 2.6 (*Continued*)  Vulnerabilities in DCS**

| S. No. | Issues | Resolving Process |
|---|---|---|
| 3 | Use of a non-dedicated communications channel for command and control | Information flow and layout of networks are essential. It is better to use a separate channel for command and control since the control messages have to reach faster which needs an immediate action. |
| 4 | The anomalous activity could be detected by the usage of tools. But it is lacking. | Implementation of certain tools to detect the malicious behavior is advisable. |
| 5 | Inadequately secured wireless communication | Deployment of authentication and authorization mechanisms for controlling the communications between the networks could be done. |
| 6 | Policies, procedures, and culture governing control system security are found to be inadequate. | Intrusion prevention system, implementing firewalls, passwords and encryption technologies could be determined in terms of policies and process to ensure security. |
| 7 | Access control is not proper during the remote access. | Forming of suitable access control policies must be implemented. |
| 8 | Applications that are installed on critical host computers are observed to be not suitable. | Before the installation of applications, careful scrutiny is required and certain permissions must be removed for the user so that applications could not be installed. |
| 9 | Scrutinizing control system software is not sufficient. | Coding standards could be followed and in addition, a precise view of operating system and application code must be performed in order to avoid interpreted solutions. |
| 10 | Authentication of command and control data is not performed. | Proper policy, procedure, and practice must be framed. Deploying multiple layers of defense in depth must be strictly followed. |
| 11 | Components that are interconnected might fail randomly. | When the components are under stress, it would fail. A data backup collected from various sources could be retrieved in case of failures. |

(*Continued*)

**Table 2.6 (*Continued*)   Vulnerabilities in DCS**

| S. No. | Issues | Resolving Process |
|--------|--------|-------------------|
| 12 | Sensors or the communication channel that carries the data from the sensors to controllers might be lost. | Maintaining the redundancy of the data and regular validations of sensors might resolve this issue. |
| 13 | Communication between control system components might be interrupted or corrupted. | Frequent monitoring is required in order to check the communication process. A backup facility is recommended to ensure the regular communication process. |

## 2.4.6  Smart Grid

A grid system is an autonomous electrical system with the hybridization of digital technologies to serve the consumers. The major components of the system are generation, transmission, distribution and consumption (Thomas & McDonald, 2015). The sources for generating electric energy are nuclear reactors, hydro and solar energy, wind and tidal energy etc. The generated energy is transmitted through high-voltage structures and further distributed by stepping down the current for ready use by consumers. Finally, the current is consumed in different ways based on its necessity for residential, commercial and industrial purposes. Since energy used in a multitude of ways, the demand is unpredictable and specific technology to analyze the demand is a challenging task. The demand can be reduced by requesting load shedding if energy consumption is analyzed or predicted earlier. The consequence of such a scenario evolved to a smart grid that integrates the electrical grid system and information communication technologies. The implementations of smart technologies are to reliably serve the consumers operating modern technology systems and consumers also monitored by smart devices.

The smart grid is a heterogeneous complex of cooperating systems (Wang & Lu, 2013) that work together to provide the functionalities based on the predicted demand. The grid has to depend on a cyber-physical system, a combination of engineered elements and cyber networks, to manage and monitor information that flows around the network drastically in the real world. The network is interconnected by millions of devices and continuous information exchange is surfacing. A smart grid offers services (Thomas & McDonald, 2015) like:

■ Renewable resources to forecast and reduce uncertainty
■ Substation digitization with less cost, time and risk

- Intelligent electronics to control devices and monitor their performance
- Monitoring and diagnostics for asset protection and life extension
- Communication infrastructure to control the devices remotely and performance visibility
- Smart metering, appliances and home control devices for utility bill savings, demand management etc.

### 2.4.6.1 Smart Grid Security

Rapid information exchange makes the grid vulnerable to attacks and security threats. It is not feasible to prevent threats and protect the system physically. It is prudent to devise algorithms and tools to secure the components from threats. The three high-level grid security objectives are availability, confidentiality, and integrity (Wang & Lu, 2013).

- Availability: This is the most important factor that ensures reliable information with rapid response to customers. The loss of data may lessen the effectiveness of the grid.
- Integrity: To safeguard information without destruction to deliver exact content and to avoid any false decision making.
- Confidentiality: To provide security of data from public access and prevent unauthorized users from intruding end users' personal information.

### 2.4.6.2 Cyberattacks in Smart Grids

Smart grid services exploit communication technologies that also ease intruders developing malicious software called malware, targeting the communication and controlling entities of the system, leading to threats to the critical infrastructure. There are possibilities for unknown threats or vulnerabilities erupting in the hardware, operating systems, and protocols, thus breaking down the working of the entire grid system (Eder-Neuhauser et al., 2017). The different types of cyberattacks are infiltration, distributed denial of service (DDOS) attacks, repurposing attacks and other malware like worms, Trojan horse, crypto-lockers etc. The attackers deploy highly persistent software utilizing only zero days, making it challenging for the defenders to trace them. To protect smart grids, it is significant to detect and analyze the operation of malware to develop and implement intelligent countermeasures in time.

The smart grid is developed to make intelligent decisions automatically and respond rationally to improve system reliability. The performance of the system depends on its services like assessing the vulnerabilities prevailing at the earliest to prevent deadly attacks, provide stable information dynamically and most importantly self-healing capability. Self-healing is to analyze and prevent the system from cyberattacks through vulnerability analysis. These global control actions are performed by cyber-physical systems that are systematically developed entities to control physical processes, monitoring and interacting with communication technologies. CPSs are developed to meet the challenges that are waiting in the dynamic domains of defense, agriculture, transportation, manufacturing, energy, healthcare, aerospace and buildings (Rajkumar et al., 2010). The possible vulnerabilities (Aloul et al., 2012) that may cause serious damage to the grid are:

- *End-user information security*: The smart meters collecting a huge amount of data from homes can leak the private details of consumers.
- *Usage of a huge number of intelligent devices*: The devices used in a smart grid are connected to communication networks and these devices involuntarily behave as entry points.
- *Physical locations of devices*: A few devices may be placed outside the premises that may be tampered by attackers.
- *Internet protocol attacks*: Attacks related to internet protocols like spoofing, DDoS etc.
- *Lifetime of devices*: Components in the grid are with different lifetimes and a few may tend to wear out and some may be incompatible with existing versions, thus leading to weak security.
- *Miscommunication between project teams*: Noncooperation in the project team may lead to decisions that will tend to make the system vulnerable.

## 2.5 Vulnerability Analysis

This section discusses the results of the beginning of the art review and synthesizes the results of the different literatures. The study reveals that most of the existing literature addressing the vulnerabilities of physical cybernetic systems is available. The results also provided a list of critical CPS threats and vulnerabilities, related factors and mitigation objectives that impact physical systems, digital systems and human elements. The study reveals several observations on the future directions needed to mitigate cyberattacks in the context of CPSs (Yeboah-ofori et al., 2018). The authors of the study examined the risks associated with these three entities, namely the fusion of physical systems, digital systems and human elements. These elements can have different inherent cyberattacks that can be triggered from anywhere in the world and when this happens, the impact can affect not only one of the elements, but also all three elements, depending on the nature of the cyberattack. We examine the elements as follows: Physical systems include the infrastructure that communicates and interacts with the real world and provides feedback. These include computers, monitors, wired and wireless networks, in-plane switching (IPS), air control centers and military command centers. Digital systems are software, embedded systems, computational programs such as programmable logic controllers (PLCs) and SCADA systems that function as intelligent agents that send and receive instructions from physical and human elements. Human elements are the components such as sensors and actuators that connect components to physical systems. Several research studies from industry reviews and articles focus on CPS design methodologies and CPS security as shown in Table 2.7.

**Table 2.7    Vulnerabilities and Mitigation Techniques for CPS**

| Cyber-Physical Systems Vulnerability | The Proposed Literature |
|---|---|
| Robles and Choi (2009) | Vulnerabilities in SCADA system were assessed and the methods were determined to resolve it. |
| Cárdenas et al. (2011) | Studied considered risks controls from technical and nontechnical areas in the CPS system design. |
| Lee and Seshia (2011) | Generic framework lacks capability to detect the diversities of attacks, threats, vulnerabilities. |
| Mo et al. (2012) | Discussed approaches to contingency analysis and detection of anomalies in the sensory system. |
| Chopade and Bikdash (2013) | Introduced a new methodological approach to comprehensively analyze the vulnerability of interdependent infrastructures. |
| Sedgewick (2014) | Proposed a framework for improving critical infrastructure cybersecurity. |
| Al Faruque et al. (2015) | Modeled attacks and countermeasure functionality using a novel security-aware functional modeling. |
| Capgemini (2015) | Identified amount stolen or spent on cyberattack as huge including the resolutions and global cost of cybercrime. |
| Humayed et al. (2017) | Generic framework lacks capability to detect the diversities of attacks, threats, vulnerabilities. |
| Yan et al. (2017) | A quantitative learning approach to identify critical attack sequences taking into account the physical behaviors of the system. |
| Coffey et al. (2018) | Critical analysis and test of SCADA equipments have been performed in order to identify the service detection on these systems. |

Risk management framework suggested by the National Institute of Standards and Technology (NIST) is to continuously monitor the components to identify possible threats and their impact. It also suggests that new information has to be gathered to gain knowledge of the attack trends. Exposure analysis (Hahn & Govindarasu, 2011) is recomputation of the risk management procedures involving security. If a security mechanism is under risk it will propagate into a privileged set, thus extracting the object under concern. The analysis will determine the increase in exposure of new units in different layers. The exposed units are reviewed for further levels of security.

Power grids exploit physical flow-based models (Correa & Yusta, 2013) for vulnerability analysis. The model identifies the components like transmitters, generators etc. in the grid that are prone to attacks. The model is designed considering game theory and min-max programming concepts that adversaries or terrorists target to trace the system components that can cause major damage or loss. Efficient methodologies will help the decision makers to forecast and prevent system damage through minimal re-dispatching the component activities to minimize the load loss. The topological structure of a grid is represented by an adjacency matrix consisting of nodes and edges that connect the nodes. The nodes are the components of the system like generators, transmitters, transformers etc., while the connecting lines are the edges. Vulnerability analysis is done considering the geodesic vulnerability index and the connectivity factor. The impairment in the network due to attacks is the result of node failures that lead to subsequent failure of nodes that are connected to them. The indices calculated from the flow graph will decide on the possible cascade failures and based on the impact of these parameters, fragmentation of the grid is suggested. The model practically does not require electric parameters for estimation or in decision making on expansion or minimizing the nodes in the network. Graph theory has proved to be a significant vulnerability assessment index to prevent intentional attacks on the critical grid infrastructure.

Vulnerability analysis is done by two methods namely structural and functional vulnerability analysis. This type of analysis is adopted for the weapons of mass destruction (WMD) attacks in SCADA networks (Chopade & Bikdash, 2013). WMD is an attack and is a threat to the USA that could be in the form of chemicals, biological viruses or nuclear weapons. Structural and functional type of analyses are crucial to protect the network. The topology of the network system and the physical properties of the nodes are also analyzed. The model will identify the shortest path to calculate the structural efficiency of the network. Followed by the analysis each smart grid node is checked for the possibility of attack and in such a scenario, the entire nodes in the path are deleted, protecting the entire smart grid. The deletion of nodes changes the topology and further analyzes for efficiency. The local power disturbances would destruct the functioning of the grid. It is dealt considering functional vulnerability analysis that includes independent and interdependent functionality models. The parameters will indicate the network robustness under WMD and failure of nodes dynamically.

Q-learning-based vulnerability analysis (Yan et al., 2017) is a reinforcement learning technique for vulnerability analysis. The methods to countermeasure the attacks are an exhaustive search, heuristic techniques, opinions from experts etc. that are all detection oriented. An approach that uses computational intelligence methods is able to learn the patterns of attacks in the sequential nodes in the network leading to cascading failure that are dynamic in nature. The intelligent technique Q-learning applies machine learning algorithms to self-tune the system to identify the attacks. The algorithm evaluates the nodes that command the actions of other neighboring nodes by acquiring a succession of actions in the network. During the trial-and-error method, the awards received by the algorithm are calculated. The learning algorithm tries to maximize the rewards or feedback from the environment and in turn identification of faulty nodes, weak nodes possible for an attack entry. The aggressiveness of the algorithm is determined by a learning parameter. Q-learning is also capable to locate the local maximum using ε-greedy method and is capable to minimize the number of lines to be attacked.

Table 2.7 summarizes the different approaches proposed in the literature to address the different CPS vulnerabilities.

## 2.6 Conclusion

This chapter provides a brief introduction to CPSs and as the key feature lies in security in CPS, related works that deal with existing vulnerability assessment in various cyber-physical systems and in addition various security frameworks models have been discussed. Vulnerabilities encountered at several layers have been presented in detail along with their countermeasures. A brief discussion on vulnerabilities in ICS, DCS, SCADA and smart grid have also been presented. The various analyses of vulnerability have also been discussed.

## 2.7 Research Challenges

Developing CPSs (Liu et al., 2017) is considered to be a challenging task since there is no unified framework that encompasses network and physical resources. Designing of a CPS must be in such a way that it must be able to resolve issues that arise due to environmental interference, collection inaccuracy, noise in measurement, and losses due to unified framework. Consequently, it is very critical to develop an abstract model in a way that the complexity of designing must be decreased and the abstracted issue must be properly maintained.

Security attacks in an uncertain environment and errors in physical devices and wireless communications massively threaten the robustness and security of the overall system (Ahola et al., 2007).

CPS systems must be designed in order to possess reliability and interoperability. Due to the cascading of various devices together, CPS must be able to resolve the errors, uncertainty errors, failures, and attacks. Thus, cascading failure could be avoided (Rohloff & Baccsar, 2008).

A high level of trust for customers along with security, safety, privacy and quality of service (QoS) must be given much focus.

# References

Ahola, T., Korpinen, P., Rakkola, J., Ramo, T., Salminen, J., & Savolainen, J. (2007). Wearable FPGA based wireless sensor platform. In *2007 29th Annual International Conference of the IEEE Engineering in Medicine and Biology Society* (pp. 2288–2291). doi:10.1109/IEMBS.2007.4352782.

Al Faruque, M., Regazzoni, F., & Pajic, M. (2015). Design methodologies for securing cyber-physical systems. In *Proceedings of the 10th International Conference on Hardware/Software Codesign and System Synthesis* (pp. 30–36). Piscataway, NJ: IEEE Press. Retrieved from http://dl.acm.org/citation.cfm?id=2830840.2830844.

Aloul, F., Al-Ali, A. R., Al-Dalky, R., Al-Mardini, M., & El-Hajj, W. (2012). Smart grid security: Threats, vulnerabilities and solutions. *International Journal of Smart Grid and Clean Energy*, (September), 1–6. doi:10.12720/sgce.1.1.1-6.

Ayaburi, E., & Sobrevinas, L. (2015). Securing supervisory control and data acquisition systems: Factors and research direction, in 21st Americas Conference on Information Systems (AMCIS).

Capgemini. (2015). Using Insurance to Mitigate Cybercrime Risk. Challenges and Recommendations for Insurers. www.capgemini.com/wpcontent/uploads/2017/07/Using_Insurance_to_Mitigate_Cybercrime_Risk.pdf.

Cárdenas, A. A., Amin, S., Lin, Z.-S., Huang, Y.-L., Huang, C.-Y., & Sastry, S. (2011). Attacks against process control systems: Risk assessment, detection, and response. In *Proceedings of the 6th ACM Symposium on Information, Computer and Communications Security* (pp. 355–366). New York: ACM. doi:10.1145/1966913.1966959.

Chopade, P., & Bikdash, M. (2013). Structural and functional vulnerability analysis for survivability of smart grid and SCADA network under severe emergencies and WMD attacks. *2013 IEEE International Conference on Technologies for Homeland Security, HST 2013*, (November 2013), 99–105. doi:10.1109/THS.2013.6698983.

Coffey, K., Smith, R., Maglaras, L., & Janicke, H. (2018). Vulnerability analysis of network scanning on SCADA systems. *Security and Communication Networks, 2018*, 1–21. doi:10.1155/2018/3794603.

Correa, G. J., & Yusta, J. M. (2013). Grid vulnerability analysis based on scale-free graphs versus power flow models. *Electric Power Systems Research, 101*, 71–79. doi:10.1016/j.epsr.2013.04.003.

David, K., O'gorman, J., Kearns, D., & Aharoni, M. (2011). *Metasploit: The Penetration Tester's Guide*. San Francisco, CA: No Starch Press.

Eder-Neuhauser, P., Zseby, T., Fabini, J., & Vormayr, G. (2017). Cyber attack models for smart grid environments. *Sustainable Energy, Grids and Networks, 12*, 10–29. doi:10.1016/j.segan.2017.08.002.

Gao, Y., Peng, Y., Xie, F., Zhao, W., Wang, D., Han, X., Tianbo, Lu, Li, Z. (2013). Analysis of security threats and vulnerability for cyber-physical systems. In *Proceedings of 2013 3rd International Conference on Computer Science and Network Technology* (pp. 50–55). IEEE. doi:10.1109/ICCSNT.2013.6967062.

Giraldo, J., Sarkar, E., Cardenas, A. A., Maniatakos, M., & Kantarcioglu, M. (2017). Security and privacy in cyber-physical systems: A survey of surveys. *IEEE Design & Test, 34*(4), 7–17. doi:10.1109/MDAT.2017.2709310.

Gonzalez, J., & Papa, M. (2008). Passive scanning in Modbus networks BT—critical infrastructure protection. In E. Goetz & S. Shenoi (Eds.), (pp. 175–187). Boston, MA: Springer.

Hahn, A., & Govindarasu, M. (2011). Cyber attack exposure evaluation framework for the smart grid. *IEEE Transactions on Smart Grid, 2*(4), 835–843. doi:10.1109/TSG.2011.2163829.

Humayed, A., Lin, J., Li, F., & Luo, B. (2017). Cyber-physical systems security—a survey. *IEEE Internet of Things Journal, 4*(6), 1802–1831. doi:10.1109/JIOT.2017.2703172.

Ledwaba, L., & Venter, H. S. (2017). A threat-vulnerability based risk analysis model for cyber physical system security. *Proceedings of the 50th Hawaii International Conference on System Sciences*, 6021–6030. doi:10.24251/HICSS.2017.720.

Lee, E. A., & Seshia, S. A. (2011). Introduction to embedded systems, a cyber-physical systems approach. Org, 6(18). http://LeeSeshia.

Liu, Y., Peng, Y., Wang, B., Yao, S., & Liu, Z. (2017). Review on cyber-physical systems. *IEEE/CAA Journal of Automatica Sinica, 4*(1), 27–40. doi:10.1109/JAS.2017.7510349.

Lu, T., Zhao, J., Zhao, L., Li, Y., & Zhang, X. (2015). Towards a framework for assuring cyber physical system security. *International Journal of Security and Its Applications, 9*(3), 25–40. doi:10.14257/ijsia.2015.9.3.04.

Mo, Y., Kim, T. H., Brancik, K., Dickinson, D., Lee, H., Perrig, A., & Sinopoli, B. (2012). Cyber–physical security of a smart grid infrastructure. *Proceedings of the IEEE, 100*(1), 195–209. doi:10.1109/JPROC.2011.2161428.

Rajkumar, R., Lee, I., Sha, L., & Stankovic, J. (2010). Cyber-physical systems: The next computing revolution. In *Design Automation Conference* (pp. 731–736). doi:10.1145/1837274.1837461.

Robles, R. J., & Choi, M. (2009). Assessment of the vulnerabilities of SCADA, control systems and critical infrastructure systems. *International Journal of Grid and Distributed Computing, 2*(2), 27–34.

Rohloff, K. R., & Baccsar, T. (2008). Deterministic and stochastic models for the detection of random constant scanning worms. *ACM Transaction Modelling Computing Simulation, 18*(2), 8:1–8:24. doi:10.1145/1346325.1346329.

Samtani, S., Yu, S., Zhu, H., Patton, M., & Chen, H. (2016). Identifying SCADA vulnerabilities using passive and active vulnerability assessment techniques. In *2016 IEEE Conference on Intelligence and Security Informatics (ISI)* (pp. 25–30). doi:10.1109/ISI.2016.7745438.

Sedgewick, A. (2014). Framework for improving critical infrastructure cybersecurity, version 1.0. *No. NIST-Cybersecurity Framework*.

Stouffer, K., Falco, J., & Scarfone, K. (2011). Guide to industrial control systems (ICS) security. *NIST Special Publication, 800*(82), 16–16.

Thomas, M. S., and McDonald, JD. (2015). *Power System SCADA and Smart Grids*. CRC Press.

Wan, J., Canedo, A., & Faruque, M. A. A. L. (2015). Security-aware functional modeling of cyber-physical systems. In *2015 IEEE 20th Conference on Emerging Technologies & Factory Automation (ETFA)* (pp. 1–4). doi:10.1109/ETFA.2015.7301644.

Wang, W., & Lu, Z. (2013). Cyber security in the smart grid: Survey and challenges. *Computer Networks, 57*(5), 1344–1371. doi:10.1016/j.comnet.2012.12.017.

Xu, Y., Bailey, M., Vander Weele, E., & Jahanian, F. (2010). CANVuS: Context-aware network vulnerability scanning BT—recent advances in intrusion detection. In S. Jha, R. Sommer, & C. Kreibich (Eds.), (pp. 138–157). Berlin, Germany: Springer.

Yan, J., He, H., Zhong, X., & Tang, Y. (2017). Q-Learning-based vulnerability analysis of smart grid against sequential topology attacks. *IEEE Transactions on Information Forensics and Security, 12*(1), 200–210. doi:10.1109/TIFS.2016.2607701.

Yeboah-ofori, A., Abdulai, J. D., & Katsriku, F. (2018). Cybercrime and risks for cyber physical systems: A review. Preprints.org. doi:10.20944/preprints201804.0066.v1.

Yu, X., & Xue, Y. (2016). Smart grids: A cyber–physical systems perspective. *Proceedings of the IEEE, 104*(5), 1058–1070. doi:10.1109/JPROC.2015.2503119.

Zhu, B., Joseph, A., & Sastry, S. (2011). A taxonomy of cyber attacks on SCADA systems. In *2011 IEEE International Conferences on Internet of Things, and Cyber, Physical and Social Computing*, 380–388.

## Chapter 3

# State Estimation-Based Attack Detection in Cyber-Physical Systems: Limitations and Solutions

Chuadhry Mujeeb Ahmed, Jianying Zhou and
Aditya P. Mathur

## Contents

In this chapter, we present a detailed case study regarding model-based attack detection procedures for cyber-physical systems (CPSs). In particular data from a real-world water treatment plant is collected and analyzed. Using this dataset and the subspace system identification technique, an input-output linear time invariant (LTI) model for the water treatment plant is obtained. This model is used to derive a Kalman filter to estimate the evolution of the system dynamics. Then, residual variables are constructed by subtracting data coming from the real-world water treatment system and the estimates obtained by using the Kalman filter. We use these residuals to evaluate the performance of statistical detectors, namely the bad-data and the cumulative sum (CUSUM) change detection procedures. First, the limitations of these model-based statistical techniques are shown. Then, an attack detection technique is proposed to improve over the threshold-based approaches. It detects data integrity attacks on sensors in CPSs. A combined fingerprint for sensor and process noise is created during the normal operation of the system. Under sensor spoofing attack, noise pattern deviates from the fingerprinted pattern enabling the proposed scheme to detect attacks. To extract the noise (the difference between expected and observed value) a representative model of the system is derived. By subtracting the state estimates from the real system states, a residual vector is obtained. It is shown that in steady state the residual vector is a function of the process and sensor noise. A set of time domain and frequency domain features is extracted from the residual vector. The feature set is provided to a machine learning algorithm to identify the sensor and process. Experiments are performed on a real-world water treatment (SWaT) facility. A class of *stealthy* attacks, designed for statistical detectors on SWaT, are detected by the proposed technique. It is shown that a multitude of sensors can be uniquely identified with an accuracy higher than 94.5% based on the noise fingerprint.

# 3.1 Introduction

A cyber-physical system (CPS) is a combination of computing elements and physical phenomenon [1–3]. In this study, a real-world water treatment testbed is considered. A CPS consists of cyber components such as programmable logic controllers (PLCs), sensors, actuators, supervisory control and data acquisition

(SCADA) workstation, and human-machine interface (HMI) elements interconnected via a communications network. The PLCs receive sensor data and control a physical process based on the sensor measurements. The advances in communication technologies resulted in the wide spread of such systems to better monitor and operate CPSs, but this connectivity also exposes physical processes to malicious entities on the cyber domain. Recent incidents of sabotage on these systems [4–6] have raised concerns on the security of CPSs [7].

CPS security poses different challenges as compared to its IT counterpart. In case of security compromise in cyber systems, the major threats are to the confidentiality, integrity and availability of the information. For a CPS, the confidentiality might not be the major concern but integrity of data may have an utmost importance [8]. The consequences in case of an attack are also different in CPS as compared to pure software systems. Attacks on a CPS might result in damage to the physical property, as a result of an explosion [9,10] or severely affecting people who depend on a critical infrastructure as was the case of recent power cutoff in Ukraine [4]. Data integrity is an important security requirement for CPSs [8], therefore the integrity of sensor data should be ensured. Sensor data can either be spoofed in the cyber (digital) domain [11] or in the physical (analog) domain [12,13]. Sensors are a bridge between the physical and cyber domains in a CPS. Traditionally, an intrusion detection system (IDS) monitors a communication network or a computing host to detect attacks. However, physical tampering with sensors or sensor spoofing in the physical/analog domain may go undetected by the legacy IDS [12].

Data integrity attacks on sensor measurement and impact of such attacks have been studied in theory, including false data injection [14], replay attacks [15], and stealthy attacks [16]. These previous studies proposed attack detection methods based on system model and statistical fault detectors [17–20] and also point out the limitations of such fault detectors against an adversarial manipulation of the sensor data. In practice, attacks on sensor measurement can be launched by analog spoofing attacks [12,21,22], or by tampering with the communication channel between a sensor and a controller by means of a classical man-in-the-middle (MiTM) attack [11].

The study reported here has the goal to look from control theoretic perspective into techniques of attack detection in a CPS. Our objective is to study these techniques on a real-time water treatment plant testbed called SWaT (details in Section 6.1). An adversary tries to modify sensor readings as a man-in-the-middle attack and based on attacker's goals would spoof a range of data values. If an adversary has the knowledge about the system and how it works, he or she can launch an intelligent attack by changing sensor values in a manner that remains within certain operating limits but yet affects the plant's performance. During such an attack, an estimator like Kalman filter would best try to track sensor measurements and remove noise. The estimator will follow the exploited values of sensor and difference (residue) between measured values, and estimation vanishes as soon as the estimator converges.

**Relevant Research**: The impact of attacks on a CPS has been reported extensively in literature besides quite a few techniques to detect and mitigate the effects of such attacks. In [23], it is concluded that perfect estimation is not feasible when half of the sensors are under attack. One major assumption here is that a local controller has complete access to the system states. In [24], the authors propose physical watermarking, where a known noise sequence is added to control input and its effect on sensor readings is evaluated. In case of an attack, since the adversary is not aware of the noise added he or she would not be able to drive sensor values to one which should have been the case because of the added noise.

In [25], false data injection attack detection in smart grid systems using Kalman filter is studied. The method shows success against random and DoS attacks but fails in case of a false data injection attack. In [26] the authors studied the power grid system for false data injection attacks. Generalized false data injection attack is introduced where an adversary adds a bias to measured data which does not change residue too much and could not be detected. In [18] the authors have used statistical detectors for attack detection on a water distribution network. A class of stealthy attacks has been designed and limitations of statistical attack detectors have been demonstrated. These mentioned works point out the limitations of estimation-based detectors against stealthy attacks using simulation models. In this article, we expose limitations of the bad-data detection schemes in detecting strategic attacks by implementing these methods in a real-world water treatment testbed. Attack vectors are systematically designed and carried out, compromising the link between PLC to SCADA. We demonstrate the success of sophisticated attacks. Furthermore, a novel solution is proposed to detect these stealthy attackers.

## 3.2 System Model-Based Attack Detection

In this section, we introduce the idea of model-based attack detection. To introduce this idea let us first consider an example of a physical process and obtain an analytical model to capture process dynamics. An example is shown in Figure 3.1.

### 3.2.1 System Model

An example of a physical process and its mathematical model is shown in Figure 3.1. In Figure 3.1, a physical phenomenon, i.e., a vehicle moving in a straight line, is considered as a toy example. A vehicle is moving in a straight line with initial position as $x_1$ at time $t_1$ and $x_2$ is the position at time $t_2$. Suppose the problem is to find the position of the vehicle at time instance $t_2$. Given initial position $x_1 = x$ and action of accelerator (force being applied) $F$ laws of physics can be used to find position $x_2$ at time $t_2$. In Figure 3.1, part (2) shows how the state $x$ of the vehicle is expressed using the first principles. In (3), a mathematical model from the first principles is presented in terms of differential equations. This representation

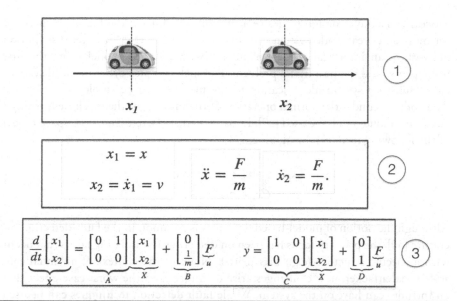

**Figure 3.1    An example of a physical process and its mathematical representation. (1) Explains the physical process (i.e., a vehicle moving in a straight line). $x_1$ is the position of the vehicle at time $t_1$ and $x_2$ is the position at time $t_2$. (2) Models this motion using the first principles. Position of the vehicle is considered as state of this physical phenomena. $F$ is the force being applied and $m$ is the mass of the vehicle. (3) A mathematical model from first principles is presented as a linear time invariant model. This representation is termed state space model in classical control theory literature.**

is called the state space model in the control theory literature [27]. Such a linear time invariant (LTI) model captures the physical process dynamics and it is useful for different kinds of system analysis including but not limited to observability and controllability of a system. In this state space model, $x \in \mathbb{R}^n$ is a system state vector, $A \in \mathbb{R}^{n \times n}$ is a state space matrix, $B \in \mathbb{R}^{n \times p}$ is the control matrix, $y \in \mathbb{R}^m$ are the measured outputs, $C \in \mathbb{R}^{m \times n}$ is a measurement matrix, $D \in \mathbb{R}^{m \times p}$ is a control matrix and $u \in \mathbb{R}^p$ denote the system control. The state space matrices $A, B, C, D$ capture the system dynamics and can be used in finding a specific system state based on an initial state. This example shows that a system model is an expressive form representing the system dynamics. It is possible to predict the system state for a physical process with a precise analytical model. This is the main idea behind the system model–based attack detection techniques. The system model can be used to model the normal behavior of a dynamic physical process, and in the case of anomalies, the system behavior deviates from the expected behavior.

The first step is to derive a system model either by the first principles or using the data-centric approaches. For the toy example shown in Figure 3.1 the physical process was simple enough to derive a system model using the first principles but for most other real-world systems, for example, the smart grid, water distribution network or an industrial process plant, deriving a system model using the first principles would not be a trivial problem. For such complex cyber-physical systems either subspace system identification technique [28] can be employed, using the data collected under the normal operation of the process, or the machine learning–based techniques might be used [29]. In particular, a discrete time state space model of the following form can be obtained:

$$\begin{cases} x_{k+1} = Ax_k + Bu_k + v_k, \\ y_k = Cx_k + \eta_k. \end{cases} \tag{3.1}$$

Although the notion of model-based detectors is common in the fault-detection literature [30], the primary focus has been on detecting and isolating faults that occur with a specific structure (e.g., bias drifts). Now, in the context of an intelligent adversarial attacker, new challenges arise to understand the worst-case effect that an intruder can have on the system. While fault detection techniques can be used to detect attacks, it is important to assess the performance of such methods against an intelligent adversary. In this work, by means of our real-world water treatment plant, we assess the performance of two model-based fault detection procedures (the bad-data and the CUSUM procedures) for a variety of attacks. These procedures rely on a state estimator (e.g., Kalman filter) to predict the evolution of the system. The estimated values are compared with sensor measurements $\bar{y}_k$ (which may have been attacked). The difference between the two should stay within a certain threshold under normal operation, otherwise, an alarm is triggered to point to a potential attack. After demonstrating the limitations of the threshold-based detectors we propose a novel attack detection technique based on the sensor and process characteristics.

### 3.2.2 Attack Detection Framework

In this section, we explain the details of our attack-detection scheme. First, we discuss the Kalman filter–based state estimation and residual generation. Then, we present our residual-based attack detection procedures (namely the CUSUM and the bad-data detectors).

### 3.2.2.1 Kalman Filter

To estimate the state of the system based on the available output $y_k$, we use a linear filter with the following structure:

$$\hat{x}_{k+1} = A\hat{x}_k + Bu_k + L_k(\bar{y}_k - C\hat{x}_k), \tag{3.2}$$

with estimated state $\hat{x}_k \in \mathbb{R}^n$, $\hat{x}_1 = E[x(t_1)]$, where $E[\cdot]$ denotes expectation, and gain matrix $L_k \in \mathbb{R}^{n \times m}$. Define the estimation error $e_k := x_k - \hat{x}_k$. In the Kalman filter, the matrix $L_k$ is designed to minimize the covariance matrix $P_k := E[e_k e_k^T]$ (in the absence of attacks). Given the system model (1) and the estimator (2), the estimation error is governed by the following difference equation:

$$e_{k+1} = (A - L_k C) e_k - L_k \eta_k - L_k \delta_k + v_k. \tag{3.3}$$

If the pair $(A, C)$ is detectable, the covariance matrix converges to steady state in the sense that $\lim_{k \to \infty} P_k = P$ exists [27]. We assume that the system has reached steady state before an attack occurs. Then, the estimation of the random sequence $x_k, k \in \mathbb{N}$ can be obtained by the estimator (2) with $P_k$ and $L_k$ in steady state. It can be verified that, if $R_2 + CPC^T$ is positive definite, the following estimator gain

$$L_k = L := (APC^T)(R_2 + CPC^T)^{-1}, \tag{3.4}$$

leads to the minimal steady state covariance matrix $P$, with $P$ given by the solution of the algebraic Riccati equation:

$$APA^T - P + R_1 = APC^T (R_2 + CPC^T)^{-1} CPA^T. \tag{3.5}$$

The reconstruction method given by (2–5) is referred to as the steady state Kalman Filter, cf. [27].

### 3.2.2.2 Residuals and Hypothesis Testing

The statistical detectors rely on a state estimator (e.g., the Kalman filter) to predict the evolution of the system. The difference between the estimated and measured values termed as the residual ($\eta_k, k \in \mathbb{N}$) can be expressed as

$$\eta_k := \bar{y}_k - C\hat{x}_k = Ce_k + \eta_k + \delta_k. \tag{3.6}$$

If there are no attacks, the mean of the residual is

$$E[\eta_{k+1}] = CE[e_{k+1}] + E[\eta_{k+1}] = \mathbf{0}_{m \times 1}. \tag{3.7}$$

where $\mathbf{0}_{m \times 1}$ denotes an $m \times 1$ matrix composed of only zeros, and the covariance is given by

$$\Sigma := E[\eta_{k+1} \eta_{k+1}^T] = CPC^T + R_2. \tag{3.8}$$

For this residual, we identify two hypothesis to be tested, $\mathcal{H}_0$ the *normal mode* (no attacks) and $\mathcal{H}_1$ the *faulty mode* (with attacks). For our particular case of study, the pressure at the nodes, flow in the pipes and the water level in the tank are the outputs of the system. Using this data along with the state estimates, we construct our residuals. Then, we have:

$$\mathcal{H}_0 : \begin{cases} E[r_k] = \mathbf{0}_{m \times 1}, \\ E[r_k r_k^T] = \Sigma, \end{cases} \quad \text{or}$$

$$\mathcal{H}_1 : \begin{cases} E[r_k] \neq \mathbf{0}_{m \times 1}, \\ E[r_k r_k^T] \neq \Sigma. \end{cases}$$

Figure 3.2 shows the approximated distributions of the residuals of the water level in the storage tank for both the attacked and the attack-free cases. In Figure 3.2, the residual under a bias injection attack (simple constant offset on the sensor measurements) is depicted. Our hypothesis can easily be verified by looking at the probability distribution of the residuals. Our null hypothesis $\mathcal{H}_0$ which follows a zero mean normal distribution with variance $\Sigma$ is also verified from plot in Figure 3.2. Similarly, for the attacked scenario $\mathcal{H}_1$, we do not have a zero mean normally distributed residual as it is shown in Figure 3.2. We can formulate the hypothesis testing in a more formal manner using existing change detection techniques based on the statistics of the residuals.

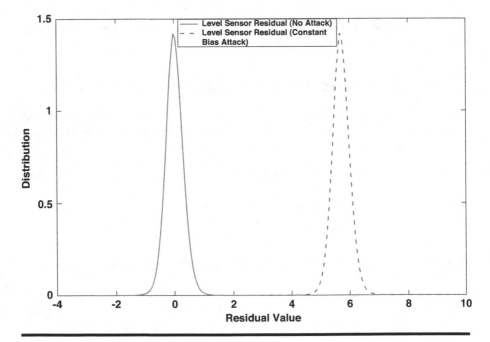

**Figure 3.2 (Left): Probability distribution of the residue for level sensor measurements without attack. (Right): Probability distribution of the residual for water level sensor measurements with bias injection attack.**

## 3.2.3 *Cumulative Sum (CUSUM) Detector*

The CUSUM procedure is driven by the residual sequences. In particular, the input to the CUSUM procedure is a *distance measure*, i.e., a measure of how deviated the estimator is from the actual system, and this measure is a function of the residuals. In this work, we assume there is a dedicated detector on each sensor (or on any sensor we want to include in the detection scheme). Throughout the rest of this article we will reserve the index $i$ to denote the sensor/detector, $i \in \mathcal{I} := \{1, 2, \ldots, m\}$. Thus, we can partition the attacked output vector as $\bar{y}_k = \text{col}(\bar{y}_{k,1}, \ldots, \bar{y}_{k,m})$ where $\bar{y}_{k,i} \in \mathbb{R}$ denotes the $i$-th entry of $\bar{y}_k \in \mathbb{R}^m$; then

$$\bar{y}_{k,i} = C_i x_k + \eta_{k,i} + \delta_{k,i}, \tag{3.9}$$

with $C_i$ being the $i$-th row of $C$ and $\eta_{k,i}$ and $\delta_{k,i}$ denoting the $i$-th entries of $\eta_k$ and $\delta_k$ respectively. Inspired by the empirical work in [31], we propose the absolute value of the entries of the residual sequence as distance measure, i.e.,

$$z_{k,i} := |r_{k,i}| = |C_i e_k + \eta_{k,i} + \delta_{k,i}|. \tag{3.10}$$

Note that, if there are no attacks, $r_{k,i} \sim \mathcal{N}(0, \sigma_i^2)$ (see Figure 3.2), where $\sigma_i^2$ denotes the $i$-th entry of the diagonal of the covariance matrix $\Sigma$. Hence, $\delta_k = 0$ implies that $|r_{k,i}|$ follows a *half-normal distribution* [32] with

$$E[|r_{k,i}|] = \frac{\sqrt{2}}{\sqrt{\pi}} \sigma_i \text{ and } \text{var}[|r_{k,i}|] = \sigma_i^2 \left(1 - \frac{2}{\pi}\right). \tag{3.11}$$

Next, having presented the notion of distance measure, we introduce the CUSUM procedure. For a given *distance measure* $z_{k,i} \in \mathbb{R}$, the CUSUM of page [33] is written as follows:

**CUSUM:** $S_{0,i} = 0, i \in \mathcal{I}$,

$$\begin{cases} S_{k,i} = \max(0, S_{k-1,i} + z_{k,i} - b_i), & \text{if } S_{k-1,i} \leq \tau_i, \\ S_{k,i} = 0 \text{ and } \tilde{k}_i = k - 1, & \text{if } S_{k-1,i} > \tau_i. \end{cases} \tag{3.12}$$

**Design parameters:** bias $b_i > 0$ and threshold $\tau_i > 0$.

**Output:** alarm time(s) $\tilde{k}_i$.

From (12), it can be seen that $S_{k,i}$ accumulates the distance measure $z_{k,i}$ over time. When this accumulation becomes greater than a certain threshold $\tau_i$ an alarm is raised. The sequence $S_{k,i}$ is reset to zero each time it becomes negative or larger than $\tau_i$. If $z_{k,i}$ is an independent nonnegative sequence (which is our case) and $b_i$ is not sufficiently large, the CUSUM sequence $S_{k,i}$ grows unbounded until the threshold $\tau_i$ is reached, no matter how large $\tau_i$ is set. In order to prevent these drifts, the bias $b_i$ must be selected properly based on the statistical properties of the distance measure. Once the bias is chosen, the threshold $\tau_i$ must be selected to fulfill a required false alarm rate $\mathcal{A}_i^*$. The occurrence of an alarm in the CUSUM when there are no attacks to the CPS is referred to as a false alarm, and $\mathcal{A}_i \in [0,1]$ denotes the *false alarm rate* for the CUSUM procedure, defined as the expected proportion of observations which are false alarms [34,35].

### 3.2.4 Bad-Data Detector

We have also implemented the bad-data detector for this case of study because it is widely used in the CPS security literature [36,37]. We also present a performance comparison between the CUSUM and the bad-data detectors. For the residual sequence $r_{k,i}$ given by (6), the bad-data detector is defined as follows.

**Bad-Data Procedure:**

$$\text{If } |r_{k,i}| > \alpha_i, \quad \tilde{k}_i = k, \ i \in \mathcal{I}. \tag{3.13}$$

**Design parameter:** threshold $\alpha_i > 0$.
**Output:** alarm time(s) $\tilde{k}_i$.

Using the bad-data detector an alarm is triggered if distance measure $|r_{k,i}|$ exceeds the threshold $\alpha_i$. Similar to the CUSUM procedure, the parameter $\alpha_i$ is selected to satisfy a required false alarm rate $\mathcal{A}_i^*$ (Figure 3.3).

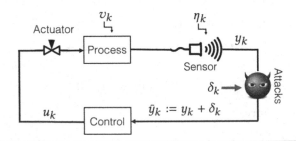

**Figure 3.3   CPS under attack.**

## 3.3 Attacker and Attack Models

In this section, we introduce the types of attacks launched on our water treatment plant. Essentially, the attacker model encompasses the attacker's intentions and its capabilities. The attacker may choose his or her goals from a set of intentions [38], including performance degradation, disturbing a physical property of the system, or damaging a component. In our experiments, three classes of attacks are modeled and executed. It is assumed that the attacker has access to $y_{k,i} = C_i y_k + \eta_{k,i}$ (i.e., the opponent has access to sensor measurements). Also, the attacker knows the system dynamics, the state space matrices, the control inputs and outputs, and the implemented detection procedure. The adversary has perfect knowledge of the Kalman filter and can modify the sensor readings to an arbitrary value.

1. *Bias Injection Attack*: First, a failure-like attack is designed. The attacker's goal is to deceive the control system by sending incorrect sensor measurements. In this scenario, the level sensor measurements are increased while the actual tank level is invariant. This makes the controller think that the attacked values are true sensor readings; and hence, the water pump keeps working until the tank is empty and the pump is burned out. The attack vector can be defined as,

$$\bar{y}_k = y_k + \eta_k + \bar{\delta}, \qquad (3.14)$$

   where $\bar{\delta}$ is the bias injected by the attacker.

2. *Stealthy Attack*: The second attack is more sophisticated and is carried out by carefully generating $\delta_k$ to drive the system to an undesired state. The objective of this attack is to maximize the damage without raising alarms. This attack is designed to deceive the detection schemes explained in Sections 2.2.3 and 2.2.4. A detailed analysis on how to design such an attack is presented in Section 3.4. This attack does not cause alarms because if the injected value and the previous steady state measurement differ only by a small amount, the residual value would not be sufficiently large to raise an alarm. By knowing the parameters of the detection procedure, it is always possible to modify the sensor values by an amount such that the residuals never cross the detection thresholds.

## 3.4 Limitations of Statistical Attack Detectors

As specified in the previous section, it is important to carefully select the parameters of the detectors. For the bad-data detector, we only have to take care of threshold $\alpha_i$ but for CUSUM, we have two parameters, the bias $b_i$ and the threshold $\tau_i$. For selecting the thresholds, it is intuitive to select them not too small or too large.

Small thresholds result in increased false alarms while large ones may result in unde-tected attacks. For the CUSUM, too-small values of bias $b_i$ lead to an unbounded growth of the CUSUM sequence while too-large $b_i$ hides the effect of the attacker. In [19], the authors present tools for selecting $b_i$ and $\tau_i$ based on the statistical prop-erties of the distance measure $z_{k,i}$. In what follows, we briefly introduce these tools.

### 3.4.1 False Alarm Rate for the Detectors

Consider the closed-loop system (1),(2–5). Assume that sensors $y_{k,i}$ are monitored for attack detection. First, for $i \in \{1,...,m\}$, let $\delta_{k,i} = 0$ and consider the CUSUM pro-cedure (12) with distance measure $z_{k,i} = |r_{k,i}|$ and residual sequence (6). According to Theorem 1 in [19], the bias $b_i$ must be selected larger than $\bar{b}_i = \sigma_i\sqrt{2/\pi}$ to ensure mean square boundedness of $S_{k,i}$ independent of the threshold $\tau_i$. The standard deviation $\sigma_i$ is given by the square root of the $i$-th entry of the residual covariance matrix $\Sigma$ given in (8). In our analysis we set $b_i = 2\bar{b}_i$. Next, for the desired false alarm rate $\mathcal{A}_i^* = 0.05(5\%)$, we compute the corresponding thresholds $\tau_i = \tau_i^*$, using Theorem 2 and Remark 2 in [19].

For the bad-data detector, we can also find the thresholds $\alpha_i$ using the tools [19]. That is, if

$$\alpha_i = \alpha_i^* := \sqrt{2}\sigma_i\mathrm{erf}^{-1}(1 - \mathcal{A}_i^*), \qquad (3.15)$$

where $\mathrm{erf}(\cdot)$ denotes the error function [39]. Then, $\mathcal{A}_i = \mathcal{A}_i^*$ for attack-free systems with $r_{k,i} \sim \mathcal{N}(0,\sigma_i^2)$, where $\mathcal{A}_i$ denotes the actual false alarm rate and $\mathcal{A}_i^*$ is the desired false alarm rate.

### 3.4.2 State Estimation under Attacks

In this section, we assess the performance of the bad-data and the CUSUM pro-cedures by quantifying the effect of the attack sequence $\delta_k$ on process dynamics when they are used to detect anomalies. In particular, we characterize for a class of stealthy attacks, the largest deviation on the estimation error due to the attack sequence. We derive upper bounds on the expectation of the estimation error given the system dynamics, the Kalman filter, the attack sequence, and the parameters of the detection procedure. For the same class of attacks, we quantify the largest deviation of the expectation for the estimation error when using the bad-data pro-cedure and then compare it with the one obtained with the CUSUM.

### 3.4.3 Design of Stealthy Attacks

We executed the two types of stealthy attacks on a SWaT testbed against the intro-duced statistical detectors.

***Stealthy Attack for Bad-Data Detector:*** This attack is designed to stay undetected by the bad-data detectors. The attacker knows the system dynamics, has
access to sensor readings, and knows the detector parameters; he or she is able to
inject false data into real-time measurements and stay undetected. Consider the
bad-data procedure and write (13) in terms of the estimated state $\hat{x}_k$,

$$|r_{k,i}| = |y_{k,i} - C_i\hat{x}_k + \delta_{k,i}| \le \alpha_i, \ i \in \mathcal{I}. \tag{3.16}$$

By assumption, the attacker has access to $y_{k,i} = C_i x_k + \eta_{k,i}$. Moreover, given his or
her perfect knowledge of the observer, the opponent can compute the estimated
output $C_i\hat{x}_k$ and then construct $y_{k,i} - C_i\hat{x}_k$. It follows that

$$\delta_{k,i} = C_i\hat{x}_k - y_{k,i} + \alpha_i - \epsilon_i, (\alpha_i > \epsilon_i) \rightarrow |r_{k,i}| = \alpha_i - \epsilon_i, \ i \in \mathcal{I}, \tag{3.17}$$

is a feasible attack sequence given the capabilities of the attacker. The constant
$\epsilon_i > 0$ is a small positive constant introduced to account for numerical precision.
These attacks maximize the damage to the CPS by immediately saturating and
maintaining $|r_{k,i}|$ at the constant $\alpha_i - \epsilon_i$. Therefore, for this attack, the sensor mea-
surements received by the controller take the form,

$$\bar{y}_{k,i} = C_i\hat{x}_k + \alpha_i - \epsilon_i. \tag{3.18}$$

***Stealthy Attack for CUSUM Detector:*** This attack is designed to stay undetected
by the CUSUM detectors. Consider the CUSUM procedure and write (12) in
terms of the estimated state $\hat{x}_k$,

$$S_{k,i} = \max(0, S_{k-1,i} + |y_{k,i} - C_i\hat{x}_k + \delta_{k,i}| - b_i), \tag{3.19}$$

if $S_{k-1,i} \le \tau_i$ and $S_{k,i} = 0$ if $S_{k-1,i} > \tau_i$. As with the bad-data procedure, we look for
attack sequences that immediately saturate and then maintain the CUSUM sta-
tistic at $S_{k,i} = \tau_i - \epsilon_i$ where $\epsilon_i$ ($\min(\tau_i, b_i) > \epsilon_i > 0$) is a small positive constant introduced to account for numerical precision. Assume that the attack starts at some
$k = k^* \ge 1$ and $S_{k^*-1,i} \le \tau_i$, i.e., the attack does not start immediately after a false
alarm. Consider the attack

$$\delta_{k,i} = \begin{cases} \tau_i - \epsilon_i + b_i - y_{k,i} + C_i\hat{x}_k - S_{k-1,i}, & k = k^*, \\ b_i - y_{k,i} + C_i\hat{x}_k, & k > k^*. \end{cases} \tag{3.20}$$

This attack accomplishes $S_{k,i} = \tau_i - \epsilon_i$ for all $k \ge k^*$ (thus, zero alarms). Note that the
attacker can only induce this sequence by exactly knowing $S_{k^*-1,i}$, i.e., the value of
the CUSUM sequence one step before the attack. This is a strong assumption since
it represents a real-time quantity that is not communicated over the communication

network. Even if the opponent has access to the parameters of the CUSUM, $(b_i, \tau_i)$, given the stochastic nature of the residuals, the attacker would need to know the complete history of observations (from when the CUSUM was started) to be able to reconstruct $S_{k^*-1,i}$ from data. This is an inherent security advantage in favor of the CUSUM over static detectors like the bad data. Nevertheless, for evaluating the worst-case scenario, we assume that the attacker has access to $S_{k^*-1,i}$. Therefore, for this attack, the sensor measurements received by the controller take the form,

$$
\bar{y}_{k,i} = \begin{cases} C_i \hat{x}_k + \tau_i - \epsilon_i + b_i - S_{k-1,i} - \epsilon_i, & k = k^*, \\ C_i \hat{x}_k + b_i, & k > k^*. \end{cases} \tag{3.21}
$$

***Residual Vector for Bias Attack and Normal Operation:*** Figure 3.4 shows a plot for a residual vector for the case of a constant bias attack on a flow meter and also shows the normal operation. In Figure 3.4 the region labeled as attacked is under attack and rest of it is in normal mode. Left hand side plot shows flow meter measurements as well as flow meter measurement estimate. On the right hand side plot the residual vector is shown. It can be observed that during the signal spoofing

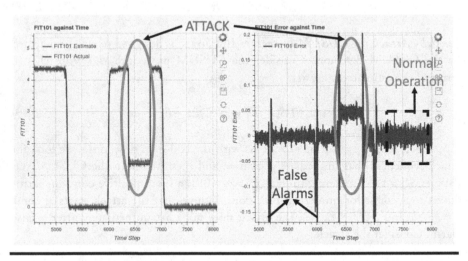

**Figure 3.4 Constant bias attack on flow meter (FIT101) in stage 1 of SWaT testbed. During the attack, a fake value is spoofed, showing a reduction in flow in the pipe. In the left-hand plot real sensor measurements under the attack are shown, and on the right-hand side residual for the sensor is shown. It can be seen that during the signal spoofing attack on the sensor, the residual deviates from the normal enabling statistics-based attack detectors to successfully detect attack. This type of attack is detected by the CUSUM, the bad-data and the NoisePrint detectors.**

attack on the sensor the residual deviates from the normal behavior enabling the statistics-based attack detectors to successfully detect attack. This type of attack is detected by the CUSUM, the bad-data and the *NoisePrint* detectors.

***Residual Vector for Stealthy Attack:*** Figure 3.5 shows a plot for the residual vector when system is under *stealthy* attack. The left-hand side plot shows the real-time data for the level sensor (LIT101) in stage 1 of the SWaT testbed, while the right-hand side plot shows the residual vector. From the design of *stealthy* attacks in the previous section, it was expected that the attacker would spoof the sensor data to stay stealthy for the statistical detectors. In an attempt to be stealthy but still be able to damage the plant [16,18,19], an attacker would inevitably modify the noise pattern in the residual vector. As we hypothesize, the randomness in the residue vector is a function of the sensor and process noise and a proof is given in the next section. The intuition for the proposed technique is based on this noise pattern in the residue vector. Sensor noise part is due to the physical structure of the sensor [40–42] and process noise is the property of the process, e.g., water sloshing in the tank [38,43].

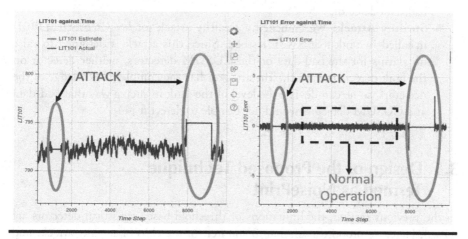

**Figure 3.5 Stealthy Attack. The right-hand plot shows the residual signal for the level sensor (LIT101) in the first stage of the SWaT testbed and the left-hand side plot is of real sensor measurements. Sensor measurements and estimates are plotted on the left and overlap each other. Two stealthy attacks are executed. The first attack executed around 1000 seconds changed the sensor value so that the injected value is chosen from the noise distribution of the sensor. This attacker could not successfully replicate the noise profile. This attack is stealthy for the CUSUM and bad-data detectors but not for the NoisePrint. For the case of the second attack, an attacker chose a value small enough to avoid detection using CUSUM or bad-data detector but as seen in the right-hand side plot the noise profile of the sensor has been changed resulting in detection by NoisePrint.**

**Table 3.1  Attack Detection Performance and Comparison Between Detectors**

| Attack Type/Detector | Bad-Data Detector | CUSUM Detector |
|---|---|---|
| Stealthy Attack | Not Detected | Not Detected |
| Constant Bias Attack | Detected | Detected |

A visual comparison of normal operation and system under attack in Figure 3.4 reveals the deviation from the normal noise pattern when the system is under attack. ***Attack Detection***: Table 3.1 shows the results for the performance of the attack detectors.

- **Constant Bias Attack:** Figure 3.4 shows the water flow at the stage 1 of the SWaT testbed, when the system is under a constant bias attack. The PLC received this attacked measurement value. In Figure 3.4, it can be seen that such an attack changes the statistics of the residual vector and could be detected using threshold-based methods, for example, the CUSUM and the bad-data detector.
- **Stealthy Attack:** We launched a stealthy attack for level sensor (LIT101) installed in tank-1 at SWaT testbed. Since this attack is designed to raise no alarms for the bad-data or the CUSUM detectors, neither detector on the tank detects the attack. The attacker has the complete knowledge of the detectors, so he can deviate the level of the tank in such a way that bad-data and CUSUM detectors would not be able to detect it [44].

## 3.5  Design of the Proposed Technique Termed as NoisePrint

In the previous section, the limitations of threshold-based statistical detectors are shown. Here, we propose a solution to counter for the class of stealthy attacks and also to authenticate the sensors in a CPS [44]. The proposed technique is termed *NoisePrint* [44]. The proposed scheme serves as a device identification framework and it can also detect a range of attacks on sensors. The proposed attack detection framework improves the limitations of model-based attack detection schemes. In general, for a complex CPS, there can be many possible attack scenarios. However, stealthy attack is a worst-case scenario for a model-based attack detection method employing a threshold-based detector. A stealthy attack exposes the limitations of threshold-based statistical attack detection methods. To be fair while making a comparison, we choose the same attack vector, namely stealthy attack. Another important thing is that the input to *NoisePrint* and reference methods is the same, i.e., a residual vector. We also executed a bias attack as an example of an attack which can be detected using CUSUM and bad-data detectors. The proposed scheme is a nonintrusive sensor and process fingerprinting method to authenticate sensors transmitting measurements to one or more PLCs. To apply this method we need to extract noise pattern for which the system model of a CPS is used. This scheme intelligently uses a model of the system in a novel way to extract noise pattern and then input that noise to NoisePrint [44]. The input to NoisePrint block is a function of sensor and process noise. Sensor noise is due to the construction of the sensor and process noise due to variations in the process, e.g., fluid sloshing in a storage tank in a process plant. Sensor noise is different from one sensor to another because of hardware imperfections during the manufacturing process [40]. Process noise is unique among different processes essentially because of different process dynamics. Sensor and process noise can be captured using a real system state (from sensor measurements) and system state estimate (from system model). These noise variations affect each device and process differently, and thus are hard to control or reproduce [45], making physical or digital spoofing of sensor noise profiles challenging.

A technique, referred to as NoisePrint, is designed to fingerprint sensor and process found in CPS. NoisePrint creates a noise fingerprint based on a set of time domain and frequency domain features that are extracted from the sensor and process noise. To extract noise pattern a system model–based method is used. A two-class support vector machine (SVM) is used to identify each sensor from a dataset, comprising a multitude of industrial sensors. According to the ground truth, one class is labeled as legitimate sensor/process and other class as illegitimate data (including attacks and data from the rest of the sensors in the plant). Experiments are performed on an operational water treatment testbed accessible for research [46,47]. A class of attacks as explained in a threat model are launched on a real water treatment testbed and results are compared with reference statistical methods. Sensor identification accuracy is observed to be at least 94.5% for a range of sensors.

The challenge in applying noise-based fingerprinting in a process plant is that the system states are dynamic. For example, for a level sensor, if the level of water stays constant we can easily extract noise fingerprint and construct a noise pattern profile for that sensor but in real processes system states keep changing; that is, fluid level in a tank keeps changing based on actuator actions. It is thus important to capture these variations as a function of control actions so that we can estimate the dynamic sensor measurements. To achieve this we have to get an analytical model for the system so that we can capture the system dynamics and be able to estimate future sensor measurements.

***System Dynamics:*** The first step is to collect the data from the real water treatment testbed called SWaT [47]. Data is collected over a period of 7 days under normal operation of the plant. This real-time dataset (RTDS) is composed of all sensor and actuator data. As explained in Section 3.2, a system model is obtained using the well-known subspace system identification techniques [28]. We used the Kalman filter to estimate the state of the system based on the available output $y_k$. The difference between real-time sensor measurement and sensor measurement estimate is the residual vector. The residual vector is a function of sensor and process noise and can be given as,

$$r_k = C \left\{ \sum_{i=0}^{k-2} (A - LC)^i (v_{k-i-1} - L\eta_{k-i-1}) \right\} + \eta_k \qquad (3.22)$$

where $r_k$ is residual at each time-instant $k \in \mathbb{N}$. $v_k \in \mathbb{R}^n$ is the process noise and $\eta_k \in \mathbb{R}^m$ is the sensor noise. $A$ and $C$ are state space matrices and $L$ is the Kalman filter gain.

This is an important intuition behind the idea of noise-based fingerprint as it can be seen that the residual vector obtained from the system model is a function of process and sensor noise. Using the system model and system state estimates it is possible to extract the sensor and process noise. Once we have obtained these residual vectors capturing sensor and process noise characteristics of the given CPS, we can proceed with pattern recognition techniques (e.g., machine learning) to fingerprint the given sensor and process.

### 3.5.1 Design of NoisePrint

In [44], the steps involved in composing a sensor and process noise fingerprint are explained. The proposed scheme begins with data collection and then divides data into smaller chunks to extract a set of time domain and frequency domain features. Features are combined and labeled with a sensor ID. A machine learning algorithm is used for sensor classification.

> *Residual Collection*: The next step after obtaining a system model for a CPS is to calculate the residual vector as explained in the previous section. Residual is collected for different types of industrial sensors present in SWaT testbed. The objective of residual collection step is to extract sensor and process noise by analyzing the residual vector. When the plant is running, an error in sensor reading is a combination of sensor noise and process noise (water sloshing etc.). The collected residual is analyzed, in time and frequency domains, to examine the noise patterns, which are found to follow a Gaussian distribution. Sensors and processes are profiled using variance and other statistical features in the noise vector. The experiment is run to obtain the sensor and process profile so that it can be used for later testing. A machine learning algorithm is used to profile sensors from fresh readings (test-data). Noise fingerprints can be generated over time or at the commissioning phase of the plant. Since these noise fingerprints are extracted from the system model, it does not matter if the process is dynamic or static.

***Feature Extraction:*** Data is collected from sensors at a sampling rate of one second. Since data is collected over time, we can use raw data to extract time domain features. We used the fast Fourier transform (FFT) algorithm [48] to convert data to the frequency domain and extract the spectral features. In total, as in Table 3.2, eight features are used to construct the fingerprint.

***Data Chunking:*** After residual collection, the next step is to create chunks of the dataset. In the following sections, it will be seen that we have performed experiments on a dataset collected over 7 days in a SWaT testbed. An important purpose of data chunking is to find out, how much is the sample size to train a well-performing machine learning model and how much data is required to make a decision about the presence or absence of an attacker.

**Table 3.2　List of Features Used. Vector $x$ is Time Domain Data from the Sensor for $N$ Elements in the Data Chunk. Vector $y$ is the Frequency Domain Feature of Sensor Data. $y_f$ is the Vector of Bin Frequencies and $y_m$ is the Magnitude of the Frequency Coefficients**

| Feature | Description |
|---|---|
| Mean | $\bar{x} = \dfrac{1}{N}\sum_{i=1}^{N} x_i$ |
| Std-Dev | $\sigma = \sqrt{\dfrac{1}{N-1}\sum_{i=1}^{N}(x_i - \bar{x}_i)^2}$ |
| Mean Avg. Dev | $D_{\bar{x}} = \dfrac{1}{N}\sum_{i=1}^{N}\lvert x_i - \bar{x}\rvert$ |
| Skewness | $\gamma = \dfrac{1}{N}\sum_{i=1}^{N}\left(\dfrac{x_i - \bar{x}}{\sigma}\right)^3$ |
| Kurtosis | $\beta = \dfrac{1}{N}\sum_{i=1}^{N}\left(\dfrac{x_i - \bar{x}}{\sigma}\right)^4 - 3$ |
| Spec. Std-Dev | $\sigma_s = \sqrt{\dfrac{\sum_{i=1}^{N}\left(y_f(i)^2\right)*y_m(i)}{\sum_{i=1}^{N}y_m(i)}}$ |
| Spec. Centroid | $C_s = \dfrac{\sum_{i=1}^{N}\left(y_f(i)\right)*y_m(}{\sum_{i=1}^{N}y_m(i)}$ |
| DC Component | $y_m(0)$ |

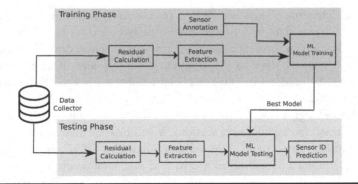

**Figure 3.6   NoisePrint framework.**

The whole residual dataset (total of $N$ readings) is divided into $m$ chunks (each chunk of $\lfloor \frac{N}{m} \rfloor$); we calculate the feature set $< F(C_i) >$ for each data chunk $i$. For each sensor, we have $m$ sets of features $< F(C_i) >_{i \in [1,m]}$. For $n$ sensors we can use $n \times m$ sets of features to train the multi-class SVM. We use a supervised learning method for sensor identification which has two phases—training and testing. For both phases, we create chunks in a similar way as explained earlier.

***Size of Training and Testing Dataset:*** It is found empirically that 2-class SVM produced highest accuracy for the chunk size of $\lfloor \frac{N}{m} \rfloor = 120$. For a total of $m$ feature sets for each sensor, at first we used half $(\frac{m}{2})$ for training and half $(\frac{m}{2})$ for testing. To analyze the accuracy of the classifier for smaller feature sets during the training phase, we began to reduce the number of feature sets starting with $\frac{m}{2}$. Classification was then carried out for the following corresponding range of feature sets for Training: $\{\frac{m}{2}, \frac{m}{3}, \frac{m}{4}, \frac{m}{5}, \frac{m}{10}\}$, and for Testing: $\{\frac{m}{2}, \frac{2m}{3}, \frac{3m}{4}, \frac{4m}{5}, \frac{9m}{10}\}$, respectively. In Section 3.7, empirical results are presented for such feature sets and the one with best performance is chosen for further analysis of the proposed scheme. For the classifier we have used a multi-class SVM library [49], as briefly described in Figure 3.6.

## 3.5.2 Support Vector Machine Classifier

SVM is a data classification technique used in many areas such as speech recognition, image recognition and so on [50]. The aim of SVM is to produce a model based on the training data and to give classification results for testing data. For a training set of instance-label pairs $(x_i, y_i), i = 1, ..., k$ where $x_i \in \mathbb{R}^n$ and $y \in \{1, -1\}^k$, SVM requires the solution of the following optimization problem:

$$\begin{aligned}
\underset{w,b,\zeta}{\text{minimize}} \quad & \frac{1}{2}w^T w + C\sum_{i=1}^{k}\zeta_i \\
\text{subject to} \quad & y_i(w^T\phi(x_i)+b) \geq 1-\zeta_i, \\
\text{where } \zeta_i \geq 0.
\end{aligned} \tag{3.23}$$

The function $\zeta$ maps the training vectors into a higher dimensional space. In this higher dimensional space a linear separating hyperplane is found by SVM, where $C > 0$ is the penalty parameter of the error term. For the kernel function in this work we use the radial basis function:

$$K(x_i, x_j) = exp(-\gamma \| x_i - x_j \|^2), \gamma > 0. \tag{3.24}$$

In our work, we have multiple sensors to classify. Therefore, multi-class SVM library LIBSVM [49] is used.

## 3.6 Case Study: Secure Water Treatment (SWaT) Testbed

In this section, a case study of a water treatment plant is presented. First, the architecture of the plant is presented, followed by the derivation of a system model and model validation. Lastly, sensor identification results are presented.

### 3.6.1 Architecture of the SWaT Testbed

SWaT is a fully operational (research facility), scaled-down water treatment plant producing 5 gallons/minute of doubly filtered water; this testbed mimics large modern plants for water treatment [47].

**Water treatment process**: The treatment process is shown in Figure 3.7. SWaT consists of six distinct stages, each controlled by an independent programmable logic controller (PLC). Control actions are taken by the PLCs using data from sensors. Stage P1 controls the inflow of water to be treated by opening or closing a motorized valve MV-101. Water from the raw water tank is pumped via a chemical dosing station (stage P2, chlorination) to another UF (ultra filtration) feed water tank in stage P3. A UF feed pump in P3 sends water via a UF unit to RO (reverse osmosis) feed water tank in stage P4. Here an RO feed pump sends water through an ultraviolet dechlorination unit controlled by a PLC in stage P4. This step is necessary to remove any free chlorine from the water prior to passing it through the reverse osmosis unit in stage P5. Sodium bisulphate ($NaHSO_3$) can be added in stage P4 to control the ORP (oxidation reduction potential). In stage P5, the dechlorinated water is passed through a two-stage RO filtration unit. The filtered

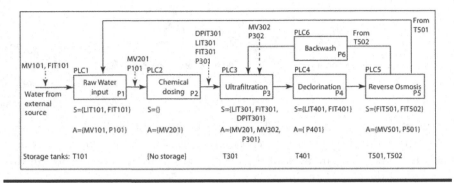

**Figure 3.7 Water treatment testbed: P1 though P6 indicate the six stages in the treatment process. Arrows denote the flow of water and of chemicals at the dosing station.**

water from the RO unit is stored in the permeate tank and the reject in the UF backwash tank. Stage P6 controls the cleaning of the membranes in the UF unit by turning on or off the UF backwash pump.

**Communications**: Each PLC obtains data from sensors associated with the corresponding stage, and controls pumps and valves in its domain. PLCs communicate with each other through a separate network. Communications among sensors, actuators, and PLCs can be via either wired or wireless links. Attacks that exploit vulnerabilities in the protocol used, and in the PLC firmware, are feasible and could compromise the communications links between sensors and PLCs, PLCs and actuators, among the PLCs, and the PLCs themselves. Having compromised one or more links, an attacker could use one of several strategies to send fake state data to one or more PLCs.

### 3.6.2 Model Validation

A system model for SWaT is obtained using the subspace system identification technique. Once we have obtained a system model, the next step is to validate the model. To validate the system model we use state space matrices obtained from system identification process and estimate the output of the system. We use the difference equation as shown in Eq. (3.2) to estimate the system state and estimate the sensor measurements. The estimate of the sensor measurements is compared with the real-time sensor measurement data. The comparison is shown in Figure 3.8. The top pane shows the sensor measurements and estimate of the sensor measurements from the obtained system model. In the middle pane we can see the difference between the real-time sensor measurements and estimate of the sensor measurements. PDF for the residual vector is plotted at the bottom pane. Here we can see that measurement estimate using the system model performs very similar to the real sensor measurements and PDF for residual vector has a very small variance. Besides visual validation

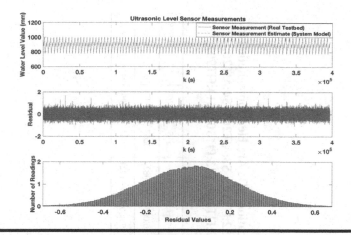

**Figure 3.8  Validating system model obtained using subspace system identification method.**

as shown in Figure 3.8, we used 1 minus the root mean square error (RMSE) as the metric of estimation accuracy or best fit of the model as shown.

$$\text{RMSE} = \sqrt{\frac{\sum_{i=1}^{n}(y_i - \hat{y}_i)^2}{n}} \qquad (3.25)$$

MSE is the difference between sensor measurement and sensor measurement estimate squared and essentially tells us the distance or how far is the estimated value from the measured value. The accuracy of the model for 18 sensors in the SWaT testbed used in this study is shown in Table 3.3. We can see that the model obtained ix very accurate with most of the accuracies as high as 99% with a couple of less accurate models. In control theory literature models with accuracies as high as 70% are considered an accurate approximation of real system dynamics [30].

## 3.6.2.1 Device Identification Performance

To see the performance of the proposed scheme we analyze the dataset in a systematic manner using machine learning and feature set as shown in Table 3.4 and the results are presented and discussed in the following. LibSVM [49] uses radial basis function (RBF) as a kernel function by default. However, SVM supports more classification functions, namely linear, polynomial (degree 3), or sigmoid. If the dataset is linearly separable then the linear model should be enough for the purpose of analysis. We performed an experiment to show which kernel function we should use to get the best accuracy models. We choose a chunk size of 120 readings and a 5 fold

**Table 3.3  Validating SWaT System Model Obtained from Subspace System Identification. FITs Are Electromagnetic Flow Meters, LITs Are Ultrasonic Level Sensors and PITs Are Pressure Sensors. S1:FIT101, S2:LIT101, S3:AIT201, S4:AIT202, S5:AIT203, S6:FIT201, S7:LIT301, S8:FIT301, S9:DPIT301, S10:LIT401, S11:FIT401, S12:FIT501, S13:PIT501, S14:FIT502, S15:PIT502, S16:FIT503, S17:PIT503, S18:FIT601**

| Sensor | S1 | S2 | S3 | S4 | S5 | S6 | S7 | S8 | S9 | S10 | S11 | S12 | S13 | S14 | S15 | S16 | S17 | S18 |
|---|---|---|---|---|---|---|---|---|---|---|---|---|---|---|---|---|---|---|
| RMSE | 0.0363 | 0.2867 | 0.0346 | 0.0113 | 0.0520 | 0.0313 | 0.2561 | 0.0200 | 0.0612 | 0.2267 | 0.0014 | 0.0096 | 0.0670 | 0.0082 | 0.0267 | 0.0037 | 0.0595 | 0.0035 |
| (1−RMSE)*100% | 96.3670 | 71.3273 | 96.5409 | 98.8675 | 94.8009 | 96.8656 | 74.3869 | 98.0032 | 93.8757 | 77.3296 | 99.8593 | 99.0377 | 93.3031 | 99.1821 | 97.3313 | 99.6251 | 94.0537 | 99.6501 |

**Table 3.4  Different Kernel Functions for SVM Classification**

| ↓ Kernel Function / Sensor → | S1 | S2 | S3 | S4 | S5 | S6 | S7 | S8 | S9 | S10 | S11 | S12 | S13 | S14 | S15 | S16 | S17 | S18 |
|---|---|---|---|---|---|---|---|---|---|---|---|---|---|---|---|---|---|---|
| Linear | 94.5721% | 94.5721% | 94.5935% | 94.5721% | 93.4229% | 94.5721% | 94.5721% | 96.4586% | 94.5721% | 100% | 94.5721% | 99.0083% | 97.3879% | 95.6213% | 94.5721% | 94.5721% | 94.5721% | 94.5721% |
| Polynomial | 42.0914% | 94.4703% | 60.6751% | 94.0697% | 75.5484% | 76.6287% | 94.4506% | 96.2649% | 94.5672% | 100% | 94.5721% | 94.4654% | 94.1141% | 95.5244% | 94.5705% | 94.408% | 93.6527% | 94.5721% |
| Radial Basis Function | 95.2125% | 95.0253% | 96.4684% | 94.6821% | 94.9547% | 96.6162% | 94.5721% | 94.5721% | 94.5721% | 100% | 94.5721% | 98.8507% | 97.2138% | 95.5884% | 94.5721% | 94.5721% | 95.8626% | 94.5721% |
| Sigmoid | 91.6809% | 89.5071% | 91.2163% | 89.517% | 91.1494% | 89.934% | 91.3706% | 89.691% | 89.824% | 100% | 89.7321% | 92.7235% | 88.7158% | 91.2967% | 90.5382% | 89.2707% | 89.2838% | 89.2691% |

cross-validation to test other classification functions. Results in Table 3.4 show that linear and RBF kernel function has high accuracy but RBF performs slightly better in all the cases, so we choose RBF as the classification function for further analysis and attack detection purposes. From these results, it can be seen that lowest sensor identification accuracy is 94%. These results support our hypothesis that sensor fingerprints exist and one can authenticate data coming from those sensors with high accuracy.

## 3.7 Conclusion

Limitations of state estimation-based attack detection schemes for CPS are discussed in this chapter. Implementation on a testbed in a realistic setting offers valuable insights. It is observed that such techniques are not effective in detecting stealthy false data injection attacks. Only random attacks similar to faults got detected, which is a well-known use of linear filters for dynamic systems. It is concluded that Kalman filter and threshold-based detection schemes cannot be employed as a defense mechanism against strategic attacks. NoisePrint is designed to overcome the limitations of threshold-based detectors. A system model for a real water treatment plant is created and validated. Higher sensor identification accuracy is the proof of fingerprint existence and it is concluded that noise-based features can be used to fingerprint sensors in cyber-physical systems.

## References

1. R. Alur. 2015. *Principles of Cyber-Physical Systems*. MIT Press.
2. E. A. Lee. 2008. Cyber physical systems: Design challenges. In *2008 11th IEEE International Symposium on Object and Component-Oriented Real-Time Distributed Computing (ISORC)*. 363–369. doi:10.1109/ISORC.2008.25.
3. NIST. 2014. Cyber-Physical Systems. https://www.nist.gov/el/cyber-physical-systems.
4. Defense Use Case. 2016. Analysis of the Cyber Attack on the Ukrainian Power Grid.
5. N. Falliere, L.O. Murchu, and E. Chien. 2011. W32 Stuxnet Dossier. Symantec, version 1.4. https://www.symantec.com/content/en/us/enterprise/media/security_response/whitepapers/w32_stuxnet_dossier.pdf.
6. J. Slay and M. Miller. 2008. *Lessons Learned from the Maroochy Water Breach*. Springer 620 US, Boston, MA, 73–82.
7. A. Cardenas, S. Amin, B. Sinopoli, A. Giani, A. Perrig, and S. Sastry. 2009. Challenges for securing cyber physical systems. In *Workshop on future directions in cyber-physical systems security*. 5.
8. D. Gollmann and M. Krotofil. 2016. *Cyber-Physical Systems Security*. Springer, Berlin, Germany, 195–204. doi:10.1007/978-3-662-49301-4_14.

9. CNN. [n. d.]. Staged Cyber Attack Reveals Vulnerability in Power Grid. http://edition.cnn.com/2007/US/09/26/power.at.risk/index.html,year=2007.

10. Wired. 2015. A Cyberattack Has Caused Confirmed Physical Damage for the Second Time Ever. https://www.wired.com/2015/01/german-steel-mill-hack-destruction/.

11. D. I. Urbina, J. A. Giraldo, A. A. Cardenas, N. O. Tippenhauer, J. Valente, M. Faisal, J. Ruths, R. Candell, and H. Sandberg. 2016. Limiting the impact of stealthy attacks on industrial control systems. In *Proceedings of the 2016 ACM SIGSAC Conference on Computer and Communications Security*. ACM, 1092–1105.

12. Y. Shoukry, P. Martin, Y. Yona, S. Diggavi, and M. Srivastava. 2015. PyCRA: Physical Challenge-Response Authentication for Active Sensors Under Spoofing Attacks. In *Proceedings of the 22Nd ACM SIGSAC Conference on Computer and Communications Security (CCS'15)*. ACM, New York, 1004–1015. doi:10.1145/2810103.2813679.

13. Y. Son, H. Shin, D. Kim, Y. Park, J. Noh, K. Choi, J. Choi, and Y. Kim. 2015. Rocking drones with intentional sound noise on gyroscopic sensors. In *Proceedings of the 24th USENIX Conference on Security Symposium (SEC'15)*. USENIX Association, Berkeley, CA, 881–896. http://dl.acm.org/citation.cfm?id=2831143.2831199.

14. Y. Mo and B. Sinopoli. 2012. Integrity attacks on cyber-physical systems. In *Proceedings of the 1st International Conference on High Confidence Networked Systems (HiCoNS'12)*. ACM, New York, 47–54. doi:10.1145/2185505.2185514.

15. Y. Mo and B. Sinopoli. 2009. Secure control against replay attacks. In *2009 47th Annual Allerton Conference on Communication, Control, and Computing (Allerton)*. 911–918. doi:10.1109/ALLERTON.2009.5394956.

16. G. Dan and H. Sandberg. 2010. Stealth attacks and protection schemes for state estimators in power systems. In *Smart Grid Communications (SmartGridComm), 2010 First IEEE International Conference on*. IEEE, 214–219.

17. C. M. Ahmed, A. Sridhar, and M. Aditya. 2016. Limitations of state estimation based cyber attack detection schemes in industrial control systems. In *IEEE Smart City Security and Privacy Workshop, CPSWeek*. IEEE.

18. C. M. Ahmed, C. Murguia, and J. Ruths. 2017. Model-based attack detection scheme for smart water distribution networks. In *Proceedings of the 2017 ACM on Asia Conference on Computer and Communications Security (ASIA CCS'17)*. ACM, New York, 101–113. doi:10.1145/3052973.3053011.

19. C. Murguia and J. Ruths. 2016. Characterization of a CUSUM model-based sensor attack detector. In *2016 IEEE 55th Conference on Decision and Control (CDC)*. 1303–1309. doi:10.1109/CDC.2016.7798446.

20. R. Qadeer, C. Murguia, C.M. Ahmed, and J. Ruths. 2017. Multistage downstream attack detection in a cyber physical system. In *CyberICPS Workshop 2017, in Conjunction with ESORICS 2017*.

21. D. F. Kune, J. Backes, S. S. Clark, D. Kramer, M. Reynolds, K. Fu, Y. Kim, and W. Xu. 2013. Ghost talk: Mitigating EMI signal injection attacks against analog sensors. In *2013 IEEE Symposium on Security and Privacy*. 145–159. doi:10.1109/SP.2013.20.

22. S. Yasser, M. Paul, T. Paulo, and S. Mani. 2013. Non-invasive spoofing attacks for anti-lock braking systems. In *CHES*, Springer Link, 8086. 55–72.

23. H. Fawzi, P. Tabuada, and S. Diggavi. 2014. Secure estimation and control for cyber-physical systems under adversarial attacks. In *IEEE Transaction Automatic Control*. 59(6), 1454–1467. doi:10.1109/TAC.2014.2303233.

24. Y. Mo, S. Weerakkody, and B. Sinopoli. 2015. Physical authentication of control systems: Designing watermarked control inputs to detect counterfeit sensor outputs. In *IEEE Control Systems Magazine* 35(1), 93–109. doi:10.1109/MCS.2014.2364724.

25. K. Manandhar, X. Cao, F. Hu, and Y. Liu. 2014. Detection of faults and attacks including false data injection attack in smart grid using Kalman filter. *IEEE Transactions on Control of Network Systems* 1(4), 370–379. doi:10.1109/TCNS.2014.2357531.

26. Y. Liu, P. Ning, and M. K. Reiter. 2011. False data injection attacks against state estimation in electric power grids. *ACM Transaction Information and System Security* 14(1), Article 13, 33. doi:10.1145/1952982.1952995.

27. K. J. Aström and B. Wittenmark. 1997. *Computer-Controlled Systems* (3rd ed.). Prentice Hall, Upper Saddle River, NJ.

28. P. Van Overschee and B. De Moor. 1996. *Subspace Identification for Linear Systems: Theory, Implementation, Applications.* Kluwer Academic Publications: Boston, MA.

29. M. Behl, A. Jain, and R. Mangharam. 2016. Data-driven modeling, control and tools for cyber-physical energy systems. In *2016 ACM/IEEE 7th International Conference on Cyber-Physical Systems (ICCPS).* 1–10. doi:10.1109/ICCPS.2016.7479093.

30. X. Wei, M. Verhaegen, and T. van Engelen. 2010. Sensor fault detection and isolation for wind turbines based on subspace identification and Kalman filter techniques. *International Journal of Adaptive Control and Signal Processing* 24(8), 687–707. doi:10.1002/acs.1162.

31. A. Cardenas, S. Amin, Z. Lin, Y. Huang, C. Huang, and S. Sastry. 2011. Attacks against process control systems: Risk assessment, detection, and response. In *6th ACM Symposium on Information, Computer and Communications Security.* 355–366.

32. M. Ross. 2006. *Introduction to Probability Models* (9th ed). Academic Press, Orlando, FL.

33. E. Page. 1954. Continuous inspection schemes. *Biometrika*, 41, 100–115.

34. B. M. Adams, W. H. Woodall, and C. A. Lowry. 1992. The use (and misuse) of false alarm probabilities in control chart design. *Frontiers in Statistical Quality Control.* 4, 155–168.

35. C.S. van Dobben de Bruyn. 1968. *Cumulative Sum Tests: Theory and Practice.* Griffin, London, UK.

36. Y. Gu, T. Liu, D. Wang, X. Guan, and Z. Xu. 2013. Bad data detection method for smart grids based on distributed estimation. In *IEEE ICC.*

37. L. Mili, T.V. Cutsen, and M.R.-Pavella. 1985. Bad Data Identification Methods in Power System State Estimation—A Comparative Study. *IEEE Trans. on Power Apparatus and Systems* (1985).

38. C. M. Ahmed and A. P. Mathur. 2017. Hardware Identification via Sensor Fingerprinting in a Cyber Physical System. In *2017 IEEE International Conference on Software Quality, Reliability and Security Companion (QRS-C).* 517–524. doi:10.1109/QRS-C.2017.89.

39. N. Lehtinen. 2010. Error functions. Stanford University, http://nlpc.stanford.edu/nleht/Science/reference/errorfun.pdf.

40. S. Dey, N. Roy, W. Xu, R. R. Choudhury, and S. Nelakuditi. 2014. Accelprint: Imperfections of accelerometers make smartphones trackable. In *Network and Distributed System Security Symposium (NDSS).*

41. C. Mujeeb Ahmed, A. Mathur, and M. Ochoa. 2017. NoiSense: Detecting data integrity attacks on sensor mseasurements using hardware based fingerprints. *ArXiv e-prints.* [arxiv]cs. CR/1712.01598.

42. J. Prakash and C. M. Ahmed. 2017. Can you see me on performance of wireless fingerprinting in a cyber physical system. In *2017 IEEE 18th International Symposium on High Assurance Systems Engineering (HASE)*. 163–170. doi:10.1109/HASE.2017.40.

43. J. Zhou, C. M. Ahmed and A. P. Mathur. 2018. Noise matters: Using sensor and process noise fingerprint to detect stealthy cyber attacks and authenticate sensors in CPS. In *2018 Annual Computer Security Applications Conference (ACSAC18)*.

44. C. M. Ahmed, M. Ochoa, J. Zhou, A. Mathur, R. Qadeer, C. Murguia, and J. Ruths. 2018. NoisePrint: Attack detection using sensor and process noise fingerprint in cyber physical systems. In *Proceedings of the 2018 ACM on Asia Conference on Computer and Communications Security (ASIA CCS'18)*. ACM. doi:10.1145/3196494.3196532.

45. R. M. Gerdes, T. E. Daniels, M. Mina, and S. F. Russell. 2006. Device identification via analog signal fingerprinting: A matched filter approach. In *NDSS*.

46. C. M. Ahmed, V. R. Palleti, and A. P. Mathur. 2017. WADI: A water distribution testbed for research in the design of secure cyber physical systems. In *Proceedings of the 3rd International Workshop on Cyber-Physical Systems for Smart Water Networks (CySWATER'17)*. ACM, New York, 25–28. doi:10.1145/3055366.3055375.

47. A. P. Mathur and N. O. Tippenhauer. 2016. SWaT: A water treatment testbed for research and training on ICS security. In *2016 International Workshop on Cyber-Physical Systems for Smart Water Networks (CySWater)*. 31–36. doi:10.1109/CySWater.2016.7469060.

48. P. Welch. 1967. The use of fast Fourier transform for the estimation of power spectra: A method based on time averaging over short, modified periodograms. *IEEE Transactions on Audio and Electroacoustics*. 15(2), 70–73.

49. C.-C. Chang and C.-J. Lin. 2011. LIBSVM: A library for support vector machines. *ACM Transactions on Intelligent Systems and Technology*. 2(2011), 27:1–27:27. Issue 3. Software available at http://www.csie.ntu.edu.tw/cjlin/libsvm.

50. Z. Akata, F. Perronnin, Z. Harchaoui, and C. Schmid. 2014. Good practice in large-scale learning for image classification. In *IEEE Transactions on Pattern Analysis and Machine Intelligence* 36(3), 507–520. doi:10.1109/TPAMI.2013.146.

# SECURITY AND PRIVACY IN CLOUD AND EMBEDDED SYSTEMS FOR CYBER-PHYSICAL SYSTEMS

# II

Chapter 4, "Towards Secure Software-Defined Networking Integrated Cyber-Physical Systems: Attacks and Countermeasures," provides an overview of software-defined networks (SDNs), smart grids and SDN-based smart grids and discusses various types of attacks related to SDN architecture and their countermeasures. It also demonstrates the applicability of SDN to provide security in smart grids. Finally, the chapter categorizes different types of attacks and their countermeasures related to SDN-based smart grids.

Chapter 5, "DDoS Defense in SDN-Based Cyber-Physical Cloud," provides an initial understanding of security challenges in SDNs by investigating how an SDN can improve network management of cyber-physical cloud computing (CPCC), additional security issues introduced over the integration of SDN and cloud computing and how SDN-based security solutions may enhance the resistance of SDN-based cloud systems against distributed denial of service (DDoS) attacks. Finally, it presents a proposed mechanism SDN-based synchronize (SYN) flooding defense in the cloud to prevent and mitigate SYN flooding attacks in the SDN-based cloud environment.

Chapter 6, "Detecting Pilot Contamination Attacks in Wireless Cyber-Physical Systems," proposes four different methods for the detection of pilot contamination attacks in wireless cyber-physical systems. The proposed methods are examined and their performance is compared.

Chapter 7, "Laboratory Exercises to Accompany Industrial Control and Embedded Systems Security Curriculum Modules," describes in detail the modules and the accompanying exercises and proposes future enhancements and extensions to these pedagogical instruments. It highlights the interaction between control and embedded systems security with Presidential Policy Directive 8, the National Preparedness Plan (NPP), cyber risk management and incident handling. Finally, it outlines the description and content of the modules in the areas of industrial control systems (ICSs); security embedded systems (SESs); and guidelines, standards and policy.

*Chapter 4*

# Towards Secure Software-Defined Networking Integrated Cyber-Physical Systems: Attacks and Countermeasures

Uttam Ghosh, Pushpita Chatterjee, Sachin S. Shetty, Charles Kamhoua and Laurent Njilla

## Contents

## 4.1 Introduction

A cyber-physical system (CPS) is an integration of the cyber world (computation and communication systems) and the man-made physical world (e.g., utility networks, vehicles, factories, etc.) as shown in Figure 4.1. CPS links the physical world with the cyber world by using sensors and actuators (Conti et al. 2012). Sensors are used to measure physical quantities and convert them into an electrical signal. This electrical signal is sampled and quantized later for computing. The cyber system calculates based on these values and sends feedback to the physical world by actuators that convert electrical signals into a physical action.

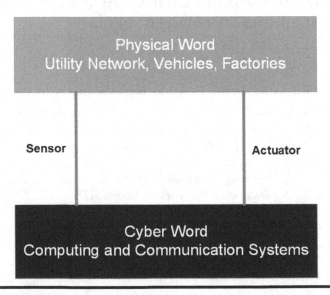

**Figure 4.1  Cyber-physical system using sensors and actuators.**

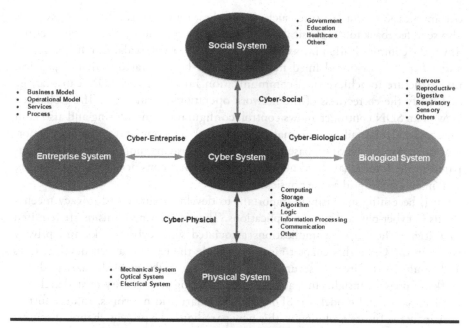

**Figure 4.2 Categorization of cyber systems based on application domains.**

CPSs combine digital and analog devices, computer systems and networks with the natural and man-made physical world. The man-made physical world includes buildings (homes, schools, offices, factories, etc.), utility networks (electricity, gas, water, etc.), transportation networks (roads, railways, airports, harbors, etc.), transportation vehicles (cars, rails, planes, etc.), healthcare systems, information technology networks, and so on. Theses physical infrastructures are integrated with cyber systems and broadly categorized into social, medical, physical and enterprise systems as depicted in Figure 4.2. Cyber systems make the physical infrastructures more smart, secure and reliable, and fully automated systems.

The emerging software-defined networking (SDN) paradigm provides flexibility to program the network centrally (logically) in controlling, managing, and dynamically reconfiguring the network. It decouples the network control and forwarding functions enabling the network control to become directly programmable and the underlying infrastructure to be abstracted for applications and network services. SDN mainly provides security and network virtualization for enhancing the overall network performance. SDN provides protection against various types of attacks by providing consistent access control, applying efficient and effective security policies, and managing and controlling the network through the use of a centralized SDN controller (Ghosh et al. 2016, 2017).

In CPSs, the deployed sensors generate a massive amount of real-time data from the physical infrastructure and send this to the cyber systems using the communication infrastructure (such as switches, routers, etc.). The cyber systems also send feedback to the physical devices using the communication infrastructure. Thus, the communication infrastructure in CPSs must be scalable, reliable, secure and efficient. Software-defined networking can be integrated with the physical infrastructure to achieve such communication infrastructure. SDN can manage and verify the correctness of the network operations at run time. The globalized view of an SDN controller allows control, configuration, monitoring and also fault (due to accidental failures and malicious attacks) detection and remediates abnormal operation in the SDN-based cyber-physical systems more efficiently as compared to the traditional based networks. This chapter considers SDN as the cyber system and smart grid as the physical system.

It is becoming increasingly important to develop security and privacy mechanisms in cyber-physical system applications. The security mechanisms are required to mitigate the negative implications associated with cyberattacks and privacy issues in the CPS. This chapter aims to provide the latest research developments and results in the areas of security and privacy for SDN-enabled smart grid networks. It presents insights into attacks in networking and security-related architectures, designs, and models for SDN-enabled smart grid networks. In an effort to anticipate the future evolution of this new paradigm, the present chapter discusses the main ongoing research efforts, frameworks, challenges, and research trends in this area. With this chapter, readers can have a more thorough understanding of SDN architecture, different security attacks and their countermeasures in SDN and SDN-enabled CPS environments.

*Chapter Organization:* The rest of the chapter is organized as follows: Section 4.2 presents a background of SDN, smart power grids and SDN-based smart grids. The attacks and countermeasures in SDN and SDN-based smart grids are presented in Sections 4.3 and 4.4 respectively. Finally, Section 4.5 concludes the chapter.

## 4.2 Background

### 4.2.1 An Overview of SDN

This section gives an overview of SDN architecture and working principles. It further discusses the importance of SDN and how SDN differs from traditional networking.

#### 4.2.1.1 SDN Architecture

Figure 4.3 presents major elements and interfaces of the SDN architecture. The SDN architecture has three layers: infrastructure layer, control layer, and the application layer.

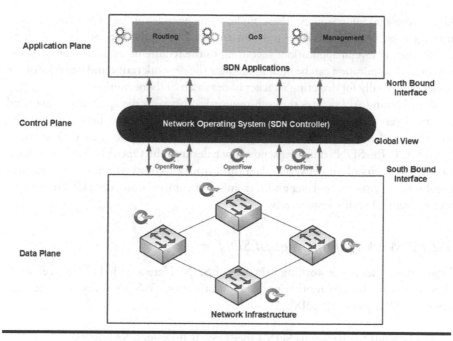

**Figure 4.3  SDN architecture.**

### 4.2.1.1.1 Infrastructure Layer

The infrastructure layer (also known as data plane) consists of a set of one or more traffic forwarding devices. These forwarding devices are known as OpenFlow (OF) switches and responsible for forwarding the data from source to destination in an SDN network based on instructions (flow rules) received from control layer.

### 4.2.1.1.2 Control Layer

The control plane consists of a set of SDN controllers. The SDN controller, also known as network operating system, is a logical entity (software program) that receives instructions or requirements from the SDN application layer and relays them to the OF switches of infrastructure layer. The controller keeps track of the network topology (global view of the network) and the statistics of the network traffic periodically. Thus, the controller is responsible for providing routing, traffic engineering or quality of service (QoS), load balancing and also security in the network.

### 4.2.1.1.3 Application Layer

The application layer comprises one or more applications (software programs) and controls the network resources with the SDN controller through the use of application programming interfaces (APIs). It collects information from the controller periodically for decision-making purposes. These applications provide routing, QoS and network management. This layer further provides an interface to the network administrator for developing several applications according to the requirements of the network. For instance, an application can be built to monitor the network traffic and behavior of the nodes periodically for detecting attacker nodes in an SDN network.

Northbound API defines the communication between the application layer and control layer whereas southbound API defines the connection between the control layer and infrastructure layer. OpenFlow protocol has been used as a southbound API. The SDN controller sends flow rules into the OpenFlow switches using OpenFlow protocol for delivering data from source to destination. OpenFlow protocol uses a secure socket layer and transmission control protocol (TCP) to provide security and reliability respectively.

### 4.2.1.2 Working Principles of SDN

Figure 4.4 presents the working principle of SDN. Here a node H2 (source) sends the packets of a flow to another node H1 (destination) in SDN using the following operations (Ghosh et al. 2018):

1. H2 sends the packet to SDN OpenFlow (OF) switch S3.
2. On receiving the packet from H2, S3 checks in its flow table for a matching flow rule. If the flow rule exists then S3 switch forwards the packet according to the flow rule towards H1. Otherwise, S3 forwards the packet (i.e., the first packet of the flow) to the SDN controller.

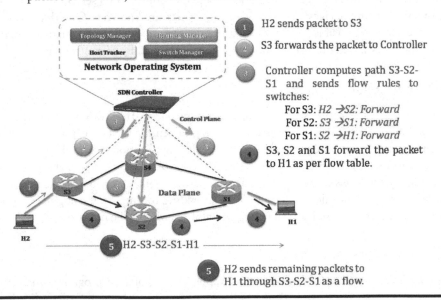

**Figure 4.4  SDN working principles.**

3. The SDN controller on receiving the packet from S3 computes the shortest path between H2 and H1 for the new flow. In this example, the shortest path is H2-S3-S2-S1-H1. In response, the SDN controller writes flow rules on all switches S3, S2 and S1 on the path. It may be noted here that the SDN controller has the global view and traffic statistics of the network. Hence, the SDN controller can choose different metrics instead of the shortest path for providing secure routing and quality of services.
4. The OF switch S3 forwards the packet to S2, S2, which forwards the packet to S1, and S1 finally forwards the packet to the actual destination H1 as per flow rule (received from the SDN controller in the previous step).
5. Thereafter, H2 sends all the packets of the flow to H1 through OF switches S3-S2-S1.

## 4.2.1.3 Traditional Networking versus SDN

In traditional or legacy networks, the forwarding devices (switches) are complicated and vendor dependent. These switches are strongly coupled between the control plane and the data plane. Thus, it is not easy to include new functionalities (applications) to the traditional networks; this fact is illustrated in Figure 4.5. The tight coupling of the control plane and data plane makes the development and deployment of new networking functionalities (e.g., routing, load balancing algorithms) very difficult. This is due to the fact that it needs a modification of the control plane of all the distributed switches in the network through the installation of new firmware and, in some cases, up-gradation of hardware. These legacy switches are distributed in a large area that makes it even more difficult to later change the network topology, configuration, and functionality. In contrast, SDN decouples the control plane from switches of the data plane and becomes

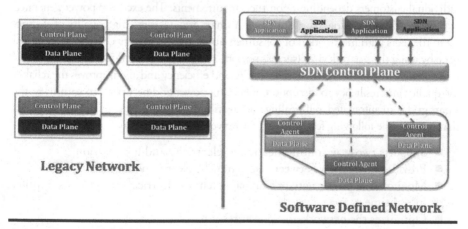

**Figure 4.5  Traditional (legacy) networking versus software-defined networking (SDN).**

a separate entity: SDN controller or network operating system (NOS). It has several advantages over traditional networks:

- An SDN controller is programmable. It is easier to include new network functionalities through programs (as applications) at the top of the SDN controller.
- An SDN controller is logically centralized. It keeps track of the network topology and statistics of the network traffic periodically. The controller is consistent and effective in taking decisions for routing, QoS and load balancing dynamically.
- A logically central SDN controller can control, configure and monitor the distributed devices of the data plane.
- A logically central SDN controller can monitor all the devices and their traffic in the data plane. The controller can run an application that can detect the malicious device whenever the device injects data falsely into the network or behaves abnormally. It can further eliminate the malicious device from the network on the fly by writing effective policies (flow rule) on the switches.

## 4.2.2 An Overview of Smart Power Grid

A power grid refers to an electric grid, which consists of generating stations that produce electrical power, high-voltage transmission lines that carry power from a few centralized generators to demand centers, and distribution lines that connect substantial numbers of individual customers. The smart grid is an up-gradation of the existing power grid with intelligent smart communication and computing devices. It forms a large-scale heterogeneous complex network between a large number of sensors, actuators, control, and data acquisition (SCADA) systems, and also smart meters located in residential and commercial premises. An architecture of a smart grid is presented in Figure 4.6 where energy can be generated seamlessly from different power sources (such as wind, solar, nuclear) and transmitted using transmission lines to the distribution center and finally distributed to the individual customers depending upon their requirements. The excessive power generated by a smart grid can be stored for future use. A smart grid delivers a high quality of electricity to the users with information of consumers about their electricity consumption in real time by using the smart devices (such as sensors, switches) to make flows of electricity and information two-way. It enhances the capacity and efficiency and also improves the reliability, quality, and resiliency to disruption of existing power grid networks. In summary, the smart grid is an automated, self-healing and distributed advanced energy delivery network that provides the following features with the power grid (Ghosh et al. 2018):

- Supports two-way communication of electricity and information
- Provides interaction between users and the electricity market
- Monitors the power network management of electrical energy consumption in real time
- Optimizes the network and power resources
- Enables integration, monitoring, control, security and maintenance
- Provides security against attacks and threats

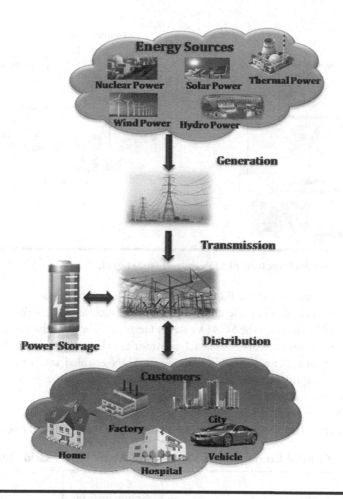

**Figure 4.6** **An architecture of smart power grid.**

## 4.2.3 *An Overview of SDN-Based Smart Grid*

An architecture of an SDN-based smart grid is presented in Figure 4.7. The architecture mainly comprises three segments: (i) a control center; (ii) communication network that consists of OpenFlow switches and links between them, and (iii) substations that consist of physical power grid devices (known as SCADA slaves in general) with sensors and actuators. The control center runs the commodity computers and servers for the SDN controller and SCADA master. The SCADA master is responsible for controlling, configuring and managing the grid devices in substations, whereas the SDN controller is responsible for configuring and managing the networking devices. The SCADA master collects the measurement data periodically from the SCADA slaves (sensors) in the substations through the communication network. It estimates the state (stability) of the smart grid by

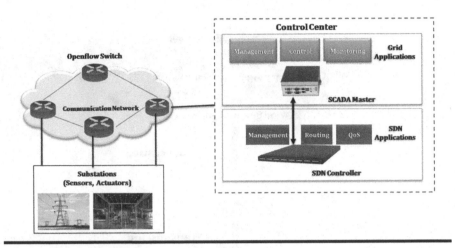

**Figure 4.7   An architecture of SDN-based smart grid.**

processing the received data. Based on the estimated state of the smart grid, the SCADA master then sends the control command (such as read, write or execute) (Lin et al. 2016a) back to the SCADA slaves (actuators) in the substations through the communication network. Table 4.1 summarizes the control commands issued by the SDN controller and SCADA master in SDN-enabled smart grid networks in order to configure and manage the network and grid devices respectively (Ghosh et al. 2017, Lin et al. 2016a).

Table 4.2 presents the comparison between the smart grid and SDN-based grid networks. The communication infrastructure must be scalable, resilient,

**Table 4.1   Control Commands in SDN-Based Smart Power Grid Networks**

| Control Commands by SCADA Master | Functionalities | Control Commands by SDN Controller | Functionalities |
|---|---|---|---|
| Read | Retrieve measurements from substations (SCADA slaves) by SCADA master | Add _Flow | Add a new flow rule to OF switches by SDN controller |
| Write | Configure smart grid devices by SCADA master | Del_Flow | Remove a flow from OF switches by SDN controller |
| Execute | Operate smart grid devices by SCADA master | Mod _Flow | Edit a flow in OF switches by SDN controller |

**Table 4.2   Smart Grid versus SDN-Based Smart Grid**

| Features | Smart Grids | SDN-Based Smart Grid |
|---|---|---|
| Programmability | Smart grids are less programmable. | SDN-based smart grids are programmable as SDN controller is programmable. |
| Protocol independency | Smart grids depend on some specific protocols. | SDN-based smart grids independent of the protocol through the use of SDN controller |
| Granularity | Fully dependent on proprietary hardware | SDN-based smart grid networks are programmable and independent of proprietary hardware; SDN controller can identify the traffic at every flow and packet level and provide QoS. |
| Resiliency | Limited resiliency to malicious attacks and failures | Resilient against malicious attacks and failures by using SDN |
| Interoperability | Smart grids depend on vendor-dependent hardware and software (vendor-lock); thus, it is difficult to configure and manage the smart grid with different vendor-specific devices and protocols. | SDN is not dependent on vendor and working on open standards; thus, it is easy to configure and manage the smart grid with different vendor-specific devices and protocols. |
| Management of Network | Managing the smart grid network is complex and time-consuming, and even sometimes manual. | SDN controller is logically centralized and it can manage the smart grid network easily and automatically. |
| Security | Several security schemes are proposed. | SDN can provide security by using controller security policies; however, it needs to develop new security schemes as an SDN controller may compromise or controller applications may get compromised. |

secure and efficient for the smart grid as it is a large-scale heterogeneous complex network that generates a massive amount of real-time data. An SDN provides such communication infrastructure in SDN-based smart grids. The SDN controller is programmable and it keeps track of the network traffic and topology periodically. Thus, it can be used for load balancing, for dynamically adjusting the routing paths for the control commands (Zhao et al. 2016), fast failure detection (Dorsch et al. 2016), security (Ghosh et al. 2017), self-healing (Lin et al. 2016b), and also for monitoring and scheduling of critical traffic flows in smart grid networks. The SDN controller can prioritize the traffic to increase the throughput and provide QoS in SDN-based smart grids. Moreover, it supports heterogeneous networks and does not depend on vendor and protocol and operates open standard (Rehmani et al. 2018).

## 4.3 Attacks and Countermeasures in Software-Defined Networking

This section presents the attacks and countermeasures in software-defined networking.

### 4.3.1 Attacks in SDN

This chapter mainly focuses on the attacks to the application, control and infrastructure layers of SDN (Kreutz et al. 2013, Shu et al. 2016). Figure 4.8 presents the attacks that can be seen in the SDN layers.

### 4.3.2 Attacks at Application Layer

In the application layer, attackers can manipulate the network configuration, steal network information, seize network resources and so on through placing malicious computer programs (such as spyware, malware, virus) in the SDN applications. As described earlier, the Northbound API provides an interface between the SDN controller and applications for managing and controlling the network. However, the lack of trust between the controller and SDN applications is a security concern as malicious SDN applications can send malicious commands to the network through the poorly designed Northbound API.

1. *Resource Exhaustion*: A malicious application can excessively use all available resources of the system that runs the SDN controller. This may lead to degradation in the performance of the SDN controller and other applications.
   a. *Memory Exhaustion*: In order to exhaust the memory of the system where the controller is running, a malicious application can allocate memory continuously.

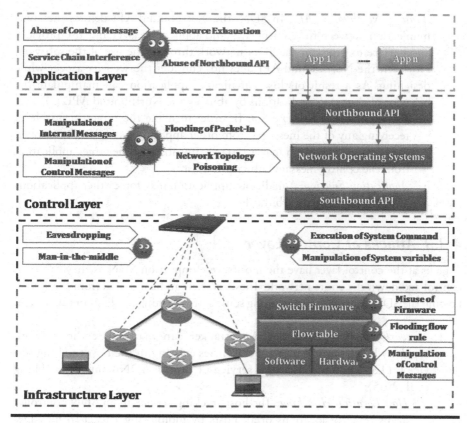

**Figure 4.8  Attacks on SDN in different layers.**

  b. *CPU Exhaustion*: A malicious application can create working threads continuously to use up all available CPU resources. As a result of this, the CPU may not be able to execute other applications.

2. *Abuse of Control Message*: An SDN application may arbitrarily issue control messages:

  a. *Flow Rule Modification*: In the flow table of an OpenFlow switch, an existing flow rule can be overwritten by a malicious application to cause unexpected network behavior.

  b. *Flow Table Clearance*: A malicious application may send the control messages to clear the flow table entries of a switch and may force termination of all the communications in the network.

3. *Service Chain Interference*: Applications of SDN with chained execution may be interfered with. For example, a malicious application can participate in a service chain and drop the control messages before the other SDN applications awaiting them. Moreover, an interference can be seen when a malicious application falls in an infinite loop to stop the chained execution of applications.

4. *Abuse of Northbound API*: In SDN, the Northbound API defines the communication between the control layer and application layer and also provides an interface to program the SDN network through the use of the software. However, the Northbound API does not consider the security aspect while designing. As a result of this, a malicious application can manipulate the behavior of the other applications by abusing the Northbound API.

   a. *Event Listener Unsubscription*: In order to make the target incapable of receiving any of the messages from other applications of SDN controller, a malicious application may arbitrarily unsubscribe the target application from the control message subscription list.

   b. *Application Eviction*: A malicious application may force other applications of SDN to terminate arbitrarily.

## 4.3.3 Attacks at Control Layer

Attacks at the control layer have the most severe impact on SDN as they can even control and down the entire network. These attacks can be initiated from the system hosting the SDN controller by exploiting some software vulnerabilities in the system.

1. *Network Topology Poisoning*: An attacker can manipulate the network topology by exploiting the vulnerabilities existing in the host tracking service (HTS) and link discovery service of various SDN controllers (Hong et al. 2015):

   a. *Host Location Hijacking*: The host tracking service in an SDN controller keeps track of locations of all hosts by monitoring Packet-In messages received by the controller from the OpenFlow switches. For instance, if a host migrates to another location, HTS can detect such migration. However, HTS does not verify the authentication of hosts. As a result, an attacker can easily impersonate (spoof) a target host and subsequently hijack the network traffic.

   b. *Link Fabrication*: The SDN controller periodically sends link layer discovery protocol (LLDP) packets to discover the links among OpenFlow switches at the infrastructure layer. However, the link discovery procedure is vulnerable as an attacker can manipulate the network topology by sending a forged or relayed LLDP packet to the controller.

2. *Flooding of Packet-In*: A Packet-In essentially represents a packet that does not match any flow rules at the infrastructure layer, and the OpenFlow protocol commands that such packets must be sent by the OpenFlow switch to the controller directly. The control layer has no built-in security mechanism to avoid the manipulation of Packet-In messages even when the OpenFlow switches are enabled with transport layer security (TLS). For instance, an attacker can flood a several Packet-In messages to place the SDN controller in an unpredictable state.

3. *Manipulation of Control Message*: An attacker can manipulate the control messages to put the control layer of SDN in an unpredictable state:

   a. *Switch Table Flooding*: An attacker can send a large number of forged control messages to flood the flow table of the OpenFlow switch. As a result of this, the switch may not be available and may cause network partitions.

   b. *Switch Identification Spoofing*: The switch identification field of a control message may be manipulated to poison the network topology and subsequently put the control layer in an unpredictable state.

   c. *Malformed Control Message*: An attacker can inject malformed control messages into the network and can cause malfunction of the control layer.

4. *Manipulation of Internal Storage*: The SDN controller shares internal storage among various SDN applications. Eventually, SDN applications can unrestrictedly access and manipulate the internal database of the SDN controller. This internal database can further be misused for many subsequent attacks, such as manipulating the network topology.

5. *Manipulation of System Variable*: System variables may be manipulated to put the SDN controller in an unpredictable state. For example, an attacker can change the timer of the hosting operating system to put the SDN controller in an unpredictable state and disconnect from the OpenFlow switches of the infrastructure layer.

6. *Manipulation of System Command*: In order to terminate the controller instance from the hosting operating system, a malicious application can execute a system exit command.

7. *Eavesdropping*: An attacker can sniff the control channel to steal sensitive information from the SDN network. For example, an attacker sniffs the ongoing control messages on the control channel to learn the network topology.

8. *Man-in-the-Middle Attack*: An attacker can actively intervene in the control channel. For example, an attacker modifies the flow rule message that is being transferred, and corrupts the behavior of the network.

## 4.3.4 Attacks at Infrastructure Layer

The infrastructure layer is located at the bottom of the SDN architecture and contains many hosts and OpenFlow switches that are interconnected with each other. These switches are responsible for forwarding packets to the end host. An attacker can attack an OpenFlow switch by simply attaching a link to a port of the switch.

*Misuse of Firmware*: An attacker can misuse the characteristics of a certain switch model. For instance, the crafted flow rules installed by an attacker may not be processed in the hardware table of a certain switch model.

*Flooding of Flow Rule*: Infrastructure layers can be in an unpredictable state when a large number of flow rules are sent to an OpenFlow switch. An attacker can capture the control channel and install numerous flow rules to the target switch to fill up the flow table. As a result, the victim switch drops the flows for authorized hosts, which leads to a denial-of-service attack.

*Manipulation of Control Message*: A malformed control message may put the infrastructure layer in an unpredictable state: An attacker injects a malformed control message to the infrastructure layer to interrupt the connection between the control layer and the infrastructure layer.         •

## 4.4 Countermeasure of Attacks in Software-Defined Networking

It can be seen from the previous section, the attackers can exploit all the layers in an SDN due to its design issues. A number of techniques have been proposed as countermeasures (Shu et al. 2016) to the attacks as presented in Table 4.3. A summary follows:

- Attacks on infrastructure layer due to limited authentication mechanisms. Message authentication code (MAC), digital signature or secure sockets layer/transport layer security (SSL/TLS) can be used for authentication and encryption to avoid eavesdropping, sniffing and spoofing of traffic. Further, the SDN controller can maintain a list of authorized switches in the access control list (ACL) and run an intrusion detection system (IDS) to mitigate attacks from the infrastructure layer.
- Forged traffic flows, man-in-the-middle and false reply attacks as traffic are sent in clear text. In order to mitigate these attacks, network traffic should be sent either with a message authentication code (MAC) or a digital signature. The timestamp can be included with the MAC or digital signature to avoid reply attacks. Further, the SDN controller can ensure that security policy is implemented for all traffic flows.
- Attacks on the control layer due to limited authentication mechanisms and open Northbound and Southbound APIs. Both Northbound and Southbound APIs can be protected by using cryptographic encryption (TLS or secure shell [SSH]). The control layer can replicate the controller periodically or can run multiple controllers to make the network more reliable and secure.
- Denial of service (DoS) or flooding attacks can be seen in any layer of SDN architecture. These attacks are possible due to OpenFlow switches having limited memory and flow table, and the SDN controller is centralized. In order to mitigate DoS attacks in switches and controller, it's necessary to limit the flow rules in switches and use either replica of the controller or multiple controllers respectively.

**Table 4.3   Attacks and Countermeasures in SDN Architecture**

| SDN Layers | Attacks | Caused by | Existing Techniques as Countermeasures |
|---|---|---|---|
| Infrastructure Layer | Man-in-the-middle attack between OpenFlow switch and SDN controller | The control link between the OpenFlow switch and SDN controller is not secure without SSL/TLS support. | FlowChecker (Al-Shaer and Al-Haj 2010) ForNOX (Porras et al. 2012) Veriflow (Khurshid et al. 2012) Controller replication (Fonseca et al. 2012) |
| | DoS attack-flooding of flow rules to overflow flow table and flow buffer | Limited authentication mechanism between OpenFlow switch and SDN Controller. OpenFlow switch has limited storage capacity for flow table and flow buffer. A large number of packets have to processed by OpenFlow switch in a short time. | FlowVisor (Sherwood et al. 2009) Virtual source Address Validation Edge (VAVE) (Yao et al. 2011) Resonance (Braga et al. 2010) |
| Control layer | Compromised controller attack | Centralized controller | FloodGuard (Wang et al. 2015) DDoS Blocking Application (Lim et al. 2014) |
| | DoS attack | Centralized controller The controller has limited computing and storage resources. A large number of flow requests have to processed by the controller in a short time. | DISCO (Phemius et al. 2014) McNettle (Voellmy and Wang 2012) HyperFlow (Tootoonchian and Ganjali 2010) |
| | Attacks from application programs | The Northbound interface is open for programming malicious applications. | FRESCO (Shin 2013) |

*(Continued)*

**Table 4.3 (*Continued*)   Attacks and Countermeasures in SDN Architecture**

| SDN Layers | Attacks | Caused by | Existing Techniques as Countermeasures |
|---|---|---|---|
| Application Layer | Illegal access to applications | No authentication mechanism Controller vulnerable as it relies on operating system | NICE (Canini 2012) Verificare (Skowyra 2013) VeriCon (Ball 2014) |
| | Conflicts in security rules and configuration for application software | The difference of access control and accountability for various application software | Flover (Son 2013) Anteater (Mai 2011) NetPlumber (Kazemian n.d.) |

## 4.5 Attacks and Countermeasures in SDN-Based Smart Power Grids

This section discusses attacks and countermeasures in SDN-based smart power grids.

### 4.5.1 Attacks in SDN-Based Smart Power Grid

The attacks in SDN-based smart power grids are presented through case studies and then the attacks are categorized in the following subsection.

### 4.5.2 Attack Case Studies

As discussed earlier, the SDN-based smart grid architecture consists of three segments: (i) a control center, (ii) communication network, and (iii) substations. Attacks can be seen in any of these three segments as illustrated in Figure 4.9. This chapter considers three attack cases in SDN-based smart power grid controller as follows (Ghosh et al. 2017, 2018):

> In the *first case*, the control center can be compromised by compromising the SCADA master or SDN controller or even by compromising their applications. The compromised SDN controller can issue malicious control-commands (such as Add_Flow, Del_Flow, Mod_Flow) to the OpenFlow switches in order to degrade the performance of the network and subsequently the smart grid. Similarly, the compromised SCADA master can issue malicious control commands (such as Read, Write, Execute) to the SCADA slaves and degrade the performance.

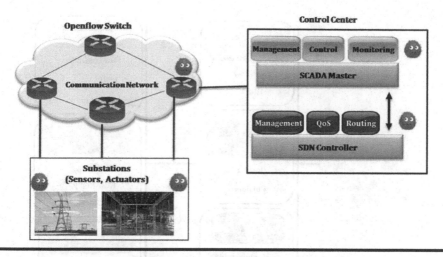

**Figure 4.9   Attacks on SDN-based smart power grid networks.**

In the *second case*, OF switches in the communication network segment may compromise and may drop or inject false packets and also delay the packets that carry measurement data/control commands from SCADA slaves/master to SCADA master/slaves. For instance, a packet that carries a critical control command, like to open a breaker of a relay, can be dropped or delayed by an intermediate compromised switch. This may cause a potential risk to the physical infrastructure of a substation in SDN-based smart grid networks.

In the *third and final case*, the SCADA slaves can be compromised and can inject malicious measurement data into the smart grid network. This is important to detect and identify bad data in measurements while estimating the state of the smart grid network correctly by the SCADA master. The failure of communication between SCADA slaves and SCADA master and the poor calibration and false injection of malicious measurement data by SCADA slaves (Liu et al. 2009) are the main sources of bad data.

## 4.5.2.1  Categorization of Attacks on SDN-Based Smart Power Grid Networks

The attacks related to the SDN-based smart grid can be classified into five different complementary categories as shown in Figure 4.10: (i) behavior-based, (ii) location-based, (iii) protocol-based, (iv) device-based, and (v) data-based.

Behavior-based attacks depend on the behavior of the attackers and their attacks' execution. It can be either passive or active. In a passive attack, the attacker can monitor (or eavesdrop) and analyze the grid data or communication

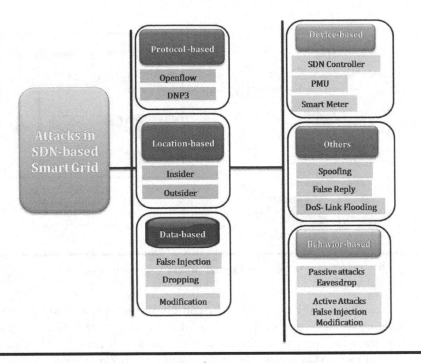

**Figure 4.10   Attack model for SDN-based smart power grids.**

data to gain meaningful information in the SDN-based smart grid network. This attack is easy to launch and it may lead to an active attack. In an active attack, the attacker can drop, modify the packets or even inject false packets to disrupt the normal operation of the SDN-based smart grid network. A passive attack is difficult to detect as the operation of the network is not affected by this attack.

An active attack is easier to detect, as the normal operation of the network may be affected seriously. Location-based attacks rely on the location of the attacker and can be either external or internal. External attacks are carried out by an attacker that does belong to the network, whereas internal attacks are carried by a compromised node which is actually a part of the network. Insider attackers are more dangerous compared to outside attackers as they have better knowledge about the secret information and internal architecture of the SDN-based smart grid network.

In protocol-based attacks, the attacker exploits the protocol (OpenFlow, distributed network protocol [DNP3], transmission control protocol/internet protocol [TCP/IP]) vulnerabilities that run in the SDN-based smart grid network. These protocols can either be associated with SDN (OpenFlow) or smart grid (DNP3). In device-based attacks, the attacker targets a specific device for maximum gains of potential malicious activities. These attacks can be either related to SDN devices (SDN controllers, OF switches) or smart grid devices (smart meters, phasor measurement units [PMUs], intelligent electrical devices [IEDs]). Lastly, in database attacks, the attacker can inject, modify and even drop the data packets. For example, the attacker can inject false meter data, prices and emergency events in the smart grid. The attacker can spread malware with the data to infect SDN controllers and SCADA devices (smart meters) so that it can steal sensitive information. These attacks may affect the smart grid financially on the electricity markets. Spoofing (IP/Hardware Address), false reply and denial of service (DoS) attacks are well known and can be seen in SDN-based smart grids too.

### 4.5.3 Countermeasures of Attacks in SDN-Based Smart Power Grid Networks

An SDN can provide security in a smart power grid by providing consistent access control, applying efficient and effective security policies, and managing and controlling the network through the use of the controller. A firewall can be developed as an application at the top of the SDN controller where efficient and effective policies can be written to provide security against attacks from the outside and even from the insider. Further, the intrusion detection system (IDS) can be developed as an application to detect attackers from inside the network. The SDN controller can perceive the entire network traffic periodically. Therefore IDS can easily detect abnormal behavior in network traffic caused by an attacker. Furthermore, the SDN controller can timely deal with new attacks as the controller is programmable. These facts are illustrated in Figure 4.11.

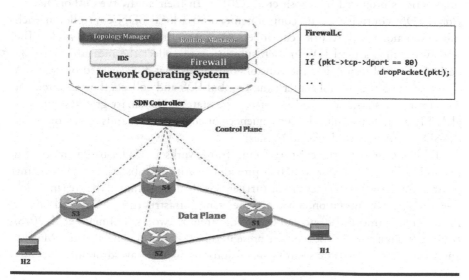

**Figure 4.11   Security using SDN.**

Recently, a number architectures (Ghosh et al. 2017, Da Silva et al. 2016, Cahn et al. 2013, Dong et al. n.d., Molina et al. 2015) have been proposed for SDN-based smart grids to provide security by utilizing the above features of SDN. In Da Silva et al. (2016), the authors proposed a network-based intrusion detection system architecture (NIDS) for SDN-based SCADA systems. One-class classification (OCC) algorithm is basically proposed in NIDS. The NIDS architecture consists of the main control center, eight distribution substations, four intermediate control center, and several field devices. The authors evaluated the performance of the proposed architecture using OpenFlow SDN controller and demonstrated that the OCC algorithm can detect the intrusion in a SCADA system with 98% accuracy.

The authors of Cahn et al. (2013) proposed a solution for power distribution subsystems. They used SDN to make the network auto-configurable, secure and reliable against possible system-inappropriate configuration. They further developed a prototype using the Ryu OpenFlow controller and evaluated in a testbed with real SCADA devices. In Molina et al. (2015), the authors presented an SDN-based architecture for a substation that follows the International Electrotechnical Commission (IEC) 61850 standard. They developed automation techniques for performing a flow-based resource management that enables features such as routing, traffic filtering, QoS, load balancing, and security.

In Dong et al. (n.d.), the authors investigated (i) how the resilience of smart grids can be enhanced against malicious attacks by an SDN, (ii) additional risks due to SDN and how to manage them, and (iii) how to evaluate and validate solutions for SDN-based smart grids. They further discussed the concrete security issues and their possible countermeasures. Another security framework for the SDN-based smart grid is proposed in Ghosh et al. (2017). In their study, the authors used a global SDN controller at the control center and a local SDN controller at each substation along with security controllers to protect the smart grid networks. The framework runs a local IDS in each substation to collect the measurement data periodically and to monitor the control commands that are executed on SCADA slaves, whereas the control center runs a global IDS and collects the measurement data from the substations and estimates the state of the smart grid system. The global IDS further verifies the consequences of control commands issued by either the SDN controller or the SCADA master.

Table 4.4 presents the existing secure frameworks for SDN-based smart grid networks. These frameworks mainly provide security in substations. The existing security frameworks/schemes can be further classified according to the area in which they are applied: substation, advanced metering infrastructure (AMI) and phasor measurement unit (PMU) networks, and different networks (Rehmani et al. 2018) as given in Figure 4.12. The schemes presented in Da Silva et al. (2015) and Maziku and Shetty (2017) provide security in substations against eavesdropping and link

**Table 4.4    Secure Architecture for SDN-Based Smart Grid Networks**

| Framework/ Architecture | First Case | | Second Case | Third Case | Security Tool/ Mechanism Used |
|---|---|---|---|---|---|
| | SCADA Master | SDN Controller Security | OF Switch Security | SCADA Slave Security | |
| Silva et al. (2016) | Not secured | Not secured | Secured | Secured | Network-based IDS (NIDS) |
| Cahn et al. (2013) | Not secured | Not secured | Secured | Not secured | SDN controller policies |
| Molina et al. (2015) | Not secured | Not secured | Secured | Not secured | sFlow collector |
| Dong et al. (n.d.) | Secured | Not secured | Secured | Not secured | Control center runs centralized IDS |
| Ghosh et al. controller Ghosh et al. (2017) | Secured | Secured | Secured | Secured | Distributed IDS: Each substation runs IDS |

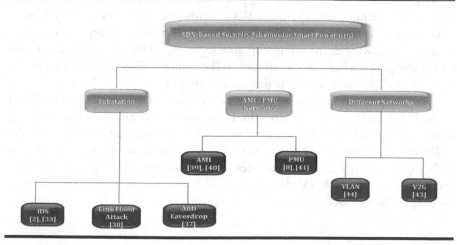

**Figure 4.12   Classification of security schemes for SDN-based smart grids according to the substation, AMI and PMU.**

flooding attacks respectively. The anti-eavesdropping scheme in Da Silva et al. (2015) achieves secure communication by using multipath routing in a SCADA system. In Maziku and Shetty (2017), the authors proposed a security score model based on SDN to protect the substation against the link flooding attack. The OpenFlow controller can easily enforce QoS policies and identify heaviest flows and busiest communications links at a real time.

A security architecture has been proposed in Irfan et al. (2015) for the protection of data of AMI networks. The proposed security architecture used Flowvisor as an SDN controller. Flowvisor provides the virtualization and slices the networks and also helps to ensure authorization, authentication, and confidentiality. Furthermore, the smart meter sends data using long-term evaluation (LTE) for the sending of smart meter data which is then compared with advanced encryption standard (AES)-128 encrypted metering data sent by the SG controller. An efficient and privacy-aware power injection (EPPI) security scheme is proposed in Zhao and Zheng (2017) for SDN-based AMI networks. It uses message authentication code (MAC) for providing security to the customers. EPPI provides security against the replay attacks. In a reply attack, the attacker captures the record of valid packets and replays it in the future.

Security schemes for SDN-based PMU networks are proposed in Lin et al. (2016b) and Jin et al. (2017). In Jin et al. (2017), the authors extended the work of Lin et al. (2016b) and integrated SDN technology with a microgrid. They mainly focused on security and reliability of the microgrid, which can be provided by SDN. The authors also developed an SDN testbed for microgrid evaluation, which is DSSnet (Hannon et al. 2016).

In Zhang et al. (2016), the authors proposed a software-defined vehicle-to-grid (SDN-V2G) architecture. The SDN-V2G architecture provides security against different types of attacks: (i) attacks on the utility server, (ii) attacks on the communication network of the utility, (iii) attacks on the SDN controller, (iv) attacks on the charging stations, and (v) attacks on the vehicles. SDN-based virtual utility network (SVUN) architecture is proposed in Kim et al. (2014) for machine-to-machine (M2M) applications in smart grids. IEEE 802.1Q is used to create virtual LANs (VLANs) in traditional networks. However, it has limitations: IEEE 802.1Q supports a limited number of devices and only one-time authentication for the M2M devices. Thus, SVUNs used SDN for creating virtual utility networks, as SDN supports a large number of devices and provides security even after the first-time authentication of M2M devices. Table 4.5 presents the attacks and countermeasures in SDN-based smart grids.

**Table 4.5  Attacks and Countermeasures in SDN-Based Smart Grids**

| Attacks Category | Attacks | Existing Frameworks/Schemes | Possible Countermeasures |
|---|---|---|---|
| Location-based | Insider | Ghosh et al. (2017), Da Silva et al. (2016), Cahn et al. (2013), Dong et al. (n.d.), Molina et al. (2015) | • IDS as an SDN application<br>• SDN controller policies |
| | Outsider | Ghosh et al. (2017), Cahn et al. (2013), Dong et al. (n.d.) | • Cryptography: Message authentication code (MAC), digital signature<br>• SDN controller policies with a list of authorized nodes |
| Behavior-based | Passive | Da Silva et al. (2015) | • Symmetric/Asymmetric cryptography for confidentiality |
| | Active | Ghosh et al. (2017), Cahn et al. (2013), Dong et al. (n.d.), Dorsch et al. (2016) | • IDS as an SDN application<br>• Cryptography: Authentication & confidentiality |
| Device-based | SDN Controller | Fonseca et al. (2012), Phemius et al. (2014), Voellmy and Wang (2012), Tootoonchian and Ganjali (2010) | • Controller replication<br>• Multiple controllers |
| | PMUs | Lin et al. (2016b), Jin et al. (2017) | Identity-based cryptographic authentication: MAC, digital signature |
| | AMIs | Zhang et al. (2016), Kim et al. (2014) | Identity-based cryptographic authentication: MAC, digital signature |

*(Continued)*

**Table 4.5 (Continued)   Attacks and Countermeasures in SDN-Based Smart Grids**

| Attacks Category | Attacks | Existing Frameworks/Schemes | Possible Countermeasures |
|---|---|---|---|
| Protocol-based | OpenFlow | Dong et al. (n.d.) | Cryptography: MAC, digital signature, secure socket layer (SSL) |
| | DNP3 | Lin et al. (2016b) | |
| Data-based | False injection | Ghosh et al. (2017), Da Silva et al. (2016), Dong et al. (n.d.) | • IDS as an SDN application<br>• Data authentication: MAC, digital signature<br>• Data confidentiality: Symmetric/Asymmetric encryption |
| | Dropping | | |
| | Modification | | |
| Others | Spoofing | Ghosh et al. (2017), Lin et al. (2016b), Da Silva et al. (2016), Irfan et al. (2015) | Identity based cryptographic authentication: MAC, digital signature |
| | False reply | | Cryptographic authentication: MAC with a timestamp, digital signature with the challenge-response scheme |
| | DoS | | Using rate limiting and packet dropping techniques |

## 4.6 Conclusion and Future Works

This chapter mainly studies different types of attacks and countermeasures in software-defined networking and SDN-based cyber-physical systems. The chapter first presented a background architecture of SDN technology, smart grids, and SDN-based smart grids. It discussed and classified different types of attacks related to SDNs and SDN-based smart grids. Existing security schemes along with their classification are then presented. The chapter further discussed possible countermeasures of various attacks that can be seen in SDN and SDN-based smart grids.

In the future, we are interested to design and develop a secure and resilient SDN-based smart grid network. In order to make the framework secure, an SDN controller can run applications like intrusion detection system (IDS), intrusion elimination system (IES), and key distribution system (KDS). IDS can periodically monitor the devices and their traffics in order to detect misbehavior nodes, whereas IES eliminates the malicious nodes from the SDN-based smart grids by using controller security policies. Key distribution system can distribute the cryptographic keys to the devices for proving authentication and confidentiality in SDN-based smart grids.

## References

Al-Shaer, E., et S. Al-Haj. FlowChecker: Configuration analysis and verification of federated openflow infrastructures. *Proceedings of the 3rd ACM Workshop on Assurable and Usable Security Configuration*, (2010): 37–44.

Ball, T., N. Bjmer, A. Gember, S. Itzhaky, A. Karbyshev, M. Sagiv, A. Valadarsky. Vericon: Towards verifying controller programs in software-defined networks. *ACM Sigplan Notices*, 49, 6 (2014): 282–293.

Braga, R., E. Mota, et A. Passito. Lightweight DDoS flooding attack detection using NOX/OpenFlow. *IEEE Local Computer Network Conference*. Denver, CO, (2010): 408–415.

Cahn, A., J. Hoyos, M. Hulse, et E. Keller. Software-defined energy communication networks: From substation automation to future smart grids. *IEEE Smart Grid Communication*, (2013): 558–563.

Canini, M., D. Venzano, P. Peresini, D. Kostic, et J. Rexford. A NICEway to test OpenFlow applications. *9th USENIX Conference on Networked Systems Design and Implementation*, (2012).

Conti, M. et al. Looking ahead in pervasive computing: Challenges and opportunities in the era of cyber-physical convergence. *Pervasive and Mobile Computing*, 8, 1 (2012): 2–21.

Dong, X., H. Lin, R. Tan, R. K. Iyer, et Z. Kalbarczyk. Software-defined networking for smart grid resilience: Opportunities and challenges. *1st ACM Workshop on CPSS*, (2015).

Dorsch, N., F. Kurtz, F. Girke, et C. Wietfeld. Enhanced fast failover for software-defined smart grid communication networks. *IEEE Global Communications Conference (GLOBECOM)*, (2016): 1–6.

Fonseca, P., R. Bennesby, E. Mota, et A. Passito. A replication component for resilient OpenFlow-based networking. *IEEE Network Operations and Management Symposium*. Maui, HI, (2012): 933–939.

Ghosh, U., P. Chatterjee, et S. Shetty. A security framework for SDN-enabled smart power grids. *2017 IEEE 37th International Conference on Distributed Computing Systems Workshops (ICDCSW)*, Atlanta, GA, (2017): 113–118.

Ghosh, U., P. Chatterjee, et S. Shetty. Securing SDN-enabled smart power grids: SDN-enabled smart grid security. Dans *Securing SDN-Enabled Smart Cyber-Physical Systems for Next Generation Networks*. IGI Global, 2018.

Ghosh, U., X. Dong, R. Tan, Z. Kalbarczyk, D. K. Yau, et R. K. Iyer. A simulation study on smart grid resilience under software-defined networking controller failures. *2nd ACM Workshop on CPSS*, (2016): 52–58.

Hannon, C., J. Yan, et D. Jin. DSSnet: A smart grid modeling platform combining electrical power distribution system simulation and software defined networking emulation. *ACM SIGSIM-PADS*. New York, (2016): 131–142.

Hong, S., L. Xu, H. Wang, et G. Gu. Poisoning network visibility in software-defined networks: New attacks and countermeasures. *Network and Distributed System Security (NDSS) Symposium (2015)*, USENIX, (2015).

Irfan, A., N. Taj, et S. A. Mahmud. A novel secure SDN/LTE based architecture for smart grid security. *IEEE PICOM*. Liverpool, (2015): 762–769.

Jin, D. et al. Toward a cyber resilient and secure microgrid using software-defined networking. *IEEE Transactions on Smart Grid*, 8, 5 (2017): 2494–2504.

Kazemian, P., M. Chan, H. Zeng, G. Varghese, N. McKeown, et S. Whyte. Real time network policy checking using header space analysis. *USENIX Symposium on Networked Systems Design and Implementation*. s.d. 99–111.

Khurshid, A., W. Zhou, M. Caesar, et P. B.Godfrey. Veriflow: Verifying network-wide invariants in real time. *ACM Proceedings of the First Workshop on Hot Topics in Software Defined*. New York, (2012): 49–54.

Kim, Y. J., K. He, M. Thottan, et J. G. Deshpande. Virtualized and self-configurable utility communications enabled by software-defined networks. *IEEE Smart Grid Communication,* (2014): 416–421.

Kreutz, D., F. M. V. Ramos, et P. Verissimo. Towards secure and dependable software-defined networks. *Proceedings of the Second ACM SIGCOMM Workshop on Hot Topics in Software Defined Networking*. Hong Kong, China, (2013): 55–60.

Lim, S., J. Ha, H. Kim, Y. Kim, et S. Yang. A SDN-oriented DDoS blocking scheme for botnet-based attacks. *2014 Sixth International Conference on Ubiquitous and Future Networks (ICUFN)*. Shanghai, China, (2014): 63–68.

Lin, H., A. Slagell, Z. Kalbarczyk, P. Sauer, et R. Iyer. Runtime semantic security analysis to detect and mitigate control-related attacks in power grids. *IEEE Transactions on Smart Grid*, (2016a).

Lin, H. et al. Self-healing attack-resilient PMU network for power system operation. *IEEE Transactions on Smart Grid*. in Print, (2016b).

Liu, Y., P. Ning, et M. K. Reiter. False data injection attacks against state estimation in electric power grid. *16th ACM Conference on Computer and Communications Security CCS*, (2009).

Mai, H., A. Khurshid, R. Agarwal, M. Caesar, P. Godfrey, et S. King. Debugging the data plane with anteater. *ACM SIGCOMM Computer Communication*, 41, 4 (2011): 290–301.

Maziku, H., et S. Shetty. Software defined networking enabled resilience for IEC 61850-based substation communication systems. *International Conference on Computing, Networking and Communications (ICNC)*, (2017): 690–694.

Molina, E., E. Jacob, J. Matias, N. Moreira, et A. Astarloa. Using software defined networking to manage and control IEC 61850-based systems. *Computers and Electrical Engineering*, 43 (2015): 142–154.

Porras, P., S. Shin, V. Yegneswaran, M. Fong, M. Tyson, et G. Gu. A security enforcement kernel for open flow networks. *Proceedings of the First Workshop on Hot Topics in Software defined networks*, (2012): 121–126.

Phemius, K., M. Bouet, et J. Leguay. DISCO: Distributed multi-domain SDN controllers. *IEEE Network Operations and Management Symposium (NOMS)*. Krakow, (2014): 1–4.

Rehmani, M. H., A. Davy, B. Jennings, et C. Assi. Software defined networks based smart grid communication: A comprehensive survey. *ArXiv*. 1801.04613, (2018).

Sherwood, R. et al. *Flow Visor: A Network Virtualization*. Deutsche Telekom Inc. R&D Lab, Stanford, Nicira Networks, (2009). Tech. Report.

Shin, S., P. Porras, V. Yegneswaran, M. Fong, G. Gu, et M. Tyson. FRESCO: Modular composable security services for software-defined networks. *20th Annual Network & Distributed System Security Symposium*, NDSS, (2013): 1–16.

Shu, Z., J. Wan, D. Li, J. Lin, A. V. Vasilakos, et M. Imran. Security in software-defined networking: Threats and countermeasures. *Springer Mobile Networks and Application*, 21 (5) (2016): 764–776.

Da Silva, E. G., A. S. D. Silva, J. A. Wickboldt, P. Smith, L. Z. Granville, et A. Schaeffer-Filho. A one-class NIDS for SDN-Based SCADA systems. *IEEE COMPSAC*. Atlanta, (2016): 303–312.

Da Silva, E. G., L. A. D. Knob, J. A. Wickboldt, L. P. Gaspary, L. Z. Granville, et A. Schaeffer-Filho. Capitalizing on SDN based SCADA systems: An anti-eavesdropping case-study. *IFIP/IEEE International Symposium on Integrated Network Management (IM)*, (2015): 165–173.

Skowyra, R., A. Lapets, A. Bestavros, et A. Kfoury. Verifiably safe software defined network for CPS. *2nd ACM International Conference on High Confidence Networked Systems*, (2013): 101–110.

Son, S., S. Shin, V. Yegneswaran, P. Porras, G. Gu. Model checking invariant security properties in OpenFlow. *International Conference on Communications ICC*, (2013): 1974–1979.

Tootoonchian, A., et Y. Ganjali. HyperFlow: A distributed control plane for OpenFlow. *Proceedings of the 2010 Internet Network Management Conference on Research on Enterprise Networking*, (2010): 3–3.

Voellmy, A., et J. Wang. Scalable software defined network controllers. *Proceedings of the ACM SIGCOMM*. New York, (2012).

Wang, H., L. Xu, et G. Gu. FloodGuard: A DoS attack prevention extension in software-defined networks. *45th Annual IEEE/IFIP International Conference on Dependable Systems and Networks*, (2015): 239–250.

Yao, G., J. Bi, et P. Xiao. Source address validation solution with OpenFlow/NOX architecture. *19th IEEE International Conference on Network Protocols (ICNP)*, (2011): 7–12.

Zhang, S., J. Wu, Q. Li, J. Li, et G. Li. A security mechanism for software-defined networking based communications in vehicle-to-grid. *IEEE Smart Energy Grid Engineering (SEGE)*, (2016): 386–391.

Zhao, J., E. Hammad, A. Farraj, et D. Kundur. *Network-Aware QoS Routing for Smart Grids Using Software Defined Networks*. Cham, Switzerland: Springer International Publishing, (2016): 384–394.

Zhao, Y. Z. J., et D. Zheng. Efficient and privacy-aware power injection over AMI and smart grid slice in future 5G networks. *Mobile Information Systems*, 17, (2017).

*Chapter 5*

# DDoS Defense in SDN-Based Cyber-Physical Cloud

Safaa Mahrach and Abdelkrim Haqiq

## Contents

## 5.1 Introduction

Industry 4.0 is the fourth industrial revolution that combines the physical world with new technologies, including the internet of things, internet of systems, and cloud computing [1,2]. The cooperation of the cyber-domain with the physical world is referred to as a cyber-physical system (CPS). CPS is a physical and engineered system whose processes are coordinated, controlled, and implemented by computer-based mechanisms [3].

As defined by the National Institute of Standards and Technology (NIST), cyber-physical cloud computing (CPCC) is a system environment that can rapidly build, modify and provision auto-scale cyber-physical systems composed of a set of cloud computing–based sensor, processing, control, and data services [4]. Due to the benefits of cloud resource provisioning capacities, such as on-demand self-service, multitenancy, elasticity, and so on, CPCC has been widely adopted for the traffic control, vehicle location detection, robotics systems, healthcare, and so on.

As cloud adoption grows, the service provider and enterprises need to create large distributed number of data centers around the world to ensure fast access to the global tenants. This places an increasing demand and usage of the cloud networking infrastructure to rapidly construct, modify and provision auto-scale networks [5,6]. However, the existing cloud networking characteristics don't meet these needs and so don't support the cloud evolution. To match these requirements a scalable, flexible, and programmable networking infrastructure is required, which may be performed while separating the control decision from the forwarding devices. In particular, software-defined networking (SDN) technology may be used to simplify the management and control of cloud networking while decoupling the control plane from the data plane.

SDN-based cloud or software-defined cloud networking is a new form of cloud networking in which SDN allows a centralized control of the network, gives a global view of the network, and provides the networking-as-a-service (NaaS) in the cloud computing environments [7–10]. However, new security issues and particularly new trends of distributed denial of service (DDoS) attacks have been introduced over the integration of SDN and cloud computing technologies [8,11–13].

The good capacities of SDN, such as software-based traffic analysis, centralized control, and dynamic network reconfiguration, may greatly enhance the prevention and mitigation of DDoS attacks in cloud computing environments [8,13–15]. Using SDN features, significant research works have been developed and proposed to defend against DDoS attacks in the enterprise networking which adopts both SDN and cloud computing [13,15–17].

Based on our study, most of the existing DDoS mitigation mechanisms are designed at the high-level SDN application plane with the involvement of the SDN controller in each operation to detect and mitigate DDoS attacks. Therefore, the communication path between the data and control planes rapidly becomes a bottleneck, which impacts the network performance and restricts its scalability and reactivity.

To address these challenges, frameworks, compilers, and programming languages [18–21] have been developed to take advantages of the SDN data plane with a way to perform dynamically specific operations (e.g., monitoring, detection, reaction, etc.) at the switch level. In this side, we designed an active DDoS mitigation mechanism which enables the SDN data plane to prevent and mitigate DDoS attacks and particularly SYN flooding attacks in the networking environment that adopts both cloud computing and SDN technologies. Our defensive mechanism exploits the capabilities of SDN data plane using the programming protocol-independent packet processor P4 [18].

In this chapter, we first discuss the SDN features which make it an appropriate technology for cloud computing–based cyber-physical systems, and then we present some proposed SDN-based frameworks for improving the cloud networking environment. In addition, we discuss the new security issues and the new trends of DDoS threats introduced over the integration of SDN and cloud computing technologies. Further, we examine and analyze some of the existing SDN-based DDoS detection and mitigation solutions. Furthermore, we present and discuss the proposed mechanism "SDN-based SYN flooding defense in cloud" which enables the SDN data plane to prevent and mitigate SYN flooding attacks in the SDN-based cloud environment.

The remainder of the chapter is structured as follows: Section 5.2 presents the SDN capabilities and their benefits for the management of cloud computing–based cyber-physical systems, and then we present some proposed SDN-based cloud solutions. In addition, we talk about the new security issues introduced over the integration of SDN and cloud computing technologies in Section 5.3. In Section 5.4, we discuss how the cloud characteristics make it more vulnerable to DDoS attacks, we examine the SDN features which improve the DDoS attack mitigation in cloud, and then we present the impact of DDoS attacks on the SDN-based cloud. In Section 5.5, we evaluate some of the existing SDN-based DDoS detection and mitigation solutions, and then we present the proposed frameworks and programming languages to address the challenges of these solutions. Further, we present and evaluate the proposed mechanism "SDN-based SYN flooding defense in cloud" in Section 5.6. Finally, we give our conclusion and perspectives in Section 5.7.

## 5.2 Background

Cyber-physical cloud computing is a cyber-physical system in which a set of sensor, processing, control, and data services rapidly and flexibly build and modify using cloud computing. Moreover, the software-defined cloud is a new form of cloud networking, in which SDN technology allows a centralized control of the network and provides a networking-as-a-service (NaaS) in cloud computing environments. In this section, firstly we present the CPS and the CPCC systems. Secondly, we discuss the SDN capabilities which make it an appropriate technology for cloud networking management, and then we present some proposed SDN-based cloud frameworks.

### 5.2.1 Cloud-Based Cyber-Physical Systems

Computing and communication powers are becoming embedded in all types of items and structures in physical systems. The cooperation of the digital-domain of computing and communications with the physical world are referred to as a cyber-physical system (CPS). CPS is a physical and engineered system whose processes are coordinated, controlled and implemented by computer-based mechanisms [3]. The increasing demand for the real-time process control, customized production, and smart decision have led to the adoption of cyber-physical systems in manufacturing, agriculture, healthcare, military, transportation, robotics, smart systems, etc.

As defined by the NIST, cyber-physical cloud computing (CPCC) is a system environment that can rapidly build, modify and provision auto-scale cyber-physical systems composed of a set of cloud computing–based sensor, processing, control, and data services [4]. Due to the benefits of cloud resource provisioning capacities [22,23], such as on-demand self-service, multitenancy, elasticity, and so on, CPCC has been widely adopted for the traffic control, vehicle location detection, robotics systems, energy, manufacturing, healthcare, etc.

### 5.2.2 Software Defined Systems

The emerging software-defined systems (SDSys) abstract the management complexities of various systems at different layers using software components. SDSys include software-defined networking (SDN), software-defined storage, software-defined servers (virtualization), software-defined datacenters (SDDs), software-defined security (SDSec), and ultimately software-defined clouds (SDClouds).

#### 5.2.2.1 Software-Defined Networking

Software-defined networking (SDN) is the most popular and used model of SDSys. SDN is considered as a key technology for improving the management and control of the large-scale networks (e.g., cloud computing). It enables IT administrators to create centralized decision-making functions for handling all control decisions while decoupling the control plane and the forwarding plane [24–26].

According to the Open Networking Foundation (ONF) (i.e., a nonprofit consortium dedicated for the development, standardization, and commercialization of SDNs) the SDN architecture consists of three layers including the data plane, control plane, and application plane [27]. The data plane consists of network devices, physical and virtual switches, which simply forward packets using the decisions (i.e., flow rules) programmed by the SDN controller. The control plane consists of a set of SDN controllers, which centrally program and control behavior of the forwarding devices through an open interface (e.g., OpenFlow). An SDN controller uses three communication interfaces to interact with the other layers: southbound, northbound and east/westbound interfaces. The OpenFlow [28] protocol has been created and considered as the standard southbound interface between the SDN controllers and the OpenFlow switches. The application plane consists of the end-user business applications that have the ability to determine specific functions (e.g., security, monitoring, virtualization, etc.) on the network devices to respond to users' dynamic requests.

## 5.2.3 SDN-Based Cloud

According to the cloud computing characteristics, such as on-demand self-service, multitenancy, elasticity, and broadband network access, IT managers face the greatest cloud challenges [29–31]. Among these challenges are availability, which is considered as a crucial security requirement in cloud since everything in cloud is defined as service [30,32] and scalability, as virtualization of computing resources, storage, etc. have been done in cloud; the network resources have to emerge to scale with cloud evolution [33,34], data security, etc.

As cloud growth, the service provider has to create an important and distributed number of data centers in the world to guarantee a quick access for the global clients. This places an increasing demand and usage of the cloud networking infrastructure [5,6]. To address these challenges, a programmable and dynamic network is needed which may be performed while separating the control decision from the forwarding devices. In particular, SDN technology may be used to simplify the management of cloud networking while decoupling the control plane from the data plane. SDN-based cloud or SD cloud networking is a new form of cloud networking, in which SDN technology allows a centralized control of networks and provides networking-as-a-service (NaaS) in cloud computing environments [7,8,10,35].

SDN-based cloud has attracted great attention recently. Cziva et al. [10] developed an SDN-based framework for live virtual machine (VM) to make easy network-server resource management over cloud data centers. The live VM migration exploits temporal network information in order to reduce the network-wide communication cost of the resulting traffic dynamics and alleviates congestion of the high cost.

Pisharody et al. [36] designed a security policy analysis framework to detect all potential conflicts over a distributed SDN-based cloud environment. The detection method is extended from the firewall rule conflict detection techniques in traditional networks.

Son et al. [9] proposed a dynamic overbooking algorithm for consolidating VMs and traffics in cloud data centers, while exploiting jointly leverages virtualization and SDN abilities. The main objective of their proposed approach is to minimize SLA violations and save energy.

Recently, Son et al. [7] from the same institution presented a taxonomy of the usage of SDN in cloud computing in various aspects including energy efficiency, performance, virtualization, and security enhancement.

## 5.3 Security Challenges of SDN-Based Cloud

Integration of SDN with cloud computing presents many advantages in comparison to the traditional networking infrastructure such as improved flexibility of the network, improved programmability of the network, etc. However, an SDN-based cloud poses several security issues. In this section, we first present the crucial security challenges defined separately in cloud computing and in SDN. Then, we discuss the new security issues introduced over the integration of SDN and cloud computing.

### 5.3.1 Cloud Security Challenges

With the high complexity of network connections on a large scale, cloud computing becomes more vulnerable to both traditional and new security issues [37,38]. The Cloud Security Alliance (CSA) is a not-for-profit organization with a mission of defining and raising awareness of best practices for offering security assurance within the cloud computing environment [39]. Recently, CSA conducted a survey of industry experts to assemble professional views on the top security challenges in cloud computing, in order to identify the greatest threats [38]. Among the critical security issues identified by CSA, we list the following:

> *Insecure interfaces and APIs*: Cloud computing providers deliver services to their clients through software user interfaces (UIs) or application programming interfaces (APIs). Provisioning and management of cloud services are all made with these interfaces. Therefore, the security and availability of cloud services are dependent upon the security level of these APIs. The UI, API

functions and web applications share a number of vulnerabilities, which may cause various attacks related to the confidentiality, availability, integrity, and accountability of cloud services. Thus, any interfaces that will connect to cloud infrastructure must be designed with strong authentication methods (e.g., transport layer security [TLS]), proper access controls, and encryption methods.

*System vulnerabilities:* No system is 100% secure: every system has vulnerabilities, which can be exploited to negatively impact confidentiality, integrity, and availability of provided services. Since virtualization is a key technology in cloud infrastructure, any vulnerability can put the security of the system and all services at significant risk. For instance, any fault and vulnerability within the hypervisor may be harnessed to launch VMs attacks (shutting down VMs) or monitor the shared resources of the present VMs.

*Service interruption/DDoS:* Since everything in the cloud is defined as service provided on-demand, the availability is the crucial security requirement in cloud computing. DoS/DDoS attacks are the main threats that can interrupt availability of cloud services. DDoS attacks may happen when an attacker forces a targeted cloud service to use excessive amounts of finite system resources like network bandwidth, memory, central processing unit (CPU), or disk space, which render services and computing resources unavailable.

*Advanced persistent threats (APT):* APT is a parasitic form of the cyberattack that infiltrates systems or networks and remains there for an extended period of time without being detected. The intention of an APT attack is usually the monitoring of the network activity and stealing valuable data rather than causing harm to the network or company. Although APT attacks can be difficult to detect and eliminate, some can be stopped with system monitoring, proactive security measures, and awareness programs.

## 5.3.2 SDN Security Challenges

Although SDN characteristics help in protecting a cloud environment against various threats, SDN itself suffers from both the present security attacks and new issues. Due to the centralized controller and the network programmability of SDN, new vulnerabilities have been introduced across SDN layers [40–43].

*Application layer:* Unauthenticated and unauthorized applications are among the serious security breaches in SDN. They may access and modify network data or reprogram the SDN components. Consequently, authorization and authentication of these applications are needed to defend network resources against malicious activities.

*Control layer:* Since the SDN controller is the brain of SDN architecture, the majority of SDN security issues are related to the control plane vulnerabilities.

Due to the separation of the data plane and the control plane within the SDN framework, the controller itself may become a target for various threats like flooding and DDoS attacks. An attacker can initiate a resource consumption attack (e.g., SYN flood attacks) on the controller to make it unavailable in response to the switch requests.

*Data layer*: The lack of secure sockets layer (SSL) or TLS adoption within controller-switch communication may cause access of unauthorized controller and so insertion of fraud flow rules in SDN OpenFlow switches. These latter suffer also from saturation attacks (i.e., flow-table overloading attacks) due to the limited storage capacity of the switch flow tables.

*Insecure interfaces and APIs*: Since all communications between the application, control and data layers, or even the communication between multiple controllers, pass through application programming interfaces (APIs) (i.e., Northbound, Southbound, and East and West Interfaces), it is paramount to secure them. The different Northbound interfaces share a number of vulnerabilities, which may cause various attacks related to the controller availability and network elements processing. As OpenFlow protocol is the standard Southbound interface of SDN; it suffers from the lack of TLS adoption by major vendors which could lead to malicious rule insertion and rule modification. In a multi-controller environment, controllers' communication passes through the East/West interfaces. Most of the time these controllers don't share a common secure channel (i.e., controllers from different vendors) between them, which could lead to sniffing important messages and exposing sensitive information.

## 5.3.3 Security Issues of SDN-Based Cloud

Although SDN-enabled cloud computing has great benefits in comparison to the traditional cloud networking, it poses several security issues [7,8,13,40] as we can see below and in Table 5.1:

*Scalability*: Scalability of the SDN-enabled cloud relies on the scalability of the networking infrastructure which is controlled by SDN technology. In the current growth networks, the SDN centralized controller may easily become a bottleneck which may disrupt cloud evolution.

*Availability*: Since everything in cloud is defined as service, availability is a crucial security requirement which is directly related to the availability of SDN infrastructure. There are two major availability issues that rely on SDN architecture: (i) in the network expansion, the SDN centralized controller may become a target for various threats like DDoS attacks; (ii) the flow tables of SDN switches suffer from saturation attacks due to their limited storage capacity.

**Table 5.1    Security Challenges of SDN-Based Cloud**

| Security Challenges | Description |
|---|---|
| Scalability | Scalability of SDN-enabled cloud relies on the scalability of the networking infrastructure which is controlled by SDN technology. In the current growth networks, the SDN centralized controller may easily become a bottleneck which may disrupt cloud evolution. |
| Availability | Since everything in cloud is defined as service, availability is a crucial security requirement which is directly related to the availability of SDN infrastructure. |
| Authorization and authentication | In the application plane, unauthenticated and unauthorized applications pose a great challenge for SDN. |
| Fraudulent flow rules | Exploiting the vulnerabilities of the Southbound interfaces, the attacker can easily insert fraudulent flow rules within the switches. |
| Flow table overloading | The flow tables of OpenFlow switches suffer from saturation attacks due to their limited storage capacity. |
| DDoS | Due to the isolation of decision from the data plane, each new flow is forwarded to the controller. Therefore, DDoS attackers may exhaust easily the controller resources while sending a great number of new flows. |

*Authorization and authentication*: In SDN architecture most of the network services are presented as third-party applications, which have access to network elements and can also handle network functions. Consequently, authorization and authentication of these applications are needed to defend network resources against malicious activities.

*Fraudulent flow rules*: Due to the isolation of decision-making functions from the SDN switches, these switches cannot identify the legitimate flow rules from fraud flow rules. Hence, an attacker may easily insert fraudulent flow rules within the switches while using vulnerabilities of Southbound interfaces.

Although security is among the major concerns in SDN-enabled cloud computing, there are a limited number of works [7,8,13,40] which analyzed and examined the security challenges of the SDN-based cloud.

Yan et al. [8] discussed the new DDoS attacks tendency and features in cloud computing and supplied a comprehensive study of SDN-based defense mechanisms against DDoS attacks. This study gives us a clear view of how to make the SDN characteristics useful to protect cloud systems from DDoS attacks and how to prevent the SDN itself from becoming a sufferer of DDoS attacks.

Son et al. [7] presented a taxonomy of the usage of SDN in cloud computing in various aspects including energy efficiency, performance, virtualization, and security enhancement.

Bhushan et al. [13] discussed SDN feasibility in the cloud environment and represent the flow table-space of a switch by using a queuing theory–based mathematical model. They presented a novel flow table sharing approach to defend the SDN-based cloud against flow table overloading attacks.

Farahmandian et al. [40] presented major security challenges in cloud, SDN, and NFV and suggested solutions using virtualization technology. The article discussed the need for a software-defined security technology for managing an integrated infrastructure platform where cloud, SDN, and NFV all play their integral parts.

# 5.4 DDoS Attacks and Mitigation in SDN-Based Cloud

With the adoption of cloud services, the rate of DDoS attacks against cloud infrastructure increases since the traditional DDoS attacks defense techniques are unable to protect the large-scale network of the cloud. Some good characteristics of an SDN approach help to defend the DDoS attacks in the cloud computing environment. However, the SDN itself may be targeted by the attackers, which raises the risk of DDoS attacks in the SDN-based cloud. In this section, we first discuss how the cloud characteristics make it more vulnerable to DDoS attacks. Then, we discuss some SDN features which improve the DDoS attack mitigation in the cloud. And finally, we present the possible DDoS attacks on SDN.

## 5.4.1 Impact of DDoS Attacks on Cloud Computing

A recent Cloud Security Alliance (CSA) survey presents that DDoS attacks are crucial threats to cloud security [31,44]. Further, a set of studies [8,12,45] show how the essential characteristics of cloud computing may be the reason to increase the rate of DDoS attacks in cloud environment, as we can see below:

> *On-Demand Self-Service*: The on-demand self-service ability allows clients to independently get cloud services (computing, storage, network, etc.); it can be easily exploited to create a powerful botnet and initiate DDoS attacks in a short time frame. Consequently, the distributed DoS attacks increased as the large-scale botnets increased in the cloud environment.

*Broad Network Access*: The broad network access feature allows customers to access cloud services through different mobile devices, such as smartphones and tablets. The lack of security on the majority of these devices can be used to launch DDoS attacks in cloud infrastructure.

*Resource Pooling*: The virtualization technology is a key for cloud computing; consequently, any virtualization vulnerability can put the security of a cloud system and all services at major risk. The multitenant model is utilized in the cloud to pool physical and virtual resources among multiple clients. The multitenant and virtualization technologies and their vulnerabilities can be exploited to easily launch DDoS attacks and also make the system more vulnerable to DDoS threats.

*Rapid Elasticity and Measured Service*: The cloud's pricing model enables its clients to pay as per their use of the cloud's services. In combination with rapid elasticity, the pricing model may be used to affect a customer financially by generating fraud bills. Such attacks can also be planned to perform an economic denial of sustainability (EDoS) attack, which is a new form of DDoS attack.

## 5.4.2 How SDN Features Enhance the DDoS Defense in Cloud

In the traditional networks, it is difficult to implement, experiment, and deploy new ideas on a large-scale network such as a cloud environment. However, the separation of the control plane from the data plane in SDNs enables experimenters to perform easily large-scale attack and defense experiments on a real environment [8]. This separation provides a programmable network in which network devices can operate and manage network traffic dynamically [14]. Hence, the programmability of SDNs allows us to flexibly implement intelligent defense algorithms against DDoS attacks in a cloud environment. A centralized control feature of an SDN gives a network-wide knowledge, which helps to build a relevant security policy for the network. Characteristics like centralized control and programmability allow SDNs to defend cloud computing against DDoS attacks [16]. In an SDN, the network traffic can be analyzed innovatively using intelligent mechanisms and various types of software tools [15]. Hence, SDNs can greatly enhance DDoS detection and mitigation abilities using the software-based traffic analysis.

## 5.4.3 DDoS Risk in the SDN-Enabled Cloud

SDN characteristics present great promise in terms of defending DDoS attacks in the cloud environment. However, the SDN itself is vulnerable to different kinds of security attacks, including unauthorized access, configuration issues, flow table attacks, malicious applications, and data leakage, as shown in Section 3.2. The SDN architecture is divided into three layers: the application layer, control layer, and data layer. All these layers and communication APIs can be targeted by the attackers to launch DDoS attacks. The controller is the brain of SDN architecture; hence, it could be seen as a single point of failure in an SDN. Moreover, it is a particularly attractive target for a DDoS attack. In the control plane DDoS attacks can target controller services, Northbound interface, Southbound interface, Eastbound interface, and Westbound interface [15]. In the data plane, OpenFlow switches suffer from the limited storage capacity of the flow tables. This limitation may be exploited by the attackers for launching the flow table, overloading attacks [13,17]. In the application plane, unauthenticated and unauthorized applications pose a great threat for SDNs. Furthermore, the problem of isolation of applications and resources is not well solved in SDNs; as a result, a DDoS attack on one application can affect other applications as well.

## 5.5 Related Work

### 5.5.1 The Current DDoS Mitigation Mechanisms in SDN-Based Cloud

An SDN-enabled cloud presents many advantages in comparison to the traditional networking infrastructure, such as improved network scalability and flexibility, and improved network programmability. However, the recent assessments [13,15–17] and our analysis in Sections 5.3 and 5.4 show the introduction of new security issues and particularly new trends of DDoS attacks over the integration of SDN and cloud computing technologies.

To the best of our knowledge, there are limited research works which address the potential challenges to mitigate DDoS attacks in the enterprise networking that adopts both cloud computing and SDN technologies, among which we present and analyze the following works with a brief description in Table 5.2.

Bhushan et al. [13] presented the SDN principle and SDN-enabled cloud. They discussed the DDoS impact in the SDN-based cloud and the existing solutions to protect the cloud environment from DDoS attacks using SDN. They presented a new approach to defense the SDN-based cloud against flow table overloading attacks. The approach utilizes the unused flow table-spaces of other switches to resist the attack at a particular switch. To increase the availability of flow tables in SDN, the proposed mechanism removes less utilized flow rules and flow rules belonging to the attack traffic. They maintain two databases for the operation of their approach: Flow Table Status and Black List. Flow Table Status records the current status of flow tables of all the switches in the network, i.e., the number of entries occupied in each flow table. The Black List database lists the IP addresses of attack sources. The objective of this approach is to enhance the resistance of the SDN against DDoS attacks while increasing the time to overload all the SDN switches (i.e., the holding time) at least up to the reaction time of DDoS defense systems.

**Table 5.2 Related Work**

| Related Work | Description |
|---|---|
| [11] | Bhushan et al. discussed the SDN feasibility in the cloud environment and represented the flow table-space of a switch by using a mathematical model based on queuing theory concepts. They presented a new approach to defend the SDN-based cloud against flow table overloading attacks. This approach utilizes the unused flow table-spaces of other switches to resist the attack at a particular switch. |
| [13] | Wang et al. studied the impact of SDN-enabled cloud on DDoS attacks defense. They remarked that if the DDoS attacks mitigation solution in SDN is designed correctly, SDN will benefit the DDoS attacks protection in a cloud computing environment. Therefore, they proposed a DDoS attacks defense architecture DaMask, which contains two modules: an anomaly-based attack detection module DaMask-D, and an attack mitigation module DaMask-M. |
| [36] | Bawany et al. made a survey of the current SDN-based DDoS attack detection and mitigation strategies. Motivated by the observed requirements, they proposed an SDN-based proactive DDoS Defense Framework (ProDefense) for a smart city data center. The approach enables the implementation of different application-security requirements and also has distributed controllers which increase the reliability and scalability of the solution. |
| [37] | Yan et al. developed a multi-queue SDN controller scheduling method to mitigate DDoS attacks in SDN. The proposed method defends the normal switches during a DDoS attack and prevents the SDN network from being unavailable by programming the flow request processing through different switches. This approach uses different time slicing strategies depending on the DDoS attack intensity. |

The experimentation showed that the holding time of SDN has been significantly improved when the proposed approach was applied and the communication between controller and switch have been reduced during the attack. However, if there are a large number of new transmission control protocol (TCP) connection attempts from different and new IP sources the switch will ask the controller for new flow rule for each connection. Hence, the SDN centralized controller may easily become a bottleneck which may disrupt the network services.

Wang et al. [15] studied the impact of the SDN-enabled cloud on DDoS attacks defense. Based on their analysis, they remarked that if the DDoS attacks mitigation solution in an SDN is designed correctly, the SDN will benefit the DDoS attacks protection in the cloud computing environment. Therefore, they proposed a DDoS attacks defense architecture (DaMask), which contains two modules: an anomaly-based attack detection module DaMask-D, and an attack mitigation module DaMask-M. For the DaMask-D module, they developed an attack detection system which is built on a probabilistic inference graphical model. The proposed detection system advances with two capabilities: an automatic feature selection to build an effective graph model and an efficient model update to address the dataset shift problem. DaMask-M is a flexible control structure which allows quick attack reaction. The authors evaluate the DaMask architecture under a private cloud and a public cloud (i.e., the Amazon Web Service [AWS]). The evaluation indicated that the proposed attack detection algorithm gives a cheaper computation cost of attack detection and a constant communication overhead as long as the link status of the network is stable. Additionally, the evaluation validated the DaMask ability to adapt to the network topology change caused by virtual machine migrations. Moreover, it exhibited that the local model update mitigates the impact of the dataset shift problems and therefore improves the detection accuracy.

Bawany et al. [16] made a comprehensive and extensive survey of the SDN-based DDoS attack detection strategies. Motivated by the key requirements of an effective DDoS attack prevention mechanism, they proposed an SDN-based proactive DDoS defense framework (ProDefense) for smart city data center. Since the smart city applications have different security requirements (i.e., catastrophic, critical, moderate), ProDefense framework uses a customized detection filter to meet the security requirements of various specific applications. The distributed control layer ability enhances the reliability and scalability of the proposed solution to support the evolution of smart city data centers. Further, the article discussed the open research challenges, future research directions, and recommendations related to SDN-based security solutions.

Yan et al. [17] developed a multi-queue SDN controller scheduling algorithm "MultiSlot" to mitigate DDoS attacks in SDNs. The proposed method defends the normal switches during a DDoS attack and prevents the SDN from being unavailable by programming the flow request processing through different switches. This approach uses different time slicing strategies depending on the DDoS attack intensity. The simulation results showed that the proposed method has better performance than the existing methods in terms of protecting the internal switches that are affected by the DDoS attacks.

### 5.5.2 Discussion

All the previously mentioned DDoS mitigation mechanisms are designed at the high-level SDN application plane with the involvement of the SDN controller in each operation to detect and mitigate DDoS attacks. Therefore, the communication path between the data and control planes rapidly becomes a bottleneck, which may impact the network performance and restrict its scalability and reactivity. In this side, the SDN community has taken into consideration the necessity to conduct specific functions directly inside the SDN Switch [46–48]. SDN data plane–based specific functions can be used to minimize the switch-controller communication and optimize the scalability and responsiveness of the defensive mechanisms.

### 5.5.2.1 Frameworks, Compilers, and Programming Languages

Frameworks, compilers, and programming languages [18,19–21,49,50] have been developed to take advantages of the SDN data plane with a way to perform dynamically specific operations (e.g., monitoring, detection, reaction) inside the switch.

Programming Protocol-Independent Packet Processors (P4) [51] is a domain-specific programming language designed for describing how packets are processed by the data plane of a programmable forwarding element, such as a hardware or software switch, router, network interface card, or network appliance. Interested P4-based security solutions have been developed. Vörös et al. [52] discussed the primary security middleware programmed and configured in P4. They proposed a stateful firewall for mitigating network flooding attacks using P4 language. Afek et al. [53] suggested a defense system against network spoofing attacks while performing selected anti-spoofing techniques in OpenFlow 1.5 and P4 (i.e., match and action rules). They also developed dynamic algorithms for automatic distribution of sophisticated rules on network switches.

As another programming language and compiler for data plane, SNAP was presented by [20] which allows stateful network-wide abstractions for packet processing. It defines some state variables and arrays to install rules and to maintain the state information of the flows. The authors in [21] have proposed the Domino compiler for packets processing in the data plane. The packet transactions are compiled by the Domino and executed on Banzai machine to allow high-level programming for line-rate switches.

## 5.6 SDN-Based SYN Flooding Defense in Cloud

### 5.6.1 System Design

In our previous work, "SDN-based SYN Flooding Defense in Cloud" [54], we proposed the design of an active mitigation mechanism which enables the SDN data plane (DP) to prevent and mitigate DDoS attacks and particularly TCP/SYN flooding attacks in the cloud environment. Our approach exploits the centralized and software-based traffic analysis of SDNs to better monitor and analyze traffic and detect flow rate anomalies. Moreover, it uses the programmability and the distributed nature of SDNs to roll security closer to the switches and uses the flexibility feature to activate the network to defend against SYN flooding attacks in a more immediate manner. The proposed work utilizes the P4 language to take advantage of the switch programmability.

The main objectives of our approach are (i) activate the data plane with customized statistical information databases which are useful for security applications to monitor traffic and detect flow rate anomalies, (ii) enable the data plane to prevent the overwhelming traffic attacks (i.e., SYN flooding attack) at an early stage using traffic anomaly detection algorithm (CUSUM), (iii) add intelligence to the data plane to defense SYN flooding attacks using SYN cookie technique, and then deploying adaptive countermeasures.

The proposed mechanism extends the data plane with three modules as we can see in Figures 5.1 and 5.3. These modules include traffic information databases (DBs), traffic anomaly detection, and inspection and mitigation.

> *Traffic information DBs*: Network traffic information is really important for any kind of security applications or appliances to monitor and analyze network traffic, detect suspicious activities, and decide further countermeasures. In this end, we propose traffic information DBs in our framework to support the previously discussed security modules in realizing the existence of flooding attacks and deciding adaptive countermeasures. Traffic information DBs are designed to easily retrieve traffic information and provide detail statistics information for upper-security applications. In this work, we use the P4 programming language to create customized DBs that can respond to our needs. We create and perform the P4 program at the switch level (Figure 5.2) to maintain information (i.e., IP address source and destination, TCP port source and destination, and TCP flags) across packets using stateful memories: counters and meters [51]. In our case, we use counters to measure the number of TCP SYN packets received by each host (server).

> *The traffic anomaly detection* activates the DP to detect if the cloud system is under overwhelming traffic attacks at an early stage. The goal of deploying this module is to reduce the pressure of the traffic to be handled by the security applications implemented above (Figure 5.3). The detection module uses the Cumulative SUM (CUSUM) [55] as a change point detection algorithm to detect the appearance of an unexpected change of the system traffic volume. It realizes this by modeling the normal traffic behavior of the network (using collected traffic information) and flagging significant deviations (i.e., when the accumulated volume of measurements surpasses some total volume threshold) from this behavioral model as anomalies. It could reduce the detection delay with a fixed false alarm rate [56] and efficiently flag the anomaly change.

> *Inspection and mitigation modules* enable the DP to inspect received packets (i.e., which do not exist in flow table) in order to realize the existence of the SYN flood attacks and activate the adaptive countermeasures, without any switch-controller

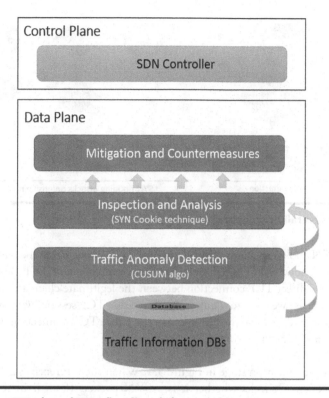

**Figure 5.1 SDN-based SYN flooding defense architecture.**

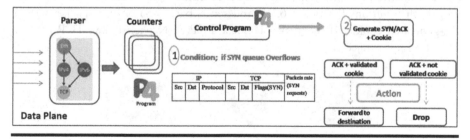

**Figure 5.2 Architecture of dynamic defense mechanism.**

communication. Inspired by the stateless SYN cookie technique [57], the inspection and mitigation modules protect both the SDN controller and cloud data center from half-open SYN attacks. When traffic flooding is detected the inspection and mitigation modules start their functions (Figure 5.3). For the first time, the DP checks if the packet exists in the flow table; if so, the packet will be immediately forwarded to the destination server. Otherwise, the DP initiates a classification phase which classifies the validated TCP sessions from failed ones (i.e., half-open SYN attacks or invalidated TCP sessions). In this phase, we select

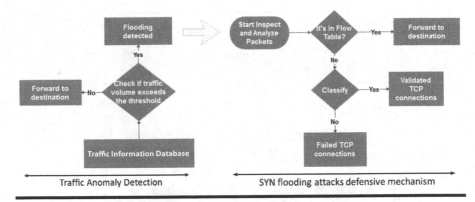

**Figure 5.3   Overall organigram.**

two SYN cookie methods: TCP-Reset and TCP-Proxy. Each one has its advantages and drawbacks [58,59]. In our case, we use as default TCP-Reset method to allow a direct TCP connection between the legitimate client and destination server. In case we want to monitor the server-side TCP session, we can use TCP-Proxy method to allow the DP to proxy the entire TCP connection, which may present a significant overhead.

We detail the classification stage in Figure 5.4. When the DP receives a TCP SYN/RST/FIN packet, it checks whether it is a SYN packet. If so, it increments the counter of the access table (i.e., access table contains information on all new TCP connection attempts) and generates a sequence number (i.e., cookie) for this packet with hash functions (Section 5.6.1.1) and sends back a SYN-ACK packet with a pre-generated cookie. If the packet is not a TCP SYN packet (i.e., TCP FIN or TCP RST), it is rejected.

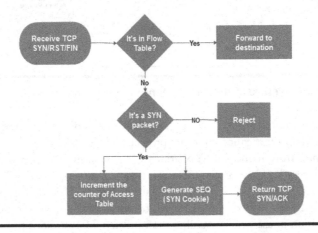

**Figure 5.4   Handling TCP SYN packets chart.**

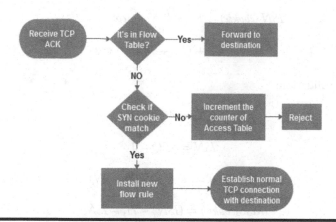

**Figure 5.5    Handling TCP ACK packets chart.**

If the DP receives an ACK packet, as shown in Figure 5.5, it checks its TCP sequence number against the cookie value that was encoded in the SYN-ACK packet (Section 5.6.1.1). If the TCP connection is validated (i.e., ACK packet contains the appropriate SYN cookie) the TCP handshake is established immediately between client and destination server. During accomplishment of the TCP handshake, the DP add the new rule in flow table to allow the future TCP requests arrived from this source. If the TCP connection is not validated, the counter of the rejected table (i.e., contains information on the failed TCP connection) is incremented and the connection is rejected.

### 5.6.1.1  Pre-generated Cookie

According to [60], the implementation of the SYN cookies must fulfill the following basic requirements:

- Cookies should contain some details of the initial SYN packet and its TCP options.
- Cookies should be unpredictable by attackers. It is recommended to use a cryptographic hashing function in order to make the decoding of the cookie more complicated. For this end, we select the recommended Linux SYN cookies method for generating and validating cookies [61].

Cookie generation:

$$H_1 = hash(K_1, IP_s, IP_d, Port_s, Port_d) \tag{5.1}$$

$$H_2 = hash(K_2, count, IP_s, IP_d, Port_s, Port_d) \tag{5.2}$$

$$ISN_d(cookie) = H_1 + ISN_s + (count \times 2^{24}) \qquad (5.3)$$

$$+ (H_2 + MSS) \bmod 2^{24}$$

Cookie validation:

$$ISN_d = ACK - 1 \qquad (5.4)$$

$$ISN_s = SEQ - 1 \qquad (5.5)$$

$$count(cookie) = (ISN_d - H_1 - ISN_s)/2^{24} \qquad (5.6)$$

$$MSS(cookie) = (ISN_d - H_1 - ISN_s) \qquad (5.7)$$

$$\bmod 2^{24} - H_2 \bmod 2^{24}$$

As we can see from the preceding list and in Table 5.3, we calculate the two hash values $H_1$ and $H_2$ (based on TCP options, secret keys $K_1$, $K_2$ and count) then we use them with initial sequence numbers (ISNs) and maximum segment size (MSS) to generate the cookie (ISNd), as it is shown in (3). For the cookie validation, there are two integrity controls: the count(cookie) and the MSS(cookie). The first one checks the age of the cookie. The second evaluates whether the value of the MSS is within the 2-bit range (0–3). If the cookie meets both integrity controls, it is considered valid, and the connection can be accepted.

**Table 5.3  Parameters of the Linux Implementation**

| Parameter | Description |
|---|---|
| $K_1$, $K_2$ | Secret keys |
| $IP_s$, $IP_d$ | Source and destination IP addresses |
| $Port_s$, $Port_d$ | Source and destination ports |
| $ISN_s$, $ISN_d$ | Source and destination initial sequence numbers |
| ACK | Acknowledgment number |
| SEQ | Sequence number |
| MSS | 2-bit index of the client's Maximum Segment Size |
| Count | 32-bit minute counter |
| Hash() | 32-bit cryptographic hash |

P4 language gives us the possibility to generate the hash values ($H_1$ and $H_2$) using the bellow function and to perform the equations (cookies generation and validation) as arithmetic operations.

```
field_list_calculation hash_value_name {
    input { fields;}
    algorithm : hash_algo;
```

## 5.6.2 Evaluation and Comparison

The main security features and strengths of our proposed framework are:

- *Useful DBs*: It exploits the stateful memories (i.e., stateful memories keeps track of the network state) capability of the SDN data plane using P4 language to realize customized traffic information DBs at the switch level. These DBs are useful for any kind of network security solutions to monitor and analyze network traffic and detect suspicious activities.
- *Early Detection*: It activates the detection of the overwhelming traffic attacks at the data plane without any external and dedicated software application or hardware appliance. It realizes this while performing the Cumulative SUM (CUSUM) algorithm which enables the switch to detect the appearance of an unexpected change of the system traffic volume. As a result, it prevents and reduces the risk of DDoS attacks in the OpenFlow switches (i.e., flow-table overloading attacks) and the SDN centralized controllers.
- *Early Classification*: It rolls SYN cookie techniques closer to the switch to analyze and classify the legitimated TCP sessions from failed ones (i.e., half-open SYN attacks or invalidated TCP sessions) without any switch-controller communication. In addition, it activates the SYN cookie technique only when the rate of TCP SYN requests reaches the defined adaptive threshold, rather than enabling them constantly.
- *TCP Session Management*: It stores information about all new TCP connection attempts including succeeded and failed connections which remain useful for detecting various attacks like network scanning attack.
- *Dynamic Network Reconfiguration*: It activates the data plane with dynamic and reactive mitigation actions (i.e., block, reject, modify, redirect traffic, and add a new rule) against TCP SYN flooding attacks.
- *Defense Against Other Protocols-Based Attacks*: It enhances the resilience against TCP SYN flooding attacks and it may also be used to defend attacks based on hypertext transfer protocol (HTTP), simple mail transfer protocol (SMTP), and others protocols. This can be done while using specific SYN cookie methods, such as HTTP Redirect and TCP Safe Reset.

Based on our study, most of the existing DDoS mitigation mechanisms [13,15–17] are designed at the high-level SDN application plane with the involvement of the SDN controller in each operation to detect and mitigate DDoS attacks. Therefore, the required communication channel between the control and data plane of an SDN creates a potential bottleneck that impacts the network performance and restricts its scalability and reactivity. In comparison to these solutions, our proposed mechanism enables the data plane to detect the network saturation attack with a minimum communication between the SDN controller and data plane. In result, the switch may prevent and reduce the risk of DDoS attacks in the OpenFlow switches (i.e., flow-table overloading attacks) and the SDN controllers at an early stage.

There are a few works [53,62] which perform SYN cookie technique at the SDN data plane to defeat TCP SYN flooding attacks. However, these approaches enable an SYN cookie all the time and for each packet which limits the responsiveness of the system. Inspired by the Linux kernel method, which automatically enables the SYN cookie only when the SYN queue is full, our framework activates the SYN cookie technique only when the rate of TCP SYN requests reaches the defined adaptive threshold, rather than enabling them constantly.

In comparison to some existing solutions, our framework implements simple method functionalities over SDN switches rather than complex methods such as machine learning or deep learning which require a high memory and processing requirements.

## 5.7 Conclusion

The growing adoption of the cyber-physical systems and the cloud technologies have highlighted the requirement for protection of computing and network infrastructure against DoS and DDoS attacks. To this end, we have made various observable contributions. In this chapter, we discussed the SDN features which make it an appropriate technology for cloud systems. For example, centralized control ability allows a global view of the network, and provides centralized network management and provisioning in the cloud computing environments. We analyzed the new security issues introduced over the integration of SDN and cloud computing technologies. Each technology has several vulnerabilities which may be exploited to launch various threats. Therefore, to ensure the security of SDN-based cloud systems, we need to fix their vulnerabilities and assure security policy that covers both SDN and cloud technologies. We examined how the cloud and SDN characteristics make them more vulnerable to DDoS attacks. Based on our analysis, we defined the challenges and opportunities raised by these new technologies. We claim that with a careful design of SDN-based DDoS mitigation, SDN will benefit the DDoS attack protection in the cloud computing environment. To prove our finding, we presented our proposed mechanism "SDN-based SYN flooding defense in cloud" to prevent and mitigate SYN flooding attacks in the SDN-based cloud environment.

In comparison to the existing solutions, our mechanism activates the detection and mitigation of TCP SYN flooding attacks at the SDN data plane without any external and dedicated appliance. Consequently, it prevents and reduces the risk of DDoS attacks in the OpenFlow switches and the SDN controllers. Our proposed solution enhances the resistance of both SDN architecture and cloud data centers against SYN flooding attacks.

For future work, we plan to conduct a study of the various cyber-physical cloud systems, such as IoT, smart cities, or smart grid, to decide which one to choose as application domain in our experiments. Moreover, further experiments and simulations will be performed to support more sophisticated attacks in the selected cyber-physical cloud system.

# References

1. Lu, Yang. Industry 4.0: A survey on technologies, applications and open research issues. *Journal of Industrial Information Integration*; 2017, vol. 6, pp. 1–10.
2. Drath, Rainer and Horch, Alexander. Industrie 4.0: Hit or hype? [industry forum]. *IEEE Industrial Electronics Magazine*; 2014, vol. 8, no. 2, pp. 56–58.
3. Rajkumar, Ragunathan and Lee, Insup and Sha, Lui and Stankovic, John. Cyber-physical systems: The next computing revolution. *Proceedings of Design Automation Conference (DAC)*, 2010 47th ACM/IEEE; 2010, pp. 731–736.
4. A Vision of Cyber-Physical Cloud Computing for Smart Networked Systems; Available from: https://ws680.nist.gov/publication/get pdf.cfm?pub id=914023.
5. Díaz, Manuel and Martín, Cristian and Rubio, Bartolomé. State-of-the-art, challenges, and open issues in the integration of Internet of Things and cloud computing. *Journal of Network and Computer Applications*; 2016, vol. 67, pp. 99–117.
6. Satyanarayanan, Mahadev. The emergence of edge computing. *Computer IEEE*; 2017, vol. 50, no. 1, pp. 30–39.
7. Son, Jungmin and Buyya, Rajkumar. A taxonomy of software-defined networking (SDN)-enabled cloud computing. *ACM Computing Surveys (CSUR)*; 2018, vol. 51, no. 3, p. 59.
8. Yan, Qiao and Yu, F Richard and Gong, Qingxiang and Li, Jianqiang. Software-defined networking (SDN) and distributed denial of service (DDoS) attacks in cloud computing environments: A survey, some research issues, and challenges. *IEEE Communications Surveys and Tutorials*; 2016, vol. 18, no. 1, pp. 602–622.
9. SLA-aware and energy-efficient dynamic overbooking in SDN-based cloud data centers. *IEEE Transactions on Sustainable Computing*; 2017, vol. 2, no. 2, pp. 76–89.
10. Cziva, Richard and Jouët, Simon and Stapleton, David and Tso, Fung Po and Pezaros, Dimitrios P. SDN-based virtual machine management for cloud data centers. *IEEE Transactions on Network and Service Management*; 2016, vol. 13, no. 2, pp. 212–225.
11. Bhushan, Kriti and Gupta, BB. Security challenges in cloud computing: State-of-art. *International Journal of Big Data Intelligence*. Inderscience Publishers (IEL); 2017, vol. 4, no. 2, pp. 81–107.
12. Somani, Gaurav and Gaur, Manoj Singh and Sanghi, Dheeraj and Conti, Mauro and Buyya, Rajkumar. DDoS attacks in cloud computing: Issues, taxonomy, and future directions. *Computer Communications*; 2017, vol. 107, pp. 30–48.

13. Bhushan, Kriti and Gupta, BB. Distributed denial of service (DDoS) attack mitigation in software defined network (SDN)-based cloud computing environment. *Journal of Ambient Intelligence and Humanized Computing.* Springer; 2018, pp. 1–13.

14. DCruze, Hubert and Wang, Ping and Sbeit, Raed Omar and Ray, Andrew. A software-defined networking (SDN) approach to mitigating DDoS attacks. *Information Technology-New Generations.* Springer; 2018, pp. 141–145.

15. Wang, Bing and Zheng, Yao and Lou, Wenjing and Hou, Y Thomas. DDoS attack protection in the era of cloud computing and software-defined networking. *Computer Networks*; 2015, vol. 81, pp. 308–319.

16. Bawany, Narmeen Zakaria and Shamsi, Jawwad A and Salah, Khaled. DDoS attack detection and mitigation using SDN: Methods, practices, and solutions. *Arabian Journal for Science and Engineering*; 2017, vol. 42, no. 2, pp. 425–441.

17. Yan, Q and Gong, Q and Yu, FR. Effective software-defined networking controller scheduling method to mitigate DDoS attacks. *Electronics Letters*; 2017, vol. 53, no. 7, pp. 469–471.

18. Bosshart, Pat and Daly, Dan and Gibb, Glen and Izzard, Martin and McKeown, Nick and Rexford, Jennifer and Schlesinger, Cole and Talayco, Dan and Vahdat, Amin and Varghese, George and others. P4: Programming protocol-independent packet processors. *ACM SIGCOMM Computer Communication Review*; 2014, vol. 44, no. 3, pp. 87–95.

19. Zhu, Shuyong and Bi, Jun and Sun, Chen and Wu, Chenhui and Hu, Hongxin. Sdpa: Enhancing stateful forwarding for software-defined networking. *Proceedings of Network Protocols (ICNP), 2015 IEEE 23rd International Conference on*; 2015, pp. 323–333.

20. Arashloo, Mina Tahmasbi and Koral, Yaron and Greenberg, Michael and Rexford, Jennifer and Walker, David. SNAP: Stateful network-wide abstractions for packet processing. *Proceedings of the 2016 ACM SIGCOMM Conference*; 2016, pp. 29–43.

21. Sivaraman, Anirudh and Cheung, Alvin and Budiu, Mihai and Kim, Changhoon and Alizadeh, Mohammad and Balakrishnan, Hari and Varghese, George and McKeown, Nick and Licking, Steve. Packet transactions: High-level programming for line-rate switches. *Proceedings of the 2016 ACM SIGCOMM Conference*; 2016, pp. 15–28.

22. Cloud Economics: Making the Business Case for Cloud. Available from: https://assets. kpmg.com/content/dam/kpmg/pdf/2015/11/cloud-economics.pdf.

23. The NIST Definition of Cloud Computing; Available from: https://nvlpubs.nist.gov/ nistpubs/Legacy/SP/nistspecialpublication800-145.pdf.

24. Jarraya, Yosr and Madi, Taous and Debbabi, Mourad. A survey and a layered taxonomy of software-defined networking. *IEEE Communications Surveys and Tutorials*; 2014, vol.16, no. 4, pp. 1955–1980.

25. Haleplidis, Evangelos and Pentikousis, Kostas and Denazis, Spyros and Salim, J Hadi and Meyer, David and Koufopavlou, Odysseas. *Software-Defined Networking (SDN): Layers and Architecture Terminology*; 2015, No. RFC 7426.

26. Kreutz, Diego and Ramos, Fernando MV and Verissimo, Paulo Esteves and Rothenberg, Christian Esteve and Azodolmolky, Siamak and Uhlig, Steve. Software-Defined networking: A comprehensive survey. *Proceedings of the IEEE*; 2015, vol. 103, no. 1, pp. 14–76.

27. The Open Networking Foundation; Available from: https://www.opennetworking.org/.

28. McKeown, Nick and Anderson, Tom and Balakrishnan, Hari and Parulkar, Guru and Peterson, Larry and Rexford, Jennifer and Shenker, Scott and Turner, Jonathan. OpenFlow: Enabling innovation in campus networks. *ACM SIGCOMM Computer Communication Review*; 2008, vol. 38, no. 2, pp. 69–74.

29. Yang, Chaowei and Huang, Qunying and Li, Zhenlong and Liu, Kai and Hu, Fei. Big data and cloud computing: Innovation opportunities and challenges. *International Journal of Digital Earth*. Taylor & Francis Group; 2017, vol. 10, no. 1, pp. 13–53.
30. Singh, Saurabh and Jeong, Young-Sik and Park, Jong Hyuk. A survey on cloud computing security: Issues, threats, and solutions. *Journal of Network and Computer Applications*; 2016, vol. 75, pp. 200–222.
31. Almorsy, Mohamed and Grundy, John and Müller, Ingo. An analysis of the cloud computing security problem. *ArXiv*; 2016. Preprint arXiv:1609.01107.
32. Rittinghouse, John W and Ransome, James F. *Cloud Computing: Implementation, Management, and Security*. CRC Press: Boca Raton, FL; 2016.
33. Varghese, Blesson and Buyya, Rajkumar. Next generation cloud computing: New trends and research directions. *Future Generation Computer Systems*; 2018, vol. 79, pp. 849–861.
34. Zhang, Haijun and Liu, Na and Chu, Xiaoli and Long, Keping and Aghvami, Abdol-Hamid and Leung, Victor CM. Network slicing based 5G and future mobile networks: Mobility, resource management, and challenges. *IEEE Communications Magazine*; 2017, vol. 55, no. 8, pp. 138–145.
35. Mayoral, Arturo and Vilalta, Ricard and Munõz, Raul and Casellas, Ramon and Martínez, Ricardo. SDN orchestration architectures and their integration with cloud computing applications. *Optical Switching and Networking*; 2017, vol. 26, pp. 2–13.
36. Pisharody, Sandeep and Natarajan, Janakarajan and Chowdhary, Ankur and Alshalan, Abdullah and Huang, Dijiang. Brew: A security policy analysis framework for distributed SDN-based cloud environments. *IEEE Transactions on Dependable and Secure Computing*; 2017.
37. Almorsy, Mohamed and Grundy, John and Müller, Ingo. An analysis of the cloud computing security problem. *arXiv*; 2016, preprint arXiv:1609.01107.
38. Treacherous 12: Top Threats to Cloud Computing (2016); Available from: https://downloads.cloudsecurityalliance.org/assets/research/top-threats/Treacherous-12 Cloud-Computing Top-Threats.pdf.
39. The Cloud Security Alliance; Available from: https://cloudsecurityalliance.org/
40. Farahmandian, Sara and Hoang, Doan B. Security for software-defined (Cloud, SDN And NFV) infrastructures–issues and challenges. *Proceedings of Eight International Conference on Network and Communications Security*; 2016.
41. Zhang, Heng and Cai, Zhiping and Liu, Qiang and Xiao, Qingjun and Li, Yangyang and Cheang, Chak Fone. A survey on security-aware measurement in SDN. *Security and Communication Networks*. Hindawi; 2018, vol. 2018.
42. Scott-Hayward, Sandra and O'Callaghan, Gemma and Sezer, Sakir. SDN security: A survey. *Proceedings of Future Networks and Services (SDN4FNS)*. IEEE; 2013, pp. 1–7.
43. Rawat, Danda B and Reddy, Swetha R. Software defined networking architecture, security and energy efficiency: A survey. *Environment*; 2017, vol. 3, no. 5, pp. 6.
44. Kalaiprasath, R and Elankavi, R and Udayakumar, Dr R and others. Cloud. security and compliance—a semantic approach in end to end security. *International Journal of Mechanical Engineering and Technology (IJMET)*; 2017, vol. 8, no. 5, pp. 482–495.
45. Deka, Rup Kumar and Bhattacharyya, Dhruba Kumar and Kalita, Jugal Kumar. DDoS attacks: Tools, mitigation approaches, and probable impact on private cloud environment. *ArXiv*; 2017. Preprint arXiv:1710.08628.
46. Dargahi, Tooska and Caponi, Alberto and Ambrosin, Moreno and Bianchi, Giuseppe and Conti, Mauro. A survey on the security of stateful SDN data planes. *IEEE Communications Surveys and Tutorials*; 2017, vol. 19, no. 3, pp. 1701–1725.

47. Shaghaghi, Arash and Kaafar, Mohamed Ali and Buyya, Rajkumar and Jha, Sanjay. Software-defined network (SDN) data plane security: Issues, solutions and future directions. *ArXiv*; 2018. Preprint arXiv:1804.00262.

48. Trois, Celio and Del Fabro, Marcos D and de Bona, Luis CE and Martinello, Magnos. A survey on SDN programming languages: Toward a taxonomy. *IEEE Communications Surveys and Tutorials*; 2016. vol. 18, no. 4, pp. 2687–2712.

49. Moshref, Masoud and Bhargava, Apoorv and Gupta, Adhip and Yu, Minlan and Govindan, Ramesh. Flow-level state transition as a new switch primitive for SDN. *Proceedings of the Third Workshop on Hot Topics in Software Defined Networking*. ACM; 2014, pp. 61–66.

50. Bianchi, Giuseppe and Bonola, Marco and Capone, Antonio and Cascone, Carmelo. OpenState: Programming platform-independent stateful OpenFlow applications inside the switch. *ACM SIGCOMM Computer Communication Review*; 2014, vol. 44, no. 2, pp. 44–51.

51. The P4 Language Specification; Available from: https://p4.org/p4-spec/p4- 14/v1.0.4/tex/p4.pdf.

52. Vörös, Péter and Kiss, Attila. Security middleware programming using P4. *Proceedings of International Conference on Human Aspects of Information Security, Privacy, and Trust*. Springer; 2016, pp. 277–287.

53. Afek, Yehuda and Bremler-Barr, Anat and Shafir, Lior. Network anti-spoofing with SDN data plane. *Proceedings of INFOCOM 2017-IEEE Conference on Computer Communications, IEEE*; 2017, pp. 1–9.

54. Mahrach, Safaa and El Mir, Iman and Haqiq, Abdelkrim and Huang, Dijiang. SDN-based SYN flooding defense in cloud. *Journal of Information Assurance and Security*; 2018, vol. 13, no. 1.

55. Granjon, Pierre. The CuSum algorithm-a small review; 2013; Available from: http://chamilo2.grenet.fr/inp/courses/ENSE3A35EMIAAZ0/document/change-detection.pdf.

56. Tartakovsky, Alexander G and Polunchenko, Aleksey S and Sokolov, Grigory. Efficient computer network anomaly detection by changepoint detection methods. In *IEEE Journal of Selected Topics in Signal Processing*; 2013, vol. 7, no. 1, pp. 4–11.

57. Fontes, Stephen M and Hind, John R and Narten, Thomas and Stockton, Marcia L. Blended SYN cookies. US Patent; 2006, 7,058,718.

58. Touitou, Dan and Pazi, Guy and Shtein, Yehiel and Tzadikario, Rephael. Using TCP to authenticate IP source addresses. US Patent; 2011, 7,979,694.

59. Touitou, Dan and Zadikario, Rafi. Upper-level protocol authentication. US Patent; 2009, 7,536,552.

60. Simpson, William Allen. TCP cookie transactions (TCPCT). RFC 6013; 2011.

61. Echevarria, Juan Jose and Garaizar, Pablo and Legarda, Jon. An experimental study on the applicability of SYN cookies to networked constrained devices. *Journal of Software: Practice and Experience*; 2018, vol. 48, no 3, pp. 740–749.

62. Shin, Seungwon and Yegneswaran, Vinod and Porras, Phillip and Gu, Guofei. AVANT-GUARD: Scalable and vigilant switch flow management in software-defined networks. *Proceedings of the 2013 ACM SIGSAC Conference on Computer and Communications Security*; 2013, pp. 413–424, ACM.

## Chapter 6

# Detecting Pilot Contamination Attacks in Wireless Cyber-Physical Systems

Dimitriya Mihaylova, Georgi Iliev
and Zlatka Valkova-Jarvis

## Contents

## 6.1 Introduction

Security is a fundamental topic in the new wireless cyber-physical systems (CPSs). Conventionally, secure communication between wireless devices is provided by different cryptographic schemes, used on the upper layers of the open systems interconnection (OSI) model. Another recently investigated strategy, which overcomes the computational complexity of crypto-algorithms, and is thus suitable for CPSs, relates to the physical properties of the wireless channel (Dohler et al., 2011).

A number of problems facing physical layer security (PLS), as emphasized in Trappe (2015), require additional research (Bash et al., 2015; Wang et al., 2015). One such weakness is the assumption that channel prediction or estimation is difficult for a malicious user to carry out. This is not valid for simple environments with poor scattering. Such a scenario is discussed in Kapetanovic et al. (2015), where the attacker overcomes the passive eavesdropping resistance of a massive multiple-input-multiple-output (MaMIMO) system by placing himself physically close to the legitimate receiver and using the correlation between the channels. Another case in Kapetanovic et al. (2014, 2015) presents an active attack, known as pilot contamination, which the eavesdropper (ED) can mount against the channel estimation procedure of a CPS. A detailed description of a pilot contamination attack (PCA) is given in Zhou et al. (2012).

In the PLS literature, it is assumed that a legitimate user (LU) knows all the channel state information (CSI). However, the ED could use a jamming attack against the channel estimate, and thus disrupt this communication (Miller et al., 2012). Different levels of CSI at the transmitter and the ED, and their influence on the privacy of a system, are discussed in Liu et al. (2015) and Bash et al. (2013).

For a time-division duplex (TDD) system, the CSI is obtained during the training phase, when the transmitter estimates the legitimate channel by means of a pilot signal sent from the receiver (Rusek et al., 2013). The pilot contamination attack consists of the eavesdropper sending pilots at the same time as the legitimate receiver does, thereby imitating the natural pilot contamination in multicell TDD systems (Jose et al., 2011). The result is an erroneous channel estimation, leading to the incorrect design of the transmitter's precoder. As a consequence, there is an improvement in the data signal sent by the transmitter to the malicious user.

In this chapter, a brief overview of existing PCA detection methods is presented, relying on either detection statistics or the use of secret keys. An analysis of the main properties of the different approaches is provided, together with commentary on their respective advantages and drawbacks. The methods are then compared, using their most prominent performance parameters. The promising features of one of the methods discussed, despite its comparatively low PCA detection probability, suggests that further in-depth research on the different attack scenarios that the method is not able to detect are worthwhile. The discussion of the weaknesses in the method's efficiency is followed by two different suggested techniques that can be used to supplement the original method in order to improve its performance. Both the techniques are analytically described, and results from simulation experiments of their use are graphically presented. As the results show, the proposed solutions successfully overcome some of the PCA detection problems of the original method, thus improving its operational efficiency.

The rest of this chapter is organized as follows: In Section 6.2, a concise survey of the literature relating to security and privacy in CPS is given, and in Section 6.3, the system model is presented. Section 6.4 describes the different solutions for PCA detection proposed in the literature; these are compared in Section 6.5. A detailed performance investigation of the 2-N-PSK method is discussed in Section 6.6, together with two newly proposed techniques which improve this method. The first concerns channel gain and the second deals with shifted constellations. Section 6.7 concludes the chapter.

## 6.2 Related Work

CPS relates to systems in which the physical characteristics of the surrounding environment are monitored and controlled through embedded intelligent cyber technologies incorporating real-time computations and communications. The cyber subsystem receives, stores and processes the data collected by small physical objects, such as sensors and radio frequency identification (RFID) tags, and issues commands to the physical devices (Lu et al., 2015). The communication between cyber and physical subsystems is usually accomplished via a wireless networking system. Direct communication between CPS devices without the need of any human interaction (Bhabad et al., 2015) is also possible by the use of IPv6 and Machine-to-Machine (M2M) applications (Weyrich et al., 2014; Mukherjee, 2015).

Different CPS architectures are proposed in the literature, assuming three (La et al., 2010; Gou et al., 2013; Zhao et al., 2013), five (Wu et al., 2010) or seven (Alguliyev et al., 2018) layers. However, as the authors in Ashibani et al. (2017) emphasize, the performance of a CPS is divided into three fundamental layers: physical, network and application. The physical layer is responsible for real-time data aggregation and incorporates devices such as sensors and actuators, barcodes, RFID tags and readers, cameras and global positioning systems (GPSs) (Ashibani et al., 2017), that collect information from the physical properties of the environment and carry out instructions from the application layer. The network layer is responsible for data transmission and routing through the devices over various network technologies, e.g., long term evolution (LTE), universal mobile telecommunications system (UMTS), Wi-Fi, Bluetooth, ZigBee, 4G and 5G (Mahmoud et al., 2015).

The data transmission between the different nodes and CPS layers is accomplished by a Gateway. Data aggregation and intelligent processing of large amounts of data are undertaken at the application layer, which is responsible for system monitoring and control by decision-making (Lee, 2008; Rajkumar et al., 2010). As a result, current CPS applications include smart homes and smart cities, various medical devices, intelligent transportation and aerospace systems, defense systems, robotic systems, smart buildings and infrastructure, meteorology and smart grid. Since for the proper implementation of most of these applications private user information is needed, enhancement of their security is of crucial importance.

The typically compact dimensions of CPS sensors and devices account for their significant hardware, memory and power constraints (Sen, 2009; Trappe et al., 2015). These devices have reduced signal processing capabilities and expose the system to new security challenges. A large number of incorporated M2M devices, together with heterogeneous radio access technologies and a dynamically changing environment (Oh et al., 2017), make it difficult to implement traditional cryptographic protocols, which require key exchange and certificate management.

Due to the resource constraints in CPS devices, security trade-offs are often necessary. One such example discussed in Pecorella et al. (2016) concerns the IEEE 802.15.4 standard for low-rate wireless personal area networks (LR-WPANs), which is one possible use of CPSs. In IEEE 802.15.4, encryption of the payload data is provided but plaintext overhead is also transmitted, containing the addresses of source and destination. As a result, a malicious user can obtain this information without the need for packet decryption. Therefore, the implementation of conventional cryptography techniques in CPSs is impeded by the specific features and limited resources of the employed devices (Suo et al., 2012; Abomhara et al., 2014). One efficient approach for secure communications in CPSs is a PLS method, which may supersede the cryptographic protocols or be used to supplement lightweight encryption algorithms, such as the one proposed in Usman et al. (2017). PLS techniques use variations in channel condition to amplify the signal received in the intended receivers, while degrading the signal strength for unauthorized users. These methods usually work independently of the radio access technology and offer embedded security that is impenetrable from the point of view of information theory.

In order to implement an overall effective solution for secure communications in the CPS, it is necessary to analyze the scenarios of different possible threats. Since most of the small CPS devices, such as sensors and RFID tags, remain unmonitored for long periods of time, a malicious user can easily access their collected data (Karygiannis et al., 2007). Different attacks, typically used against wireless sensor networks (WSNs), which are closely related to CPSs, are presented in Padmavathi et al. (2009), Sen (2009), Virmani et al. (2014), and Shahzad et al. (2016), together with some possible measures for counteracting them. The security principles involved at the different layers of CPS architecture are discussed in Mahmoud et al. (2015), while in Zhang et al. (2014) and Conti et al. (2018), the authors present an analysis

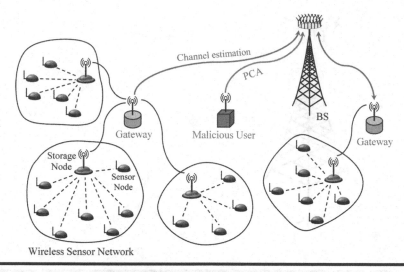

**Figure 6.1 PCA representation in the training phase of a WSN.**

of the opportunities and challenges related to CPS security. In Mukherjee (2015), several PLS techniques that can be used for security provisioning of CPSs are considered, with the various strategies being compared in terms of their complexity, energy efficiency, and flexibility, which are important to CPS applications.

Wireless communication systems, to which CPSs belong, are vulnerable to two main types of attack on the physical layer—jamming and eavesdropping (Zou et al., 2016). The eavesdropping attack represents an unauthorized user trying to intercept the legitimate data, which is easily achieved when the ED is located inside the coverage area of an LU. One strategy that the ED could employ to improve the strength and quality of the eavesdropped signal concerns the channel identification procedure between an LU and a base station (BS), and represents the so-called pilot contamination attack. A schematic representation of a PCA initiated by a malicious user in a WSN is illustrated in Figure 6.1, where each gateway acts as an LU, and BS performs channel estimation.

The PCA undermines security at the physical layer and can have a detrimental effect on the communication privacy of a CPS. Hence the significance of introducing schemes for the detection of pilot contamination attacks and the need to interrupt the communication if one is uncovered.

# 6.3 System Model

This chapter focuses on studies based on different PCA detection methods. The most common scenario, used for simplicity, comprises a single cell—in order that the multicell interference can be ignored, channels with additive white Gaussian

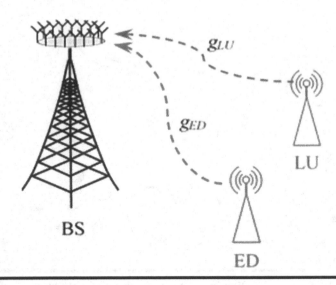

**Figure 6.2 System model.**

noise (AWGN) and no user mobility. In addition, the model examines the situation depicted in Figure 6.2, where a BS with M multiple antennae communicates with only one single-antenna LU, while one single-antenna ED tries to eavesdrop on the information exchange during the training phase.

A TDD system with reciprocity between uplink and downlink channels is assumed. During the uplink training phase both the LU and the ED synchronically send their pilot signals to the BS, which undertakes channel estimation. After obtaining the CSI the BS computes its precoder to match the characteristics of the estimated channel and performs beamforming in the downlink TDD phase.

The uplink channel from the LU to the BS is denoted as $g_{LU} = \sqrt{P_{LU}d_{LU}}\,h_{LU}$ and includes the influence of the transmit power of the $LU - P_{LU}$, the large-scale fading $d_{LU}$ which is a scalar, and the small-scale fading $h_{LU}$ which is an $M \times 1$ vector. Likewise, the channel from the ED to the BS is represented as $g_{ED} = \sqrt{P_{ED}d_{ED}}\,h_{ED}$, where $P_{ED}$ is the ED's transmit power, $d_{ED}$ is the large-scale fading and $h_{ED}$ is an $M \times 1$ vector for the small-scale fading.

## 6.4 Detection Methods

The PCA detection methods can be classified into two groups: detection statistic methods and secret keys (SKs) confirmation methods. In this chapter three detection statistic schemes will be presented as well as one SKs confirmation scheme.

### 6.4.1 Detection Statistic Schemes

The first group of methods is based on the detection statistic of the signal received at the BS. As the channel estimation of all three methods unified in this group is undertaken at the BS, they are resistant to attacks by jamming the LU of a CPS.

#### 6.4.1.1 Two Random N-PSK Pilots Detection Scheme (2-N-PSK)

One of the methods proposed in the literature is based on sending two random N-PSK symbols during the channel estimation phase (Kapetanovic et al., 2013). A geometric constellation of the 8-PSK is depicted in Figure 6.3, where a vector representation of a complex number $q$ by its module and phase is shown.

The pilot symbols are publicly known, so the ED can send the same pilot sequence as the LU, and its behavior would not be recognized by the BS during the process of obtaining the CSI. For that reason, to enable detection of the malicious user at the BS, an analysis is made of the scalar product of the correlation between the two received N-PSK vectors. The signals received at the BS for each of the two training periods are given by Eq. (6.1) (Kapetanovic et al., 2013):

$$
\begin{aligned}
y_1 &= g_{LU}\, p_1^{LU} + g_{ED}\, p_1^{ED} + n_1 \\
&= \sqrt{P_{LU} d_{LU}}\, h_{LU}\, p_1^{LU} + \sqrt{P_{ED} d_{ED}}\, h_{ED}\, p_1^{ED} + n_1, \\
y_2 &= g_{LU}\, p_2^{LU} + g_{ED}\, p_2^{ED} + n_2 \\
&= \sqrt{P_{LU} d_{LU}}\, h_{LU}\, p_2^{LU} + \sqrt{P_{ED} d_{ED}}\, h_{ED}\, p_2^{ED} + n_2,
\end{aligned}
\tag{6.1}
$$

where $p_1^{LU}$ and $p_1^{ED}$ are the pilots sent from the LU and the ED respectively during the first training slot, and $p_2^{LU}$ and $p_2^{ED}$ are the pilots from the second training interval; $n_1$ and $n_2$ denote the AWGN in the first and second slot respectively.

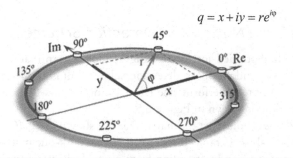

**Figure 6.3** **Geometric representation of a complex number.**

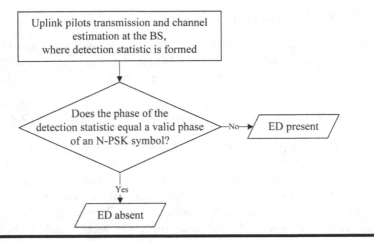

**Figure 6.4   Flowchart of the 2-N-PSK method.**

The correlation follows Eq. (6.2) (Kapetanovic et al., 2013), which forms the detection statistic $z$ as the phase of $y_1^H y_2$; $(\cdot)^H$ stands for Hermitian matrix and $n_{12}$ is the noise result:

$$z_{12} = \frac{y_1^H y_2}{M} = \frac{1}{M}\left(\sqrt{P_{LU}d_{LU}}\,h_{LU}\,p_1^{LU} + \sqrt{P_{ED}d_{ED}}\,h_{ED}\,p_1^{ED}\right)^H$$
$$\times \left(\sqrt{P_{LU}d_{LU}}\,h_{LU}\,p_2^{LU} + \sqrt{P_{ED}d_{ED}}\,h_{ED}\,p_2^{ED}\right) + n_{12} \qquad (6.2)$$

Depending on the correlation result and the angle of its vector, the presence of the ED can be determined. If the angle of $z_{12}$ does not converge to an angle of a valid N-PSK symbol, the non-legitimate user is present in both the training slots. Otherwise, it is concluded that the ED is absent. The approach is summarized in Figure 6.4.

### 6.4.1.2 L Random N-PSK Pilots Detection Scheme (L-N-PSK)

The second method studied for discovering a pilot contamination attack repeats the logic of the two random pilots' detection scheme but is applied to $L$ number of pilots, aiming to improve performance (Kapetanovic et al., 2015). The block diagram of this method is shown in Figure 6.5.

The main idea is to construct the matrix $R$ as it is shown in Eq. (6.3) (Kapetanovic et al., 2015) and analyze its rank. If $R$ converges to a rank-one matrix, then the ED is absent. When there is an intervention from a non-legitimate user, $R$ converges to a full-rank matrix.

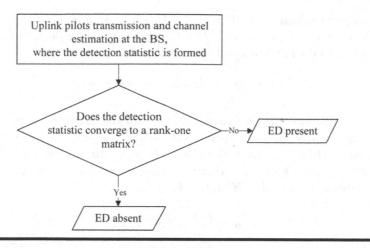

**Figure 6.5  Flowchart of the L-N-PSK method.**

$$R = \frac{y^H y}{M} + W, \tag{6.3}$$

where $y = [y_1, y_2, ..., y_L]$ is a $1 \times L$ row-vector composed of the signals received at the BS, and $W$ is an $L \times L$ noise matrix.

### 6.4.1.3 Generalized Likelihood Ratio Test (GLRT) Scheme

Another method, introduced in Im et al. (2013a, 2013b) and expanded in Im et al. (2015), can be applied in the case of a system with multiple legitimate receivers. This technique employs the GLRT to distinguish between two models, each of them having no unknown parameters. Therefore, the model is applicable only when the large-scale fading of every LU is determined in advance. On the other hand, the large-scale fading of the ED is not revealed to the BS and the LUs, and for the purposes of GLRT its influence must be replaced with its maximum-likelihood estimate (MLE).

During the uplink phase all the LUs, $K$ in number, simultaneously send their training sequences $\psi$ to the BS at the beginning of each coherence block. The pilot sequences of the users are orthonormal to one another and thus enable individual channel estimation at the BS. Each training sequence consists of $L$ number of pilots and represents a $1 \times L$ vector that for the $k$-th user is denoted as $\sqrt{L}\psi_k$.

In the case where an ED contaminates the pilot sequence of the $l$-th LU, the signal received at the BS is (Im et al., 2013b):

$$Y = \sum_{k=1}^{K} \sqrt{P_{LU} d_{LU_k} L} h_{LU_k} \psi_k + \sqrt{P_{ED} d_{ED} L} h_{ED} \psi_l + W, \tag{6.4}$$

where $P_{LU}$ is the transmit power of each LU and $W$ is an $M \times L$ noise matrix.

The signal received at the BS for the $l$-th user is given by Eq. (6.5) (Im et al., 2013b), where $c_l = P_{LU} d_{LU_l} L$, $\omega = \sqrt{P_{ED} d_{ED} / P_{LU} d_{LU_l}}$ is the degree of direction steering toward the ED, and $w_l = W \psi_l^H$ is the noise vector:

$$y_l = Y \psi_l^H = \sqrt{c_l} \left( h_{LU_l} + \omega h_{ED} \right) + w_l \tag{6.5}$$

The GLRT algorithm defines two models: the null and alternative hypotheses, $H_0$ and $H_1$, respectively. The null hypothesis considers that the ED is absent, the alternative that the ED is present. Both the hypotheses are formulated by the detection statistic at the BS—$y_l$, which indicates the energy variance in the two cases. A larger variance is observed in $H_1$ (Im et al., 2013b):

$$H_0 : y_l = \sqrt{c_l} h_{LU_l} + w_l;$$
$$H_1 : y_l = \sqrt{c_l} \left( h_{LU_l} + \omega h_{ED} \right) + w_l. \tag{6.6}$$

As $\omega$ is dependent on the ED's large-scale fading, which is an unknown parameter, its value in the GLRT equation must be replaced by its MLE. By doing this, the final decision for validating one of the hypotheses follows Eq. (6.7) (Im et al., 2013b), where $\lambda$ is the sensitivity of the detector:

$$\ln \Lambda_1 (y_l) = \frac{\|y_l\|^2}{(1+c_l) M} - \ln \left\{ \frac{\|y_l\|^2}{(1+c_l) M} \right\} - 1 \underset{H_0}{\overset{H_1}{\underset{<}{>}}} \frac{\ln \lambda}{M}. \tag{6.7}$$

A brief overview of the method can be seen in Figure 6.6.

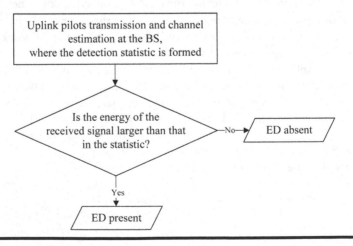

**Figure 6.6 Flowchart of the GLRT method.**

## 6.4.2 Secret Keys (SKs) Confirmation Scheme

Another method, explored in Tomasin et al. (2016), reveals the presence of the ED by means of bilateral channel estimation. In the first training phase, the BS sends publicly known pilots to the LU, which obtains the CSI of $g_{LU}$. During the second training phase, the LU sends pilots to the BS, where another assessment of the legitimate channel is made. Both estimations are then compared following a key-confirmation procedure, explained briefly in Figure 6.7.

From their estimation results both the BS and the LU extract SKs in the form of $N$-bit numbers—$a$ and $b$ respectively. Thereafter the BS generates a random sequence of bits $r$ in the same length, which is added modulo-2 to the BS's SK and the result $x = r + a$ is sent to the LU. The LU decrypts the message, applying modulo-2 sum with its SK, and gets $r' = x + b$, which is given to the input of an invertible non-identity function $f(r')$. The encrypted result $y = f(r') + b$ is then transmitted to the BS. At the BS the random bit sequence $r$ is processed with the same function $f(r)$ and the mapped value

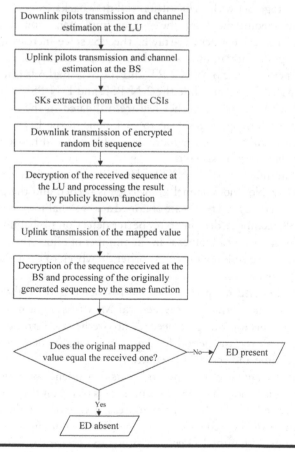

**Figure 6.7   Flowchart of the SKs method.**

is compared to the decrypted message received from the $LU - z = y + a$. If, apart from noise, both the values coincide, i.e., $f(r) = z$, the SKs of the BS and the LU are the same, the BS concludes that there is no active ED contaminating the pilots. Conversely, if the ED is present in one of the training periods, both the CSIs differ more than the noise level, and the SKs of the BS and the LU are different.

In a scenario when the ED attacks both of the pilot phases, the SKs do not coincide because the obtained CSIs at the BS and the LU are influenced by two different channels—the channel between the ED and the BS and that between the ED and the LU.

## 6.5 Comparison of the Reviewed PCA Detection Schemes

A comparison of the PCA detection schemes already discussed will clarify some of their advantages, together with their substantial drawbacks.

One common requirement for the proper implementation of all the techniques discussed is a large antenna array at the BS, since increasing the number of its antennae significantly improves the detection probability of all the methods. However, good performance of the GLRT and the SKs methods is feasible even in a conventional MIMO system, while the 2-N-PSK and L-N-PSK schemes are reliable in a MaMIMO scenario. Another feature, validated in the literature, is that the four reviewed solutions are resistant to variations in the ED's transmit power and the more powerful the ED, the more detectible it is and hence the better the performance of the detection schemes.

One main advantage of the 2-N-PSK detection strategy over the others is its reduced complexity. No prior channel knowledge is needed and the performance is robust against noise, thus good results are obtained at low signal-to-noise ratios (SNRs).

The L-N-PSK technique demonstrates better detection probability at moderate to high SNRs, at the cost of an increase in the number of pilots, the complexity and the time needed for the training phase. Another drawback of the L-N-PSK scheme is its high sensitivity to noise.

Among the discussed methods, the GLRT scheme is the only one that does not require any special changes to the general MIMO system model. Moreover, although prior channel knowledge is needed, no overhead is introduced during the training phase. However, the period for training is long, which means that more resources such as time and energy are required.

One main disadvantage of this method is its dependence on large-scale fading coefficients. The conclusion that the ED is present is based on the energy allocation of the signal received at the BS, i.e., the increased variance of the contaminated signal compared to the non-contaminated one. The authors in Kapetanovic et al. (2015) offer a strategy that the ED could employ to deceive any detection scheme, reaching a decision purely via estimation of the large-scale fading enlargement.

The posited behavior of the ED is that it imitates the natural channel improvement. That is to say, the ED starts sending pilots at low power and increases them gradually over the separate coherence intervals of the large-scale fading. As the value of $d_{LU}$ changes slowly over time and frequency, the BS cannot detect the intervention of the ED and concludes that no pilot contamination attack is occurring.

Apart from the GLRT technique, none of the other three methods discussed in this chapter utilizes knowledge of the large-scale fading coefficients to reveal the ED's presence.

A drawback common to all three schemes that obtain the CSI by detection statistics is that their decision rules mainly use the phase information of the signal received at the BS. In consequence, they are vulnerable to phase noise induced due to hardware imperfections. In contrast, the SK technique is based only on theoretical principles and exhibits strong phase noise resistance.

Looking closely at the SKs method, its main benefit is reliable detection without the need for any prior knowledge of the channel. Since the technique does not rely on signal processing mechanisms, it is resistant to noise and interference. Its main disadvantages include increased complexity, a long training period, large overhead, and vulnerability to jamming of the signal at the LU due to its participation in the channel estimation procedure.

For comparison purposes, the most important features of the methods discussed are given in Table 6.1, where the degree of relation between each technique

**Table 6.1  Comparison of the Detection Methods**

| | 2-N-PSK | L-N-PSK | GLRT | SKs |
|---|---|---|---|---|
| SNR dependence | L | H | H | L |
| Knowledge of large-scale fading necessary | N | N | H | N |
| Phase noise dependence | H | H | H | N |
| Number of antennae dependence | H | H | H | H |
| MaMIMO necessary | H | H | N | N |
| Prior channel knowledge necessary | N | N | H | N |
| System model changes necessary | H | H | N | H |
| Complexity | L | L | H | H |
| Long training period | L | L | H | H |
| Overhead induced | L | L | N | H |
| Robustness to ED's power variations | H | H | H | H |
| Robustness to jamming of the LU | H | H | H | N |

**Table 6.2   System Performance of the Detection Methods**

|  | 2-N-PSK | L-N-PSK | GLRT | SKs |
|---|---|---|---|---|
| Crucial parameter | SNR = 0dB | SNR = 0dB | $\omega^2 = -10$dB | $\zeta = 1$ |
| Performance parameter | $p_{DP} \to 0.5$ | $p_{DP} \to 1$ | $p_{DP} \to 1$ | $p_{OUT} \to 10^{-2}$ |

and the given properties is denoted by *L* for *low* and *H* for *high*. *N* is used for *none* to indicate that the method does not depend on the current feature.

The effectiveness of the different solutions can be compared by the basic performance parameters of each detection scheme. The performance parameter for the three techniques that are based on the statistics of the contaminated training is the probability of detection of the ED's presence, which is denoted as $p_{DP}$. While the execution of both the N-PSK schemes is strongly dependent on the SNR, the crucial parameter for the GLRT method is the degree of direction steering toward the ED—$\omega$, which is related to the transmission power and the large-scale fading of the LU and the ED. Since the fundamental principles of the SKs algorithm are based more on information from theoretical approaches than on signal processing, the performance of the system can be analyzed by the secrecy outage probability (SOP)—$p_{OUT}$. The SOP is defined as the probability that the ED knows something about the secret message or key, and it is affected by the degree of correlation between the legitimate and non-legitimate channels, measured by the correlation factor—$\zeta$.

Some interesting values of the different schemes' performance parameters are given in Table 6.2.

An overview of the data in Table 6.2 shows that for SNR = 0°dB the detection probability of the 2-N-PSK scheme converges to 0.5°, while at the same level of SNR the probability of successfully discovering the presence of an ED with the L-N-PSK scheme converges to 1. However, at low SNRs in the order of −10 dB, the behavior of the 2-N-PSK is significantly better than that of the L-N-PSK technique. The detection probability of the GLRT scheme converges to 1° when the square of the steering toward the ED is −10°dB. When the value of the critical parameter for the SKs method, i.e., the correlation factor, is high the outage probability of the model converges to 0.01, leading to comparatively weak system performance.

# 6.6  New Techniques to Improve the Operation of the 2-N-PSK Detection Scheme

## 6.6.1  Performance Analysis of the 2-N-PSK Detection Scheme

The useful properties of the 2-N-PSK method warrant some additional analysis of its low detection probability. Four different situations exist where the non-legitimate pilots have values forming a correlation result that equals an angle from the original

N-PSK constellation (Mihaylova et al., 2017a, 2017c). The detection problems of the 2-N-PSK are listed below, where $\varphi(\cdot)$ denotes the angles of the pilots.

1. The pilots of the adversary in both the training intervals equal the corresponding legitimate pilots:

$$\varphi\left(p_1^{ED}\right)=\varphi\left(p_1^{LU}\right) \quad \text{and} \quad \varphi\left(p_2^{ED}\right)=\varphi\left(p_2^{LU}\right)$$

2. The pilots of the adversary in both the training intervals are reciprocal to the corresponding legitimate pilots, or one is reciprocal and the other is equal:

$$\varphi\left(p_1^{ED}\right)=\varphi\left(p_1^{LU}\right) \quad \text{and} \quad \varphi\left(p_2^{ED}\right)=\varphi\left(p_2^{LU}\right)+180°$$

$$\varphi\left(p_1^{ED}\right)=\varphi\left(p_1^{LU}\right)+180° \quad \text{and} \quad \varphi\left(p_2^{ED}\right)=\varphi\left(p_2^{LU}\right)$$

$$\varphi\left(p_1^{ED}\right)=\varphi\left(p_1^{LU}\right)+180° \quad \text{and} \quad \varphi\left(p_2^{ED}\right)=\varphi\left(p_2^{LU}\right)+180°$$

3. The adversary initiates the attack during the second training interval and its pilot equals the second legitimate pilot or its reciprocal:

$$p_1^{ED}=0 \quad \text{and} \quad \varphi\left(p_2^{ED}\right)=\varphi\left(p_2^{LU}\right)$$

$$p_1^{ED}=0 \quad \text{and} \quad \varphi\left(p_2^{ED}\right)=\varphi\left(p_2^{LU}\right)+180°$$

4. The pilots of the adversary differ from the corresponding legitimate pilots by the same angle:

$$\varphi\left(p_1^{ED}\right)=\varphi\left(p_1^{LU}\right)+x° \quad \text{and} \quad \varphi\left(p_2^{ED}\right)=\varphi\left(p_2^{LU}\right)+x°$$

Of all four of the detection problems of this method, a solution is proposed in the literature only for case 3, where the ED starts the attack together with a second legitimate pilot of a pair and manages to duplicate its angle or its reciprocal. If the correlation result has a valid N-PSK angle, a revision of the amplitudes of the received signals can be implemented. Since the received power of the contaminated pilot differs from the power of the non-contaminated pilot, the authors (Kapetanovic et al., 2015) suggest a comparison of the ratio of the received signals by certain thresholds. In the case where the ratio is outside the range defined by the thresholds, it is assumed that the ED is present in only one of the pilot sessions. Otherwise, the power difference of the received signals is assumed to be due to noise influence and channel imperfections. However, some previous channel knowledge is needed for setting the threshold values, making this technique unreliable.

## 6.6.2 A Technique Based on Channel Gain

In Mihaylova et al. (2017a, 2017b), a new technique omitting the use of thresholds is proposed to solve the problems listed in case 3 of the original 2-N-PSK method—the relevant flowchart is shown in Figure 6.8.

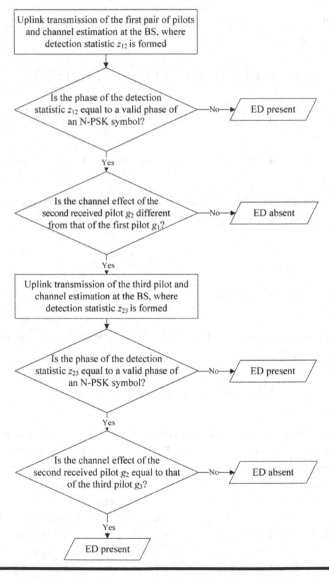

**Figure 6.8   Flowchart of the improved 2-N-PSK technique based on channel gain.**

Initially, the concept of the original 2-N-PSK method is followed. If the correlation result of the first pair of pilots has an N-PSK angle, the channel gain values of both the pilots are computed and compared in accordance with Eq. (6.8):

$$g_1 = \frac{y_1}{p_1^{LU}}, \quad g_2 = \frac{y_2}{p_2^{LU}}. \tag{6.8}$$

If the channel gain during the second pilot transmission differs from that of the first pilot, i.e., $g_1 \neq g_2$, the ED is assumed to be present only in the second training interval. However, this difference might be a consequence of noise and, in order to eliminate the use of thresholds, a third pilot is then transmitted. The correlation between the second and third received signals is calculated at the BS. In the worst case, when its angle equals one from the N-PSK constellation, a revision of the channel gain of the third pilot session is conducted.

$$g_3 = \frac{y_3}{p_3^{LU}}. \tag{6.9}$$

If the value of $g_2$ continues to appear during the next pilot transmission, i.e., $g_2 = g_3$, the conclusion of the proposed 2-N-PSK supplement is that a pilot contamination attack is present in the CPS.

The improved technique based on channel gain is illustrated in Figure 6.9 (Mihaylova et al., 2017a, 2017b).

Although the improved technique based on channel gain is able to reveal attacks of type 3 initiated in a CPS, its successful performance depends on the values of ED's pilots. When the second and third pilots of ED are either both equal

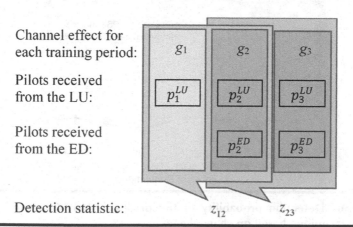

**Figure 6.9 Channel gain comparison for the received pilots.**

or both reciprocal to the corresponding legitimate pilots, malicious intervention is detected. On the other hand, an intruder whose second pilot is the same as the corresponding legitimate pilot and whose third pilot is reciprocal to it, or vice versa, cannot be detected.

An additional benefit of the use of this proposed technique is its ability to identify an adversary who initiates an attack of type 2 with the first pilot equal, and the second and third reciprocal, to the corresponding legitimate pilots.

The detection probability of the original method and the improved technique is studied (Mihaylova et al., 2017a) via a large number of simulations. At first, the pilots of both the LU and ED are generated randomly so that all the four problem cases of the original method can be observed. Then another attempt is conducted that investigates the detection probability of the methods only for attacks of type 3, again simulated with random conditions. The simulations are separated into groups depending on their number, and the averaged results of 10 independent experiments for each of the groups are presented in Figures 6.10 and 6.11 (Mihaylova et al., 2017a).

As the graphs show, the use of the channel gain technique increases the PCA detection probability of the original method by approximately 5% when all types of attacks are simulated, and by around 21% when the attacks begin with the second pilot transmission.

An important consideration for the implementation of the improved technique is that its performance is reliable when the channels are static and a good estimate of the noise is available, so that its influence can be taken into account. Should the impact of noise in the channels change over time, the channel gain values will alter the different training intervals, which will prevent successful detection.

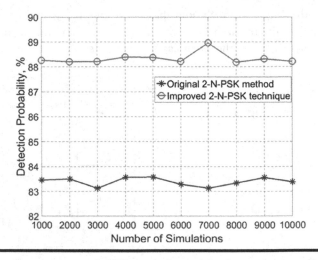

**Figure 6.10 Detection probability of the original 2-N-PSK method and the improved technique based on channel gain.**

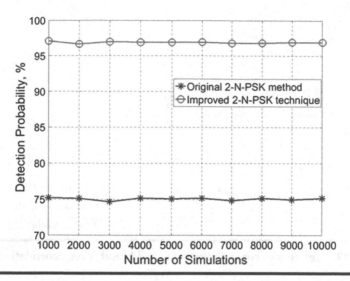

**Figure 6.11** **Detection probability of the original 2-N-PSK method and the improved technique based on channel gain when ED misses the first pilot transmission.**

## 6.6.3 Shifted 2-N-PSK Detection Technique

Another technique proposed to improve the PCA detection capabilities of the original 2-N-PSK method in a CPS is the so-called Shifted 2-N-PSK method, suggested in Mihaylova et al. (2017d) and evaluated in Mihaylova et al. (2017c).

The improved method works with two pilots, shifted from the N-PSK constellation by previously specified degrees. In a common scenario, every odd legitimate pilot is shifted by one angle, $x_1$, and every even pilot has another offset value, $x_2$. As with the original 2-N-PSK method, the correlation of the received pilots is computed at the BS and its angle is analyzed. An attack on the CPS is surmised when the correlation result angle does not coincide with an angle from a reference constellation, obtained in the absence of an intrusion. In contrast to the 2-N-PSK, the Shifted 2-N-PSK method operates with two reference constellations that differ from the original N-PSK constellation by certain degrees. One constellation is used for the odd correlations, such as $z_{12}$ between the first and the second received signals, and another for the even correlations, for instance $z_{23}$ between the second and the third pilots. The reference angles for odd correlations are derived by adding $x_2 - x_1$ degrees to every N-PSK angle, while for even correlations $x_1 - x_2$ is added. An example of the two reference constellations is given in Figure 6.12 (Mihaylova et al., 2017d). The block diagram of the Shifted 2-N-PSK method is shown in Figure 6.13.

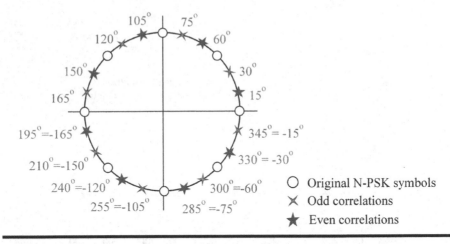

**Figure 6.12 Reference constellations for odd and even correlations when $x_1 = 20°$ and $x_2 = 5°$.**

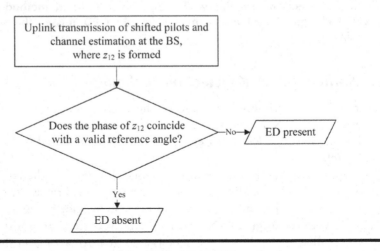

**Figure 6.13 Block diagram of the Shifted 2-N-PSK method.**

The successful performance of the Shifted 2-N-PSK method depends on the shift values for the odd and even legitimate pilots. The different scenarios investigated are described below.

1. *Neither the legitimate constellations' shift values nor the difference between them equal an N-PSK angle: $x_1 \neq x_2 \neq \varphi(N\text{-}PSK)$ and $|x_1 - x_2| \neq \varphi(N\text{-}PSK)$*

   When the offsets of both the pilots of a pair are chosen such that none of the legitimate and reference constellations coincides with the original N-PSK constellation, all detection problems of the original 2-N-PSK method are successfully solved.

2. *The shift value of either the odd or the even legitimate pilots equals an N-PSK angle: $x_1 = \varphi(N\text{-}PSK)$ or $x_2 = \varphi(N\text{-}PSK)$*

   In the case where only one of the legitimate pilots of the pair is shifted from the original N-PSK constellation, an attack on the CPS that the Shifted 2-N-PSK method is unable to detect may occur. The method reveals attacks of type 3 only when the second legitimate pilot of the pair belongs to a shifted constellation. Hence, if only the second pilot of the pair is shifted from the original N-PSK constellation, every even correlation is vulnerable to attacks initiated in the second training interval, and if only the first pilot of the pair is shifted, the odd correlations are susceptible. Nonetheless, all the other types of attack that the original method is unable to reveal are efficiently detected when only one of the legitimate constellations is shifted.

3. *The shift values of odd and even legitimate pilots differ by an N-PSK angle: $|x_1 - x_2| = \varphi(N\text{-}PSK)$*

   Equal offsets in both the pilots of the pair resulted in a reference constellation for the odd and even correlations, which coincide with the original N-PSK symbols. Although the use of a single constellation simplifies the analysis at the BS, when the shift values of all the pilots are the same only problems 2 and 3 of the original method are overcome by the Shifted 2-N-PSK.

An assessment of the detection capabilities of the Shifted 2-N-PSK is given in Mihaylova et al. (2017c) for both noiseless and noisy scenarios, and the results are compared to those of the original method. The three different Shifted 2-N-PSK variations, described as *A*, *B* and *C* in the current section, are examined. In a similar way to the experiments of the improved technique based on channel gain in Section 6.2, the arithmetic means values of 10 independent observations with channels free of noise are graphically presented in Figure 6.14.

The results depicted in the figure confirm the expectation that perfect detection probability in the absence of noise is achieved by the Shifted 2-N-PSK when the pilots of the legitimate pair differ according to the rules described in *A*, while in scenarios *B* and *C* the detection probability is lower. The implementation of cases *A*, *B* and *C* of this method improves the original 2-N-PSK's probability of successful PCA detection by about 17%, 15% and 6% respectively.

When the influence of noise is also considered, the performance of the 2-N-PSK and the Shifted 2-N-PSK deteriorates as the detection regions, in which the attack is taken to be absent, broaden around the angles of the reference constellation. The distance for constructing the detection region around the reference angles is obtained by Eq. (6.10), suggested by the authors of Kapetanovic et al. (2013), where $N_0$ is the noise power and $c$ is a scaling constant:

$$r(c_B) = c \frac{\sqrt{N_0 \left( MN_0 + 2\dfrac{P_B d_B \|h_B\|^2}{M} \right)}}{M}. \tag{6.10}$$

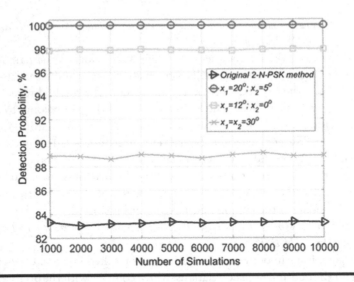

**Figure 6.14** Detection probability of the original 2-N-PSK method and the Shifted 2-N-PSK in the noiseless case.

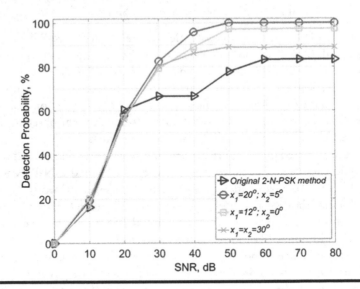

**Figure 6.15** Detection probability vs. SNR of the original 2-N-PSK and the Shifted 2-N-PSK method.

The simulation results of the detection probability of the Shifted 2-N-PSK and original 2-N-PSK methods as a function of SNR, presented in Mihaylova et al. (2017c), can be observed in Figure 6.15.

When SNR is low, the behavior of the method based on shifted constellations is broadly similar to that of the original method, although an improved detection probability is observed when the SNR is greater than 20°dB. Perfect detection is demonstrated by the Shifted 2-N-PSK when the SNR is 50°dB or more and the pilots belong to two different constellations, both shifted from the original.

## 6.7 Conclusion

This chapter reviews four methods for the detection of pilot contamination attacks. All the solutions discussed are for a single-cell model, AWGN and with no user mobility. A future study could be aimed at a more realistic scenario with a multicell system where user mobility is included and the noise has a complex distribution. Other research could investigate to what extent increasing the number of pilots in the L-N-PSK scheme achieves a related improvement in the PCA detection performance of the CPS.

Two new techniques to improve the detection probability of the original 2-N-PSK method are also examined. The channel gain technique is developed to reveal pilot contamination attacks initiated during the transmission of the second pilot from the legitimate training sequence. Although its implementation increases the PCA detection probability of the system, an investigation of the channel gain solution in the presence of noise is still an open issue for future research. The second 2-N-PSK improvement—the Shifted-2-N-PSK method—is studied for three different relationships between the offset values of the first and second legitimate pilots. As a result of the analysis of analytical expressions and simulation experiments, it is demonstrated that the best performance of Shifted-2-N-PSK is observed when neither the legitimate nor the reference constellations coincide with the original N-PSK constellation. In such a scenario, and when the influence of noise is discounted, the PCA detection probability of Shifted-2-N-PSK approaches 100%. Since the detection capabilities of the Shifted-2-N-PSK method depend on the relation between the legitimate pilots' offsets, further study can be directed toward investigating whether optimal values for the shifted angles of the constellations exist.

## Acknowledgment

This work was supported by European Regional Development Fund and the Operational Program "Science and Education for Smart Growth" under contract UNITe № BG05M2OP001-1.001-0004-C01 (2018–2023).

## References

Abomhara M. and G. M. Koien, "Security and privacy in the Internet of Things: Current status and open issues," *Proceedings of the IEEE International Conference on Privacy and Security in Mobile Systems*, Aalborg, Denmark, 2014: pp. 1–8.

Alguliyev R., Y. Imamverdiyev, L. Sukhostat, "Cyber-physical systems and their security issues," *Comput. Ind.*, 100 (2018): 212–223.

Ashibani Y., Q. H. Mahmoud, "Cyber physical systems security: Analysis, challenges and solutions," *Comput. Secur.*, 68 (2017): 81–97.

Bash B. A., D. Goeckel, D. Towsley, "Limits of reliable communication with low probability of detection on AWGN channels," *IEEE J. Sel. Areas Commun.*, 31, no. 9 (2013): 1921–1930.

Bash B. A., D. Goeckel, D. Towsley et al., "Hiding information in noise: Fundamental limits of covert wireless communication," *IEEE Commun. Mag.*, 53, no. 12 (2015): 26–31.

Bhabad M. A., P. G. Scholar, "Internet of Things: Architecture, security issues and countermeasures," *Int. J. Comput. Appl.*, 125, no. 14 (2015): 1–4.

Conti M., A. Dehghantanha, K. Franke et al., "Internet of Things security and forensics: Challenges and opportunities," *Future Gener. Comp. Syst.*, 78, no. 2 (2018): 544–546.

Dohler M., R. W. Heath, A. Lozano et al., "Is the PHY layer dead?" *IEEE Commun. Mag.*, 49, no. 4 (2011): 159–163.

Gou Q., L. Yan, Y. Liu et al., "Construction and strategies in IoT security system," *Proceedings of Green Computing and Communications (GreenCom) 2013 IEEE and Internet of Things (iThings/CPSCom) IEEE International Conference on and IEEE Cyber Physical and Social Computing (CPSCom)*, Beijing, China, 2013: pp. 1129–1132.

Im S., H. Jeon, J. Choi et al., "Robustness of secret key agreement protocol with massive MIMO under pilot contamination attack," *Proceedings of the ICTC*, Jeju, South Korea, 2013a: pp. 1053–1058.

Im S., H. Jeon, J. Choi et al., "Secret key agreement under an active attack in MU-TDD systems with large antenna arrays," *Proceedings of the Globecom*, Atlanta, GA, 2013b: pp. 1849–1855.

Im S., H. Jeon, J. Choi et al., "Secret key agreement with large antenna arrays under the pilot contamination attack," *IEEE Trans. Wireless Commun.*, 14, no. 12 (2015): 6579–6594.

Jose J., A. Ashikhmin, T. L. Marzetta et al., "Pilot contamination and precoding in multicell TDD systems," *IEEE Trans. Wireless Commun.*, 10, no. 8 (2011): 2640–2651.

Kapetanovic D., A. Al-Nahari, A. Stojanovic et al., "Detection of active eavesdroppers in massive MIMO," *Proceedings of the IEEE International Symposium on Personal, Indoor and Mobile Radio Communications (PIMRC)*, Washington, DC, 2014: pp. 585–589.

Kapetanovic D., G. Zheng, and F. Rusek, "Physical layer security for massive MIMO: An overview on passive eavesdropping and active attacks," *IEEE Commun. Mag.*, 53, no. 6 (2015): 21–27.

Kapetanovic D., G. Zheng, K.-K. Wong et al., "Detection of pilot contamination attack using random training and massive MIMO," *Proceedings of the IEEE International Symposium on Personal, Indoor and Mobile Radio Communications (PIMRC)*, London, UK, 2013: pp. 13–18.

Karygiannis T., B. Eydt, G. Barber et al., "Guidelines for securing radio frequency identification (RFID) systems," *National Institute for Standards and Technology (NIST) Special publication 800-98* (2007): 1–154.

La H. J., and S. D. Kim, "A service-based approach to designing cyber physical systems," *Proceedings of IEEE/ACIS 9th International Conference on Computer and Information Science (ICIS)*, Yamagata, Japan, 2010: pp. 895–900.

Lee E. A., "Cyber physical systems: Design challenges," *Object Oriented Real-Time Distributed Computing (ISORC) 2008 11th IEEE International Symposium on*, 2008: pp. 363–369.

Liu T.-Y., P.-H. Lin, S.-C. Lin et al., "To avoid or not to avoid CSI leakage in physical layer secret communication systems," *IEEE Commun. Mag.*, 53, no. 12 (2015): 19–25.

Lu T., J. Lin, L. Zhao et al. "A security architecture in cyber-physical systems: Security theories, analysis, simulation and application fields," *IJSIA*, 9, no.7 (2015): 1–16.

Mahmoud R., T. Yousuf, F. Aloul et al., "Internet of Things (IoT) security: Current status, challenges and prospective measures," *Proceedings of 10th International Conference for Internet Technology and Secured Transactions (ICITST)*, London, UK, 2015: pp. 336–341.

Mihaylova D., Z. Valkova-Jarvis, and G. Iliev, "A new technique to improve the 2-N-PSK method for detecting wireless pilot contamination attacks," *WSEAS Trans. Commun.*, 16, 2017a: 176–183.

Mihaylova D., Z. Valkova-Jarvis, and G. Iliev, "An improved technique for the detection of pilot contamination attacks in TDD wireless communication systems," *Proceedings of CSCC 2017*, Crete, Greece, 2017b.

Mihaylova D., Z. Valkova-Jarvis, and G. Iliev, "Detection capabilities of a shifted constellation-based method against pilot contamination attacks," *Proceedings of RTUWO 2017*, Riga, Latvia, 2017c.

Mihaylova D., Z. Valkova-Jarvis, G. Iliev et al., "Shifted constellation-based detection of pilot contamination attacks," *Proceedings of GWS 2017*, Cape Town, South Africa, 2017d.

Miller R., and W. Trappe, "On the vulnerabilities of CSI in MIMO wireless communication systems," *IEEE Trans. Mobile Comput.*, 11, no. 8 (2012): 1386–1398.

Mukherjee A., "Physical-layer security in the Internet of Things: Sensing and communication confidentiality under resource constraints," *Proc. IEEE*, 103, no. 10 (2015): 1747–1761.

Oh S.-R. and Y.-G. Kim, "Security requirements analysis for the IoT," *Proceedings of the 2017 International Conference on Platform Technology and Service (PlatCon)*, Busan, South Korea, 2017: pp. 1–6.

Padmavathi G. and D. Shanmugapriya, "A survey of attacks, security mechanisms and challenges in wireless sensor networks," *Int. J. Comput. Sci. Inf. Secur.*, 4, no. 1 & 2 (2009): 1–9.

Pecorella T., L. Brilli, L. Mucchi, "The role of physical layer security in IoT: A novel perspective," *Information*, 7, no. 3 (2016): 49.

Rajkumar R., I. Lee, L. Sha et al., "Cyber-physical systems: The next computing revolution," *Proceedings of the 47th Design Automation Conference, DAC'10*, Anaheim, CA, 2010: pp. 731–736.

Rusek F., D. Persson, B. K. Lau et al., "Scaling up MIMO," *IEEE Signal Process. Mag.*, 30, no. 1 (2013): 40–60.

Sen J., "A survey on wireless sensor network security," *Int. J. Comput. Sci. Inf. Secur.*, 1, no. 2 (2009): 55–78.

Shahzad F., M. Pasha, and A. Ahmad, "A survey of active attacks on wireless sensor networks and their countermeasures," *Int. J. Comput. Sci. Inf. Secur.*, 14, no. 12 (2016): 54–65.

Suo H., J. Wan, C. Zou et al., "Security in the Internet of Things: A review," *Proceedings of the 2012 International Conference on Computer Science and Electronics Engineering*, Hangzhou, China, 2012: pp. 648–651.

Tomasin S., I. Land and F. Gabry, "Pilot contamination attack detection by key-confirmation in secure MIMO systems," *Proceedings of IEEE Globecom*, Washington, DC, 2016.

Trappe W., "The challenges facing physical layer security," *IEEE Commun. Mag.*, 53, no. 6 (2015): 16–20.

Trappe W., R. Howard, R. S. Moore, "Low-energy security: Limits and opportunities in the Internet of Things," *IEEE Secur. Priv.*, 13, no. 1 (2015): 14–21.

Usman M., I. Ahmed, M. I. Aslam et al., "SIT: A lightweight encryption algorithm for secure Internet of Things," *Int. J. Comput. Sci. Inf. Secur.*, 8, no. 1 (2017): 402–411.

Virmani D., A. Soni, S. Chandel et al., "Routing attacks in wireless sensor networks: A survey," *preprint, arXiv:1407.3987* (2014).

Wang H.-M., and X.-G. Xia, "Enhancing wireless security via cooperation: Signal design and optimization," *IEEE Commun. Mag.*, 53, no. 12 (2015): 47–53.

Weyrich M., J.-P. Schmidt, and C. Ebert, "Machine-to-machine communication," *IEEE Software*, 31, no. 4 (2014): 19–23.

Wu M., T. Lu, F. Ling et al., "Research on the architecture of Internet of Things," *Proceedings of the 3rd International Conference on Advanced Computer Theory and Engineering (ICACTE)*, Chengdu, China, 2010: pp. 20–22.

Zhang Z.-K., M. C. Yi Cho, C. W. Wang et al., "IoT security: Ongoing challenges and research opportunities," *Proceedings of the IEEE 7th International Conference on Service-Oriented Computing and Applications*, Matsue, Japan, 2014.

Zhao K. and L. Ge, "A survey on the Internet of Things security," *Proceedings of the 9th International Conference on Computational Intelligence and Security (CIS)*, Leshan, China, 2013: pp. 663–667.

Zhou X., B. Maham, and A. Hjorungnes, "Pilot contamination for active eavesdropping," *IEEE Trans. Wireless Commun.*, 11, no. 3 (2012): 903–907.

Zou Y., J. Zhu, X. Wang et al., "A survey on wireless security: Technical challenges, recent advances, and future trends," *Proc. IEEE*, 104, no. 9 (2016): 1727–1765.

## Chapter 7

# Laboratory Exercises to Accompany Industrial Control and Embedded Systems Security Curriculum Modules

Guillermo A. Francia, III, Jay Snellen
and Gretchen Richards

## Contents

## 7.1 Introduction

In June 2017, the National Institute of Standards and Technology (NIST) published the first revision to the NIST SP 800-12 document, which contains guidelines that addresses the assessment and analysis of security control effectiveness and security posture of an organization (Nieles et al. 2017; NIST SP800-12r1).

In early 2015, the U.S. Department of Energy's Office of Electricity Delivery and Energy Reliability published a document titled "Energy Sector Cybersecurity Framework Implementation Guidance" (DOE, 2015) in response to NIST's Framework for Improving Critical Infrastructure Cybersecurity (NIST, 2014). Almost invariably, the North American Electric Reliability Corporation (NERC) continues to update and enforce a suite of critical infrastructure protection (CIP) (NERC, 2015) standards related to the reliability of cybersecurity.

These guidelines and standards underscore the importance of protecting our critical infrastructures which are mostly operating through automated controlled systems (Solomon, 2016). On an almost daily basis, this national need for cybersecurity-related CIP becomes more pronounced in light of intrusions and attempted attacks on Internet-facing control systems. As documented in ICS-CERT (2015), the Industrial Control Systems Cyber Emergency Response Team (ICS-CERT) responded to 245 incidents reported by asset owners and industry partners in 2014.

The rest of the paper is organized into five parts. First, we present background materials and the motivation behind this work. Second, we provide details on the design and implementation of embedded systems (ESs) and industrial control systems (ICSs) security curriculum resources. Third, we describe the laboratory setup and the associated hardware wherein the exercises are conducted. Fourth, we describe in detail the laboratory activities pertinent to each module and present the results of the initial evaluation of the pedagogical materials. Finally, we provide concluding remarks and present directions to possible extensions to this work.

## 7.2 Background

Recognizing the need to protect our critical infrastructures from intended or unintended harm, we embark on an ES and ICS security curricula enhancement project with the following objectives:

■ To develop and enhance curriculum modules focused on embedded and industrial control systems security,

■ To develop hands-on exercises to support the ES and ICS security learning modules,

■ To present lessons learned at various information security conferences and to offer mini-training workshops to widely disseminate the learning module to the Center for Academic Excellence (CAE) community, and

■ To evaluate teaching and learning effectiveness of the ES and ICS security curricula.

A professional development workshop on ICS security was conducted in June 2017 for college instructors. Overall, the pre- and post-workshop surveys indicate that the topics for the workshop were well chosen and well delivered, and the inexpensive control systems hardware was rated as excellent. The results highlight that ICS security is a topic that is not well covered in information assurance/cybersecurity curricula, and the workshop, as intended, highlighted the importance of that and other aspects of cybersecurity and provided instructors with tools and knowledge to integrate ICS security into their courses. Interesting remarks, coming from two of the participants, indicate that our workshop is far superior to the recently attended programmable logic controller (PLC) training workshop that was offered by one of the biggest industrial control manufacturers in the country. The ongoing project will build on the success of the recently concluded ICS workshop to effectively fill a void in cybersecurity training for the CAE community and the Department of Defense (DoD) training personnel across the nation. It will have significant contributions to the Cybersecurity National Action Plan (CNAP) on addressing the expansion of the national cybersecurity workforce. As previously mentioned, ES and ICS security education, which is critical to our national interest, has lagged behind other cyber-related curricula due to the limited access to ICS/SCADA (supervisory control and data acquisition) equipment and testbed facility. Obviously, ICS security hands-on training programs cannot be performed on operational ICS without disrupting normal operating processes and thus, the need for hands-on activities that are adequately supplemented by an inexpensive embedded and control system hardware similar to that used in this project is paramount.

## 7.2.1 Prior and Similar Works

There have been similar efforts to address the need for enhancing control systems security. Prior and notable related works that this project builds upon are found in Francia et al. (2012, 2016a, 2016b), Francia and Snellen (2014), Thornton et al. (2012), and Francia and Francia (2014). A national SCADA testbed program has been established by the Department of Energy (INL, 2014). A primary goal of the program is to provide control system security training through workshops. Although these workshops provided the necessary training for various individuals and groups, the training materials are not directly adaptable to the rapid training of DoD personnel. The Cyber Security Education Consortium (CSEC) has created centers of excellence in automation and control systems to provide training on SCADA and control systems security (CSEC, 2014). The courses that were created for this security curriculum are excellent training tools to upgrade the security skills of operators. However, widespread adoption is restricted by the high cost and the lack of hardware resources to support the courses in a portable and affordable setting. The SANS Institute offers a course on industrial control systems and SCADA security (SANS, 2014) which targets that personnel who are directly involved with the operation of industrial controls. The exorbitant registration cost for the course makes it impractical for training workshop adoption. Our proposed capacity-building project offers freely available course modules using affordable resources that can deliver hands-on and realistic control systems security training and education.

## 7.3 Security Curriculum Modules

### 7.3.1 Industrial Control Systems (ICS) Security Curriculum Modules

Given the constraint that ICS curriculum modules must be designed to be self-contained as much as possible, we strived to cover the four basic areas of control systems application and security: control system networks and protocols, programmable logic controller (PLC) programming, human-machine interface (HMI) and system historian development and security, control system vulnerability assessment and penetration testing, and defensive techniques and incident response for control systems. These modules are detailed in Table 7.1.

**Table 7.1   The Control System Security Curriculum Modules**

| | |
|---|---|
| **Module Name:** Control System Networks and Protocols; Python Programming<br>Learning Objectives: To understand control system networking concepts and communication protocols. To be able to write Python scripts for security applications<br>Prerequisite: Basic knowledge of computer networks.<br>Topic Outline:<br>• Control systems and networks (SCADA, DCS,[1] ICS[2])<br>• Human-Machine Interfaces (HMI)<br>• Communication Protocols: Modbus, Profibus, OPC,[3] DNP3,[4] EtherNet/IP[5]<br>• Deep Packet Inspection of Control Packets<br>• Python scripting<br><br>Associated Problem-Based Laboratory Exercises:<br>• Control system packet capture and analysis<br>• Deep packet inspection<br>• Python scripts for log analysis and reverse engineering | **Module Name:** PLC Programming and HMI Development and Security<br>Learning Objectives: To understand the basic functions and programming of PLCs; to be able to design and implement a control system HMI; to understand HMI security.<br>Prerequisite: Basic knowledge of control devices and associated protocols.<br>Topic Outline:<br>• PLC programming using Ladder Logic<br>• Secure programming of control systems<br>• HMI design and implementation<br>• HMI vulnerability analysis and penetration testing<br><br>Associated Problem-Based Laboratory Exercises:<br>• PLC programming<br>• Creating a control system Human-Machine Interface (HMI) (see sample HMI in Figure 7.1)<br>• Customizing the toolkit |

(Continued)

**Table 7.1 (*Continued*)   The Control System Security Curriculum Modules**

| | |
|---|---|
| **Module Name:** Defensive Techniques and Incident Response for Control Systems<br>Learning Objectives: To understand attack methodologies, defensive techniques and incident response for control systems.<br>Prerequisite: Basic knowledge of computer networks, control system protocols, and security principles.<br>Topic Outline:<br>• Understanding basic firewall rule configuration (Authentication, Authorization, and Accounting)<br>• Intrusion Detection and Prevention Systems on control systems<br>• Indicators of compromise on control systems<br>• Event investigation and data analysis<br>• Incident response policy and plans on control systems<br>• Evidence handling and administration<br><br>Associated Problem-Based Laboratory Exercises:<br>• Configure an IDS for a control system environment<br>• Configure and test a firewall configuration for the toolkit<br>• Design a modular firewall policy; critique a given firewall policy<br>• Perform a behavioral analysis of a compromised control system | **Module Name:** Control System Vulnerability Assessment and Penetration Testing<br>Learning Objectives: To understand control system vulnerability assessment; to be able to perform penetration testing of control systems; to be able to recommend remedial actions for control system hardening.<br>Prerequisite: Basic knowledge of control system and network protocols.<br>Topic Outline:<br>• Attack surfaces of control systems<br>• Vulnerability assessment and tools<br>• Penetration testing and tools<br><br>Associated Problem-Based Laboratory Exercises:<br>• Control system reconnaissance and mapping<br>• Vulnerability assessment of control systems<br>• Penetration testing of control system networks<br>• Maintaining exploits and backdoors<br>• Data exfiltration detection |

[1] Distributed Control System
[2] Industrial Control System
[3] Object Linking and Embedding for Process Control
[4] Distributed Network Protocol 3
[5] Ethernet Industrial Protocol

## 7.3.2 Embedded Systems (ES) Curriculum Modules

The following ES security curriculum modules are also designed to be self-contained: Secure Firmware Development and Embedded System Authentication. These modules are detailed in Table 7.2.

## 7.3.3 Guidelines, Standards, and Policy Curriculum Modules

The protection of control systems operating our nation's critical infrastructures should also include a basic understanding of the managerial, legal, and physical aspects pertinent to those systems. Thus, additional curriculum modules specific to industrial control systems security were developed and made available as supplementary materials. These modules include, but are not limited to, Guidelines and Standards, Regulations and Compliance, Security Policies and Procedures, Management and Operational Controls, Risk Management, and Physical Security. The proposed modules covered the managerial, legal, regulatory, and physical aspects of critical infrastructure protection are shown in Table 7.3.

**Table 7.2    The Embedded Systems Security Curriculum Modules**

| | |
|---|---|
| **Module Name:** Secure Firmware Development<br>Learning Objective: To understand the secure coding of firmware on embedded systems<br>Prerequisite: Basic knowledge of computer programming.<br>Topic Outline:<br>• Understanding embedded systems<br>• Secure firmware coding<br><br>Associated Problem-Based Laboratory Exercises:<br>• Exploiting a firmware<br>• Reverse engineering a firmware<br>• Secure coding | **Module Name:** Embedded System Authentication<br>Learning Objectives: To understand basic cryptographic techniques; to be able to design and implement a lightweight encryption system for authentication; to understand the limitations of ES security.<br>Prerequisite: Basic knowledge of discrete math and programming.<br>Topic Outline:<br>• Cryptography<br>• Tiny Encryption Algorithm (TEA)<br>• Networking fundamentals<br><br>Associated Problem-Based Laboratory Exercises:<br>• Design of a basic authentication system for an embedded system<br>• Implementation of the TEA encryption system on the client and embedded system |

**Table 7.3 Guidelines, Standards, and Policy Curriculum Modules**

| | |
|---|---|
| **Module Name:** National Preparedness Plan and Cyber<br>Learning Objective: To understand the National Preparedness Guidelines in PPD-8 as it relates to cybersecurity and critical infrastructure<br>Prerequisite: Basic knowledge of laws and regulations.<br>Topic Outline:<br>• Review of PPD-8<br>• Examination of the cybersecurity and critical infrastructure plan<br>• Understand the laws and regulations<br>• Case study<br><br>Associated Problem-Based Laboratory Exercises:<br>• Discussion on the plan and how they can implement increased cybersecurity protocols. | **Module Name:** Risk, Response, Recovery, and the Command Center<br>Learning Objectives: To understand how to assess risk to build response and recovery plans in an ICS model<br>Prerequisite: Basic knowledge of risk response, and recovery.<br>Topic Outline:<br>• Defining Risk<br>• Identifying Risk Assessment Methods<br>• Transitioning Risk to Response and Recovery Planning<br>• The role of Command Center in Response and Recovery<br><br>Associated Problem-Based Laboratory Exercises:<br>• Scenario-based group activity will walk them through the risk assessment, framing a response and recovery plan based on the ICS model. |
| **Module Name:** Protecting Systems: Physical and Virtual Security and Policy Design<br>Learning Objective: To understand the role of physical and virtual security rely on policy design<br>Prerequisite: Basic knowledge of critical infrastructure and security practices.<br>Topic Outline:<br>• Overview of physical security methods<br>• Review virtual security measures<br>• Examination how policies can enhance or detract from those measures<br><br>Associated Problem-Based Laboratory Exercises:<br>• Conduct a strengths, weaknesses, opportunities, threats (SWOT) analysis of their current security measures and policies | **Module Name:** Managing Operations and Controls<br>Learning Objectives: To understand techniques effective strategies in managing operations and controls<br>Prerequisite: Basic knowledge of operations and controls.<br>Topic Outline:<br>• Defining operations and controls<br>• Identifying management strategies<br>• Change Management<br><br>Associated Problem-Based Laboratory Exercises:<br>• Building upon the SWOT analysis, participants will discuss how to implement increased protocols to provide increased security measures as it relates to operations and controls. |

### 7.3.4 Development and Implementation Processes

A subset of the course modules was used in a course titled "Embedded and Control Systems Security" in the Spring semester of 2018. The goals are to expose the students in that course to the problem-based learning approach and to measure the effectiveness of the learning modules.

We believe that this collection of curriculum content is appropriate for the level of expertise that we expect from the students at the Center for Academic Excellence 2 year(CAE-2Y), Center of Academic Excellence-Cyber Defense (CAE-CD), Center of Academic Excellence-Research (CAE-R) and Center of Academic Excellence-Cyber Operations (CAE-CO). Further, for each module, we provide multiple hands-on laboratory projects that introduce the problem based learning (PBL) approach (Hung et al., 2008; Lieb, 1991) (Hung et al., 2008; Lieb, 1991) to learning and enable the learners to practice the technique in order to gain a better understanding of the concepts involved. The subjects were arranged in a natural progression from baseline processes to a complete in-depth analysis and recovery of ICS systems after an incident. To enable widespread dissemination, we provide an accompanying videocast for each module and hands-on activity and an easy access to the pedagogical materials through a website. The curriculum modules are embodied as living documents, which will be continuously enhanced and expanded in subsequent years.

## 7.4 Laboratory Setup and Associated Hardware

To facilitate active learning, each module is accompanied by one or more laboratory activities that reinforce the concepts that were taught in the lectures. A typically computer laboratory is augmented by two very inexpensive devices: a PLC toolkit and a development board. These devices are equipped with Ethernet ports and can easily be attached to the laboratory's network switch/router. The total cost of implementing these embedded systems and industrial control system curricula modules is approximately $700: $200 for the development board and $500 for the PLC toolkit. The affordability of the accompanying hardware makes these modules appealing for widespread adoption.

The Do-more® H2 Series PLC starter kit with embedded 10/100 Base-T Ethernet, as shown in Figure 7.1, includes an H2-DM1E CPU, a 3-slot base, an input simulator, an output module, a USB port for programming, and a free development software. Total memory space is 262 kBytes and capable of 65K instruction words. It also includes an RS-232 port for Modbus RTU master and slave connections.

The BIG8051 Development System, shown in Figure 7.2, from MikroElektronika is a full-featured platform for embedded systems programming. It is based on the Silicon Labs C8051F040, a highly integrated microcontroller derived from the popular and mature Intel 8051 architecture. It features a rich variety of integrated peripheral modules, including an MMC/SD card slot, a serial Ethernet module, a USB communication interface, and numerous input and output ports. This makes the

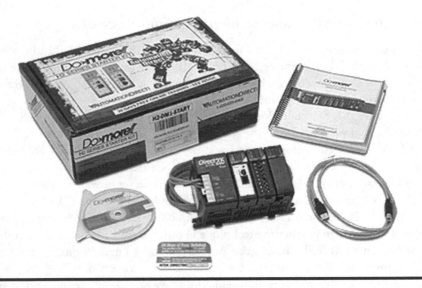

**Figure 7.1   PLC starter kit (AutomationDirect).**

**Figure 7.2   BIG8051 development board (MikroElectronika).**

BIG8051 ideal for device prototyping, and for exploring microcontroller programming and Internet of Things applications in a classroom or laboratory environment.

## 7.5 Details of the Laboratory Projects

The development of the hands-on exercises that were used in the ICS laboratory projects is based on the 10 curriculum modules described above.

### 7.5.1 ICS Network Protocols

#### 7.5.1.1 Description

Industrial control system protocols range from wired to wireless. Wired protocols include Ethernet Industrial Protocol (Ethernet/IP), Common Industrial Protocol (CIP), Modbus, Modbus Transmission Control Protocol (Modbus/TCP), Distributed Network Protocol version 3 (DNP3), Process Field Bus (Profibus), DeviceNet, Controller Area Network (CAN), and Ethernet for Control Automation Technology (EtherCAT). With the ever-increasing risk that ICS is being subjected to, it is imperative that cybersecurity professionals gain a good understanding of the communication protocols with which these systems operate and the threats that exist in securing them. The laboratory exercises are focused on the analysis of network packets of various ICS protocols and the development of Python-based utility tools.

#### 7.5.1.2 Laboratory Exercises

**Exercise 1:** Perform a deep packet analysis of captured DNP3 network packets.
**Exercise 2:** Perform a deep packet analysis of captured IEC 60870-5-104 network packets.
**Exercise 3:** Perform a deep packet analysis of captured Ethernet/IP network packets.
**Exercises 4–6:** Various introductory and security-related Python programming projects.

### 7.5.2 PLC Programming and HMI Development

#### 7.5.2.1 Description

A programmable logic controller consists of a CPU, a storage space, and input/output circuits. It is a self-contained computing device that is geared mostly for industrial control. It is programmed to realize the various functions required by industrial processes such as robot controls, equipment operations, status diagnostics, etc. The PLC instruction set may include logic, timing, counting, communication, math, and input/output (I/O) control.

A human-machine interface provides an intuitive interface to facilitate the interaction between the human operator and the control devices. Further, it enables effective process automation and monitoring by depicting process data and status using graphical displays.

The laboratory activities in this module provide the hands-on experiences for students to learn PLC programming using ladder logic diagram and to implement the associated HMI for each of the PLC programs.

### 7.5.2.2 Laboratory Exercises

**Exercise 1:** Implementation of a standard up-counter in a PLC utilizing ladder logic diagram.

**Exercise 2:** Implementation of a two-way light switching system for a two-storey building utilizing a ladder logic diagram.

**Exercise 3:** Implementation of a timer down PLC component to realize an automatic baking oven utilizing a ladder logic diagram.

**Exercises 4–6:** Implementation of corresponding HMIs for each of the PLC applications described in Exercises 1–3.

## 7.5.3 Defensive Techniques and Incident Response for ICS

### 7.5.3.1 Description

The purpose of a cybersecurity incident response plan is to provide the necessary mechanism to detect and respond to cybersecurity incidents as well as to protect critical data, assets, and systems to prevent incidents from happening.

The concept of a firewall originated from the wall that separates sections of a building having the purpose of preventing fire from spreading from one section to another. In network security, a firewall is a device that determines whether to allow or discard a packet that goes through it, depending on certain preset policies.

The lab exercises associated with this module involve analyzing and expanding a default firewall rule set, developing incident handling checklists, and configuring a Snort-base intrusion detection system that is specific to ICS. To make the firewall exercises as universally applicable as possible, we have chosen DD-WRT, an open-source firewall firmware solution which is supported by many commercial routers.

### 7.5.3.2 Laboratory Exercises (Firewall)

**Exercise 1:** Prepare the laboratory scenario by deploying an HMI application and PLC firmware. Both are designed to communicate using the Modbus/TCP protocol; in this lab, they are used to test Modbus/TCP connectivity within and between local area networks.

**Exercise 2:** Prepare a commercial router for use with control systems networks. This includes enabling secure shell access for remote management and enabling logging.

**Exercise 3:** Prepare the lab workstation for subsequent activities. If the workstation is not already equipped with a secure shell (SSH) client, this lab exercise guides the participant through the installation and configuration of the client and concludes by testing connectivity with the secure shell server in the router.

**Exercise 4:** Configure the firewall. After critiquing the default firewall configuration, the participant is guided through the implementation of several layered firewall policies. These range from open Modbus/TCP access to more restricted access; at each stage, Modbus/TCP connectivity is tested between the HMI, PLC, and the lab workstations. Finally, the participant is guided through the process of monitoring the firewall, checking the firewall logs, and permanently committing the completed and tested firewall policy.

### 7.5.3.3 Laboratory Exercises (Incident Response)

**Exercise 1:** Develop an incident handling checklist similar to checklist in the Generalized Incident Handling Guide, NIST SP 800-61r2, that is specific to a data exfiltration incident. Describe the purpose of each checklist item.

**Exercise 2:** Develop an incident handling checklist similar to the checklist in the Generalized Incident Handling Guide, NIST SP 800-61r2, that is specific to a malicious code incident. Describe the purpose of each checklist item.

### 7.5.3.4 Laboratory Exercises (Intrusion Detection System)

**Exercise 1:** Deploy an IDS for a Modbus device that will monitor critical activities in a Modbus protocol system.

## 7.5.4 Control System Vulnerability Assessment and Penetration Testing

### 7.5.4.1 Description

Vulnerability assessment (VA) is the process of identifying, documenting, and analyzing the vulnerabilities of a system. This process yields a list of vulnerabilities which are prioritized based on their criticality to impact the business objectives of the enterprise. Further, the results of this process, including the remediation steps taken to resolve the discovered vulnerabilities, can often be used to satisfy certain regulatory audit or compliance requirements.

ICS network reconnaissance is the process of scanning the network for the purpose of discovering the ICS devices attached to the network and their vital characteristics such ports opened and closed, IP addresses, services offered, and operating systems. Although this process seems to be passive and nonintrusive, it could have a harmful effect on an ICS.

Penetration testing is the process of testing a system, network, web interfaces, or applications with the intent of discovering vulnerabilities that an adversary may be able to exploit. The process can be performed manually, automated with software tools, or carried out using a technique that combines both. In some sense, penetration testing is a simulated adversarial attack.

### 7.5.4.2 Laboratory Exercises

**Exercises 1 and 2:** Perform scanning and enumeration techniques utilizing Zenmap to discover control devices on an internal network. Perform a system reconnaissance using Zenmap and Shodan to discover PLC devices on the local network and on the Internet, respectively.

**Exercise 3:** Perform vulnerability assessment of a control device on the internal network utilizing OpenVas and generate a detailed report of the discovered vulnerabilities.

**Exercises 4 and 5:** Utilizing Kali Linux, Metasploit, Armitage, and Modbusclient, exploit an HMI device with an attack originating from the external network.

## 7.5.5 Secure Firmware Development

### 7.5.5.1 Description

Secure firmware development explores embedded systems from a programming perspective, with an emphasis on secure applications. Starting with an introduction to embedded systems in general, and microcontrollers in particular, it proceeds to explore the various applications of embedded systems and the unique challenges faced by embedded system designers. The labs introduce the participant to embedded system development tools and present hands-on exercises which include the use of an in-circuit debugger for real-time memory inspection and troubleshooting. The module concludes with a detailed discussion of secure coding standards and defensive programming practices and presents the participant with the challenge of applying these practices to resolve a security-related flaw in an example embedded system.

### 7.5.5.2 Laboratory Exercises

**Exercise 1:** Install and configure the suite of microcontroller development tools provided by Silicon Labs and configure the BIG8051 Development System for first-time use.

**Exercise 2:** Use the in-circuit debugger, and the memory inspection tools of the Silicon Labs IDE, to control and monitor a firmware program during execution. Activities include single-stepping through program instructions, viewing and modifying the contents of memory during program execution, and programming the microcontroller's input/output ports.

**Exercise 3:** Apply the defensive programming ideas discussed in the lessons to discover and repair a security-related flaw. The participants are given a precompiled firmware binary for a complete embedded system which deliberately includes a serious security flaw. In Part One, the participants must discover a consistent exploit for this flaw using black-box testing techniques, and in Part Two, the participants are given the source code and must reverse-engineer it, using the debugging tools and the programming practices discussed earlier. Once the participants have identified the source of the security flaw and the reasons the flaw has the effect that it does in the system, they are asked to resolve the flaw, without affecting the functionality of the system.

## 7.5.6 Embedded System Authentication

### 7.5.6.1 Description

Embedded system authentication explores the application of cryptographic techniques to embedded systems. It begins with an introduction to cryptography, the major applications of cryptographic algorithms, and the characteristics of strong cryptosystems. It then discusses the various problems of implementing strong cryptography within the constraints of embedded systems, and presents one cryptographic algorithm, the Tiny Encryption Algorithm (TEA), as an ideal example of small and efficient embedded systems programming which nonetheless provides a reasonable level of protection. It then introduces the user to the Java programming language and platform, exploring input/output in Java and the use of the RXTX communication libraries for data exchange with embedded systems.

### 7.5.6.2 Laboratory Exercises

**Exercise 1:** An entry-level exercise, intended to introduce the use of serial ports and serial networks for embedded system communications. This exercise involves configuring the BIG8051 development system to enable data exchange with the PC; this involves using the virtual serial port on the BIG8051, and the Java Development Kit (JDK) and the RXTX communication libraries on the PC.

**Exercise 2:** Explore lightweight cryptography for embedded systems by implementing the TEA on two platforms: on the microcontroller platform using C, and on the PC platform using Java. The participant is challenged to use this algorithm to implement secure communications between the two platforms, encrypting on one platform and decrypting on the other.

**Exercise 3:** Expand the encryption algorithm implementation completed in the previous lab to add the ability to encrypt and decrypt 8 byte alphanumeric blocks. Again, the participant must implement the algorithm on both platforms; in the process, the participant explores such cross-platform development issues as integer precision and bitwise manipulation of signed numbers.

**Exercise 4:** The final exercise explores the security of lightweight cryptography by staging a brute-force attack on the TEA algorithm. Using the encryption tools developed in the earlier labs, the participant is presented with a brute-force password cracking tool (or is asked to develop this tool themselves, at the instructor's discretion), and is challenged to crack the encryption using keys of varying degrees of complexity.

### 7.5.7 National Preparedness Plan and Cyber

#### 7.5.7.1 Description

The National Preparedness Plan and Cyber modules introduced and expanded the participants understanding of policy decisions and government role in protecting the critical infrastructure and cyberspace. Few are aware that Presidential Policy Directive-8 (Obama, 2015), known as the National Preparedness Plan, includes cybersecurity in addition to natural and other man-made disasters.

#### 7.5.7.2 Laboratory Exercises

**Exercise 1:** Cyber Hygiene Kit introduced a connection between the *Blueprint for a Secure Cyber Future* (DHS, 2011) and the protection of the critical infrastructure. Participants in this lab are asked to evaluate their level of protection and maintenance on their home systems. The checklist will verify the number of systems, peripheral devices, authorized users, maintenance schedule, and patch levels to ensure the system is properly maintained and protected. Then participants are asked to reflect on how the practice on their systems at home translates to the CI and ICS practices.

**Exercise 2:** Participants are asked to read the "Electric Power Generation and Transmission: Creating a Sustainable NERC CIP Compliance Program" (Lockheed Martin, 2014) case study with a software solution and *Cybersecurity of Power Grid: State-of-the-Art* (Sun et al., 2018), which reviews "(1) a survey of the state-of-the-art smart grid technologies, (2) power industry practices and standards, (3) solutions that address cybersecurity issues, (4) a review of existing cyber physical systems (CPS) testbeds for cybersecurity research, and (5) unsolved cybersecurity problems." Based on the review of the laws, presidential directives, regulations, and policies, participants will be asked to compare the information in the article concerning unsolved cybersecurity problems to identify best practices that could be used to detect and deter attacks against the power grid.

## 7.5.8 Protecting Systems: Physical and Virtual Security and Policy Design

### 7.5.8.1 Description

The Protecting Systems: Physical and Virtual Security and Policy Design module has participants explore the steps of physical and cyber vulnerability threat assessments. A discussion on the information obtained in the vulnerability threat assessments can be used to create, enhance, or remove the policy. Participants are asked to retain their lab information for future labs.

### 7.5.8.2 Laboratory Exercises

**Exercise 1:** In this lab, participants will assess the potential hazards at their workplace. The identified hazards will be categorized as low, moderate, and high according to impact and probability. Participants are given websites containing statistics on active shooters, hazardous waste contamination, hurricanes, power outages, tornadoes, and wildfires to determine the probability in their area. In addition to the natural and man-made disasters, participants are asked to speak to their systems administrator or information technology (IT) staff to determine the number of malware attacks, denial of service, and other intrusions. This information will provide insight to events that have the highest impact and probability, which should be addressed first. This information is used in future modules to create cyber incident risk assessment, CIRT response and IT disaster recovery plan, and post-incident handling labs.

**Exercise 2:** The self-assessments: Vulnerability management asks participants to identify the vulnerabilities on their home system. The areas to be reviewed include the tools used to identify vulnerabilities such as scheduled patching for systems, updated software applications, and the ability to detect malicious code on home systems and mobile devices. Participants will be asked to list the identified vulnerabilities, categorize and prioritize the vulnerability, determine if the vulnerability is actively discovered, and discuss the relevance of the vulnerability of their home system. Maintaining a list of vulnerabilities and actions is recommended.

**Exercise 3:** Conduct a second SWOT analysis using the materials discussed in the article titled "Transforming Power Operations and Maintenance Efficiency with Advanced Asset Information Management" (Smith, 2018).

Identify and list additional or modified best practices to add to your organization based on the content in the article and content within this module.

## 7.5.9 Risk, Response, Recovery, and the Command Center

### 7.5.9.1 Description

Risk Management, Response, Recovery, and the Command Center is predicated on the content and labs from Modules 7 and 8. To effectively determine risk, response, recovery, and command center operations, the identification of mitigating circumstances is paramount. The first five labs are the assessment steps in the cyber incident risk management (CIRM) plan. The confluence of data from physical and virtual vulnerability threat assessments, all hazards models, and cyber incident risk management coalesces into a comprehensive IT disaster recovery, cyber incident response, and a unified command center. At the end of Module 9, participants in the final assessment and lab combination will be engaged in a decision-tree cyber incident scenario to test their understanding of how to respond to a cyberattack. The first five lab exercises will walk participants through a cyber incident risk management assessment to build a plan.

### 7.5.9.2 Laboratory Exercises

**Exercise 1:** Participants are asked to conduct a cyber risk assessment worksheet will identify potential risks at either their workplace or home systems.

**Exercise 2:** The Risk Impact Scale is the next step to rank the identified issues in Lab 1 based on severity and area of risk. This process teaches the participants how to prioritize risk based on the mission, strategies, financial, regulatory, and compliance guidelines.

**Exercise 3:** The Opportunity Impact Scale will convey to participants that some risks can present opportunities. Participants will need to determine out of the identified areas of risk and what opportunities if any are present within that identified area.

**Exercise 4:** Likelihood Scale Lab instructs the participants to rank the possibility of occurrence based on the identified areas of risk and/or opportunity. To score the risk or opportunity, participants consider the impact and probability based on low, low-medium, medium, medium-high, and high for both categories. Once a decision is made, the item is plotted on the chart.

**Exercise 5:** Risk/Opportunity Evaluation and Response Lab will have participants determine the risk/opportunity response strategy, response plan, cost estimate, other resources needed, and the target completion date. This is the last step to determine which risks and opportunities will be addressed based on priority, the time frame of resolution, the resolution, and by whom the risk/opportunity will be addressed.

**Exercise 6:** The Cyber Scenario is the final lab for Module 9 and will be a decision tree cyber scenario that uses a cyberattack method discussed in the module. Based on the information the participants have been given in this module and other external resources, the participants will have to decide the type of attack, and the correct order of handling the incident, before power is restored to the region. Participants will use a variety of techniques such as multiple choice and drag and drop to determine the best method of solving the incident. Once the scenario is completed, participants are asked to print a copy of their scenario. This information will then be used to complete the Capstone at the end of Module 10.

## 7.5.10 Managing Operations and Controls

### 7.5.10.1 Description

Although managing operations, controls, and compliance is a daily operational task, these areas are critical during a post-incident. The recovery from a cyberattack and restoring compliance and regulatory standards is a process. Therefore, participants are asked to conduct labs and assessments to regain regulatory and compliance standards with minimal impact to the organization.

### 7.5.10.2 Laboratory Exercises

**Exercise 1:** ICS Systems lab asks participants to search the Internet to find consumer products that are or could be classified as supervisory control and data acquisition (SCADA), distributed control systems (DCSs), and process control systems (PCSs).

**Exercise 2:** Participants will search the Internet to find cyber laws that are associated with industrial control systems and/or the critical infrastructure. This lab will demonstrate the number of laws and regulations that are in place to protect ICS and CI.

**Exercise 3:** Participants are also asked to review the 10 Self-Assessment domains in the Cyber Resilience Review (CRR) to determine if the primary role of the domain is under managing operations, controls, or compliance and to give explanations for their decision. This exercise will demonstrate the purpose of CRR and how that framework guides or governs the actions of ICS and CIP.

**Exercise 4:** The participants' last lab is a capstone, which refers back to the cyber scenario in Module 9. In the capstone, the participants will be asked to identify from prior assessments and labs the areas where their organization could permit such an attack at their location. Then the participants will determine which laws, policies, regulations, and compliance standards have been violated.

The final step is an action plan or post-incident handling plan on how the organization can regain their "good standing" with the laws, regulations, operations, and compliance guidelines that were breached or violated during the attack.

## 7.6 Lab Assessments

The labs were created with supporting documentation and examples to demonstrate the correct protocol or processes. Content assessments were included throughout the modules to confirm the participants' level of understanding in the information given. For instance, to gauge the effectiveness of the materials used in the Embedded System Security and Industrial Control System Security curriculum modules, the lessons and laboratory exercises were piloted in two sections of an Embedded and Control Systems Security course during the Spring 2018 semester.

Two pilot surveys were given to the students in both sections. The first survey was distributed at the beginning of the semester; a total of 36 students ($n = 36$) participated. The second survey was given upon completion of the course. The post-survey had 35 students ($n = 35$) participate. Although the students that participated in the pre-survey were to take the post-survey, the pilot research design was disrupted. In Spring 2018 an EF3 tornado struck the university campus and the greater community area. Due to this unusual event occurring close to the close of the Spring semester, students were provided completion options that included (1) complete the course, (2) accept the current grade, or (3) receive an Incomplete. Therefore, the pilot pre- and post-surveys will be redistributed in future semesters to establish reliability and validity.

The six questions on the pre-semester survey were as follows:

1. Rate your own level of experience with project-oriented microcontroller kits, such as the Arduino.
2. Rate your own level of experience with Programmable Logic Controllers (PLCs), such as those used in Industrial Control Systems.
3. What is your present level of experience with using ladder logic to develop firmware for Programmable Logic Controllers (PLCs)?
4. What is your present level of experience with using human-machine interface (HMI) tools to develop applications for monitoring and controlling Programmable Logic Controllers (PLCs)?
5. How would you rate your present level of awareness of Control Systems Security issues?
6. How would you rate your present level of awareness of Embedded Systems Security issues?

The corresponding six questions on the post-semester survey included the following:

1. How would you rate your new level of understanding of microcontrollers and microcontroller programming?

2. How would you rate your new level of understanding of Programmable Logic Controllers (PLCs)?
3. How would you rate your new level of understanding of ladder logic programming?
4. How would you rate your new level of understanding of human-machine interface (HMI) programming?
5. How would you rate your new level of awareness of Control Systems Security issues?
6. How would you rate your new level of awareness of Embedded Systems Security issues?

The pre-semester survey questions were designed to assess (1) the students' previous experience with embedded systems and (2) their level of awareness of embedded systems security issues, and (3) to establish a baseline of measure to compare the post-semester survey. When the post-semester survey was distributed, students were asked if the work in the course had increased or decreased their level of interest in pursuing further study of embedded systems by selecting the indicator that best described their level of agreement:

- "My experience in this class has made me more comfortable with the idea of working with embedded systems (including microcontrollers and PLCs) as development platforms."
- "My experience in this class has made me more interested in pursuing projects which make use of embedded systems, either as a career or for personal enjoyment."
- "My experience in this class has helped me to better understand the role of embedded systems in everyday life, including their use in devices that I regularly use or rely on."
- "My experience in this class has helped me to better understand the security-related challenges of embedded systems design."
- "My experience in this class has helped me to better understand the approaches and tools that programmers can take to solving the aforementioned security problems."

In addition to the questions listed above, students were asked to rate the following on a 1–10 scale, and the findings are displayed in Table 7.4, Pre-and Post-Semester Assessment Results. The four areas students were asked to rank included the following:

- The effectiveness of the course materials
- The development tools and equipment used in the course
- Their own confidence in undertaking ES-ICS security activities
- How much the course increased or decreased their interest in information security in general, and embedded systems security in particular

**Table 7.4  Pre- and Post-Semester Assessment Results**

| Question/Response Options | Pre-(n = 36) | Post-(n = 35) |
|---|---|---|
| **Question 1** | | |
| Rate your level of understanding (or experience) of microcontrollers and microcontroller programming. | | |
| • I never used them/Very Unfamiliar | 83.33% | 0% |
| • I've used them only once or twice/Somewhat Unfamiliar | 16.67% | 2.86% |
| • I've used them occasionally/Somewhat Familiar | 0% | 57.14% |
| • I've used them extensively/Very Familiar | 0% | 40% |
| **Question 2** | | |
| Rate your level of understanding (or experience) with Programmable Logic Controllers (PLCs). | | |
| • I never used them/Very Unfamiliar | 88.89% | 0 |
| • I've used them only once or twice/Somewhat Unfamiliar | 11.11% | 2.86% |
| • I've used them occasionally/Somewhat Familiar | 0% | 60% |
| • I've used them extensively/Very Familiar | 0% | 37.14% |
| **Question 3** | | |
| Rate your level of understanding (or experience) with ladder logic programming. | | |
| • I never used them/Very Unfamiliar | 88.89% | 0% |
| • I've used them only once or twice/Somewhat Unfamiliar | 11.11% | 11.43% |
| • I've used them occasionally/Somewhat Familiar | 0% | 57.14% |
| • I've used them extensively/Very Familiar | 0% | 14.29% |
| • N/A (did not participate due to the tornado) | | 17.14% |

*(Continued)*

**Table 7.4 (*Continued*)    Pre- and Post-Semester Assessment Results**

| Question/Response Options | Pre-(n = 36) | Post-(n = 35) |
|---|---|---|
| **Question 4** | | |
| Rate your level of understanding (or experience) with HMI programming. | | |
| • I never used them/Very Unfamiliar | 88.89% | 0% |
| • I've used them only once or twice/Somewhat Unfamiliar | 11.11% | 11.43% |
| • I've used them occasionally/Somewhat Familiar | 0% | 57.14% |
| • I've used them extensively/Very Familiar | 0% | 14.29% |
| • N/A (did not participate due to the tornado) | | 17.14% |
| **Question 5** | | |
| What is your present level of awareness of Control Systems Security issues? | | |
| • I never used them/Very Unfamiliar | 50% | 0% |
| • I've used them only once or twice/Somewhat Unfamiliar | 30.56% | 5.72% |
| • I've used them occasionally/Somewhat Familiar | 19.44% | 45.71% |
| • I've used them extensively/Very Familiar | 0% | 48.57% |
| **Question 6** | | |
| What is your present level of awareness of Embedded Systems Security issues? | | |
| • I never used them/Very Unfamiliar | 55.56% | 0% |
| • I've used them only once or twice/Somewhat Unfamiliar | 30.55% | 8.57% |
| • I've used them occasionally/Somewhat Familiar | 13.89% | 40% |
| • I've used them extensively/Very Familiar | 0% | 51.43% |

The summary of the post-semester assessment findings is shown in Table 7.5 (Figure 7.3).

In the assessments for embedded systems, the modules contained additional assessments to gauge the participant's understanding of the content. The lab assignments can connect to a database or learning management system (LMS) to track the participants' progression, which would allow further analysis. The culmination of surveys, successful completion of the labs, and assessments in the module are validating the effectiveness of the module content and the accompanying laboratory exercises.

**Table 7.5   Post-Semester Assessment Results**

| Questions | Results (n = 35) | |
|---|---|---|
| My experience in this class has made me more comfortable with the idea of working with embedded systems (including microcontrollers and PLCs) as development platforms. | Strongly disagree: | 0% |
| | Somewhat disagree: | 2.86% |
| | Somewhat agree: | 8.57% |
| | Agree: | 45.71% |
| | Strongly agree: | 42.86% |
| My experience in this class has made me more interested in pursuing projects which make use of embedded systems, either as a career or for personal enjoyment. | Strongly disagree: | 0% |
| | Somewhat disagree: | 5.71% |
| | Somewhat agree: | 11.43% |
| | Agree: | 48.57% |
| | Strongly agree: | 34.29% |
| My experience in this class has helped me to better understand the role of embedded systems in everyday life, including their use in devices that I regularly use or rely on. | Strongly disagree: | 0% |
| | Somewhat disagree: | 0% |
| | Somewhat agree: | 5.72% |
| | Agree: | 25.71% |
| | Strongly agree: | 68.57% |
| My experience in this class has helped me to better understand the security-related challenges of embedded systems design. | Strongly disagree: | 0% |
| | Somewhat disagree: | 0% |
| | Somewhat agree: | 2.86% |
| | Agree: | 40% |
| | Strongly agree: | 57.14% |

*(Continued)*

**Table 7.5 (*Continued*)   Post-Semester Assessment Results**

| Questions | Results (n = 35) | |
|---|---|---|
| My experience in this class has helped me to better understand the approaches and tools that programmers can take to solving the aforementioned security problems. | Strongly disagree: | 0% |
| | Somewhat disagree: | 2.86% |
| | Somewhat agree: | 11.43% |
| | Agree: | 45.71% |
| | Strongly agree: | 40% |
| On a scale of 1–10 (in which 10 is "very confident" and 1 is "no confidence at all"), rate your confidence in undertaking Embedded Systems and Industrial Control Systems (ES-ICS) security activities. | 10 (very confident): | 0 |
| | 9: | 5 |
| | 8: | 11 |
| | 7: | 10 |
| | 6: | 3 |
| | 5: | 5 |
| | 4: | 1 |
| | 3: | 0 |
| | 2: | 0 |
| | 1 (no confidence at all): | 0 |
| On a scale of 1–10 (in which 10 is "very effective" and 1 is "not effective at all"), rate the effectiveness of the course materials used in learning about ES-ICS security. | 10 (very effective): | 15 |
| | 9: | 7 |
| | 8: | 7 |
| | 7: | 3 |
| | 6: | 1 |
| | 5: | 1 |
| | 4: | 0 |
| | 3: | 0 |
| | 2: | 0 |
| | 1 (not effective at all): | 0 |

(*Continued*)

**Table 7.5 (*Continued*)   Post-Semester Assessment Results**

| Questions | Results (n = 35) | |
|---|---|---|
| On a scale of 1–10 (in which 10 is "very effective" and 1 is "not effective at all"), rate the effectiveness of the tools used in learning about ES-ICS security. | 10 (very effective): | 17 |
| | 9: | 6 |
| | 8: | 8 |
| | 7: | 3 |
| | 6: | 1 |
| | 5: | 0 |
| | 4: | 0 |
| | 3: | 0 |
| | 2: | 0 |
| | 1 (not effective at all): | 0 |
| On a scale of 1–10 (in which 10 is "greatest increase," 5 is "neutral," and 1 is "greatest decrease"), rate how much the course increased or decreased your interest in ES-ICS security, or Information Security in general. | 10 (greatest increase): | 9 |
| | 9: | 5 |
| | 8: | 10 |
| | 7: | 4 |
| | 6: | 2 |
| | 5 (neutral): | 5 |
| | 4: | 0 |
| | 3: | 0 |
| | 2: | 0 |
| | 1 (greatest decrease): | 0 |

# 7.7 Conclusion and Future Plans

In this paper, we presented the design and implementation of Embedded Systems (ES) and Industrial Control Systems (ICS) security curriculum modules. We also described the laboratory setup and the associated hands-on exercises that are pertinent to each module. We believe that the problem-based approach to learning is enhanced by the carefully designed activities that accompany the lecture modules. The capstone exercise on applying the various laws, policies, regulations, and compliance standards to specific attack scenarios provides a realistic tabletop exercise

**Figure 7.3    The ES-ICS security workbench.**

for participants to be able to apply their newly acquired knowledge on ICS security and critical infrastructure protection.

Future plans, connected with these curriculum modules and activities, are the following:

- The continuous evaluation of the effectiveness of the curriculum modules and laboratory exercises;
- The enhancement of the table-top exercises with additional real-world scenarios; and
- The development of additional curriculum modules in the areas of threat intelligence, machine learning, indicators of compromise, and attack attribution pertaining to ICS and ES security.

## Acknowledgments

This work is supported in part by a Center for Academic Excellence (CAE) Cyber Security Research Program grant (Award # H98230-17-1-0326) from the National Security Agency (NSA) and a National Science Foundation (NSF) grant (Award # 1515636). Opinions expressed are those of the author and not necessarily of the granting agencies. The United States government is authorized to reproduce and distribute reprints notwithstanding any copyright notation herein.

# References

Cyber Security Education Consortium (CSEC). (2014). CSEC advances cybersecurity & homeland defense. https://atecenters.org/st/csec/. Accessed August 10, 2018.

U.S. Department of Homeland Security (DHS), (2011). Blueprint for a Secure Cyber Future, The Cybersecurity Strategy for the Homeland Security Enterprise. November 2011. https://www.dhs.gov/xlibrary/assets/nppd/blueprint-for-a-secure-cyber-future.pdf. Accessed February 21, 2019.

U.S. Department of Energy (DOE). (2015). Office of Electricity Delivery and Energy Reliability, Energy Sector Cybersecurity Framework Implementation Guidance. January 2015. http://www.energy.gov/sites/prod/files/2015/01/f19/Energy%20Sector%20Cybersecurity%20Framework%20Implementation%20Guidance_FINAL_01-05-15.pdf. Accessed August 10, 2018.

Francia, G. A., Bekhouche, N., Marbut, T. M., & Neuman, C. (2012). Portable SCADA security toolkits. *International Journal of Information & Network Security* (*IJINS*), 1(4), 265–274.

Francia, G. A. & Francia, X. P. (2014). Critical infrastructure protection and security benchmarks. In M. Khrosrow-Pour (Ed.), *Encyclopedia of Information Science and Technology* (3rd ed., pp. 4267–4278). Hershey, PA: IGI-Global Publishing.

Francia, G. A., & Snellen, J. (2014). Embedded and control systems security projects. *Information Security Education Journal* (*ISEJ*), 1(2), 77–84.

Francia, G. A., Randall, G., & Snellen, J. (2016a). Pedagogical resources for industrial control systems security: Design, implementation, conveyance, and evaluation. *Journal of Cybersecurity Education, Research and Practice*, 2016(4). https://digitalcommons.kennesaw.edu/ccerp/2016/Academic/4/.

Francia, G. A., Francia, X. P., & Pruitt, A. M. (2016b). Towards an in-depth understanding of deep packet inspection using a suite of industrial control systems protocol packets. *Journal of Cybersecurity Education, Research and Practice*, 2016(2), Article 2. http://digitalcommons.kennesaw.edu/jcerp/vol2016/iss2/2.

Hung, W., Jonassen, D. H., & Liu, R. (2008). Problem-based learning. In J. M. Spector, J. G. van Merriënboer, M. D. Merrill, & M. Driscoll (Eds.), *Handbook of Research on Educational Communications and Technology* (3rd ed., pp. 485–506). Mahwah, NJ: Erlbaum.

Idaho National Laboratory (INL). (2014). National SCADA test bed program. http://www.inl.gov/scada/. Accessed November 8, 2015.

Industrial Control Systems Cyber Emergency Response Team (ICS-CERT). (2015). Incident response/vulnerability coordination in 2014. ICS-CERT Monitor, September 2014–February 2015. https://ics-cert.us-cert.gov/sites/default/files/Monitors/ICS-CERT_Monitor_Sep2014-Feb2015.pdf.

Lieb, S. (1991). *Principles of Adult Learning*. Honolulu, HI: Honolulu Community College. http://design2learn.ch/downloads/principles_of_adult_learning_lieb.pdf. Accessed July 1, 2016.

Lockheed, M. (2014, October). *Electric Power Generation & Transmission Creating a Sustainable NERC CIP Compliance Program*. Penn Energy White Papers. Retrieved July 27, 2018.

National Institute of Standards and Technology (NIST). (2014). Framework for improving critical infrastructure cybersecurity. February 12, 2014. http://www.nist.gov/cyberframework/upload/cybersecurity-framework-021214.pdf. Accessed March 30, 2016.

Nieles, M., Dempsey, K., & Pillitteri, V. Y. (2017, June). NIST special publication 800-12 revision 1. from National Institute of Standards and Technology: https://nvlpubs.nist. gov/nistpubs/SpecialPublications/NIST.SP.800-12r1.pdf. Accessed August 7, 2018.

North American Electric Reliability Corporation (NERC). (2015). Critical infrastructure protection (CIP) standards. http://www.nerc.com/pa/Stand/Pages/CIPStandards. aspx. Accessed April 6, 2016.

SANS. (2014). ICS410: ICS/SCADA security essentials. http://www.sans.org/course/ics-scada-cyber-security-essentials. Accessed November 5, 2015.

Smith, C. (2018, June). *Transforming Power Operations and Maintenance Efficiency with Advanced Asset Information Management.* Bentley White Paper, 1–8. Retrieved July 27, 2018.

Solomon, H. Over 90 percent of ICS devices exposed to Internet are vulnerable, says Kaspersky. IT World Canada, July 11, 2016. https://www.itworldcanada.com/ article/over-90-per-cent-of-ics-devices-exposed-to-internet-are-vulnerable-says-kaspersky/384856.

Sun, C., Hahn, A., & Liu, C. (2018). Cyber security of a power grid: State-of-the-art. *International Journal of Electrical Power & Energy Systems*, 99, 45–56. doi:10.1016/j. ijepes.2017.12.020.

Thornton, D. C., Francia, G. A., & Brookshire, T. (2012). Cyberattacks on SCADA systems (pp. 9–14). *Proceedings of the 16th Colloquium for Information Systems Security Education.* Lake Buena Vista, FL, June 11–13, 2012 (Best Paper in CISSE Conference Award).

# SECURITY AND PRIVACY IN BIG DATA CYBER-PHYSICAL SYSTEMS

# III

Chapter 8, "Security and Privacy in Big Data Cyber-Physical Systems," presents the existing privacy-preserving data publishing and privacy-preserving data mining techniques for big data cyber-physical systems.

Chapter 9, "Big Data Technologies–Supported Generic Visualization System in an Enterprise Cyber-Physical Environment," presents a security visualization design which provides a generic structure to enable the use of multiple data sources and a standardized structure to enable the use of multiple JavaScript-based visualization systems. It provides expert evaluation results and the user requirement traceability information as a part of the validation efforts.

Chapter 10, "Searching for IoT Resources in Intelligent Transportation Cyberspace (T-CPS)—Requirements, Use-Cases and Security Aspects," analyzes the features and challenges of IoT search technologies. Then, it proposes a distributed search model, Transportation Cyber-Search (TCS), and defines its properties. It addresses the fundamental strategies that are to be considered when developing and implementing TCS engines for IoT-aided Transportation-CPS (T-CPS). And also it holistically analyzes the privacy, threats and forensics, challenges, and mechanisms inherent in TCS design, while highlighting potential synergies and avenues of collaboration.

## Chapter 8

# Security and Privacy in Big Data Cyber-Physical Systems

L. Josephine Usha and J. Jesu Vedha Nayahi

## Contents

# 8.1 Introduction

Today, knowingly or unknowingly, people get connected with each other in one virtual world known as *cyber society* (Mueller 2006). The interaction of people with cyber society components, such as social media, search engines, blogs, and websites with their services causes generation of the enormous amount of data (Alguliyev et al. 2018). Generally, the size of the data is from exabyte to zettabyte. Similarly, a cyber-physical system, which is the connection between physical devices and the cyber system, generates a large volume of continuous flowing data that are required to be processed in a secured manner. Almost every industry is trying to explore this huge amount of data for making critical decisions and planned business moves. This will add more challenges while storing and processing data (Cardenas et al. 2008). Traditional methods of analysis are not suitable for analyzing big data because of its size and complexity. Therefore, new techniques ensuring privacy and security are to be developed to store and analyze data in the real-time cyber-physical system.

The personal information collected from individuals is governed by the privacy law of that country and hence ensuring big data privacy is a vital need. *Privacy-preserving data publishing* (PPDP) is a concept used in cyber-physical systems (Mitchell and Chen 2014; Sanislav et al. 2014; Mourtzis and Vlachou 2016; Song et al. 2017) that provides various tools and techniques for preserving data privacy while publishing the data over the Internet. In this chapter, we discuss various approaches to PPDP, evaluate their differences in a systematic way, and clarify the differences and requirements that distinguish PPDP from other related problems. We have identified the challenges of the existing techniques and propose future scope for the researchers.

The chapter is organized as follows: Section 8.2 discusses the overview of big data and privacy and security concerns in big data. Section 8.3 discusses the various approaches for ensuring the privacy of big data. Section 8.4 discusses various privacy-preservation methods and their comparisons. Section 8.5 discusses limitations and challenges in big data security. Finally, Section 8.6 provides the concluding remarks.

## 8.2 Background

"Cyber-physical system" (CPS) (Altawy and Youssef 2016) refers to the integration of computation and physical processing which combines both cyber and physical components. In the term "cyber-physical system," the term "cyber" refers to the integration of computation, communication and control, and the term "physical" refers to the natural or human-made systems that are operated in a controlled environment. A CPS uses embedded computers and network to compute, communicate, and control the physical processes and collect feedback about how the physical processes are affected based on the changing operating conditions and vice versa. Figure 8.1 shows the basic architecture of a cyber-physical system.

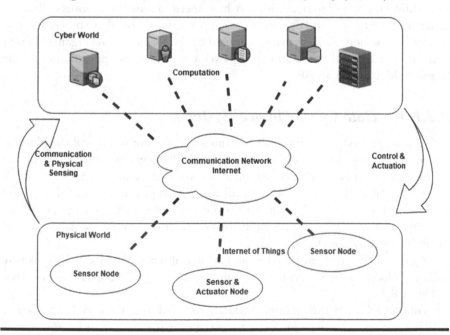

**Figure 8.1   Cyber-physical system.**

One of the recent developments in cyber-physical systems is a cloud-based cyber-physical system (CBCPS) (Reddy 2014), which takes the benefits of both cloud and physical systems. The key aspects of CBCPS include cloud computing (Neela and Saravanan 2013), Internet of Things (IoT), big data analytics, and cyber-physical systems. The cyber-physical system emerges in various areas, including aerospace, automotive, chemical production, civil infrastructure, energy, healthcare, manufacturing, materials, and transportation (Giraldo et al. 2017).

CPS systems are vulnerable to both physical attacks as well as cyberattacks, so as to threaten the human lives. For example, transportation cyber-physical systems (TCPSs) produce and exchange vast amounts of security-critical, safety-critical, and personal sensitive information, which leads to various attacks by cybercriminals (Kenyon 2018) Ensuring security of CPS is important in order to protect the information from vulnerabilities and safeguard the system against intruders.

The amount of data in the world is growing day by day due to recent technological development. The introduction of big data cyber-physical systems in banking, finance, retail industry, healthcare, smart city, smart grid, social media, and IT sector, has started gaining importance along with many research challenges such as heterogeneity, data lifecycle management, data processing, scalability, security and privacy, and data visualization and interpretation. The rapid growth of the cyber-physical system in various applications generates a huge volume of data originated from different physical sources with very high speed in numerous formats. The data generated from the physical devices can be gathered and analyzed in a timely manner, to obtain the actionable ideas that can be used by the businesses and organizations to improve their internal decision-making power and create new opportunities for the business (Matturdi et al. 2014).

## 8.2.1 Big Data Cyber-Physical Systems

Cyber-physical systems (CPSs) can continuously monitor or control the physical systems by using computer-based algorithms. The physical component in the cyber-physical system is operated in various domains and generates a huge volume of data. This will lead to significant challenges in terms of design and management of CPS in different aspects such as performance, energy efficiency, security, privacy, reliability, sustainability, fault tolerance, scalability, and flexibility (Reddy 2014).

Big data in cyber-physical systems and in general is described using 5 Vs, namely volume, velocity, variety, veracity, and value, used to describe big data as described in Figure 8.2.

Volume refers to the amount of data generated and Velocity is the rate at which new data is generated. Variety means that the data is in heterogeneous formats like text, audio, image, or videos collected from heterogeneous sources.

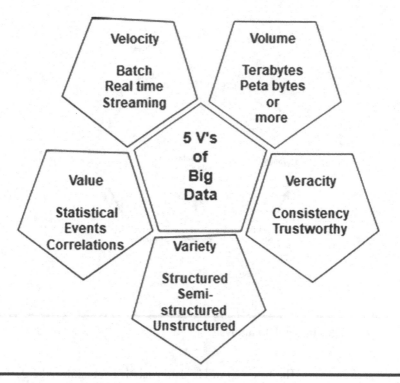

**Figure 8.2    5 V's of big data.**

Big data veracity refers to the biases, noise and abnormality in data, i.e., it tells about the consistency and trustworthiness of data, and Value that includes a large volume and variety of data that is easy to access and delivers quality analytics that enables informed decisions. It provides outputs for gains from large data sets.

## 8.2.2 Stages in Big Data Cyber-Physical Systems

In cyber-physical systems, big data analytics is defined as the process of gathering, structuring and analyzing a large volume of data to extract useful information for making a better decision about the future. Any organizations can use the concept of big data analytics to identify the useful patterns and extract valuable information from the large amounts of data. Various software tools used in conjunction with advanced analytics disciplines like predictive analytics, data mining concepts, text analytics, and statistical analysis are used for analyzing the big data (Karthikeyan et al. 2015; Chaudhari and Srivastava 2016). There are certain steps to be followed in the life cycle of big data processing such as data gathering, processing, and visualization. Figure 8.3 shows the steps in big data processing.

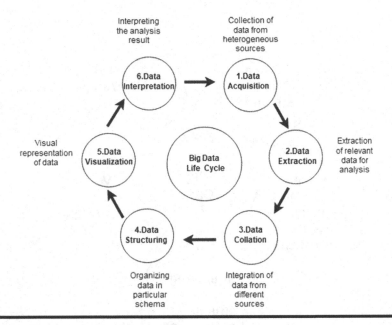

**Figure 8.3  Stages in big data analysis.**

- *Data Acquisition*: The first step in big data analysis is obtaining the data itself. With the advanced technological revolution, the data generation rate is rising exponentially. The smart devices equipped with a wide array of sensors generate data continuously. Most of this data is not useful and can be discarded. However, due to the unstructured nature of big data, selectively discarding the data presents a challenge situation. This data becomes valuable information when it is merged with other valuable data and superimposed. Due to the inconsistent nature of devices over the World Wide Web, data is widely collected and stored in the cloud.

- *Data Extraction*: After collecting the data from different sources, the next step is to extract relevant data for the purpose of analysis, because the collected data may have redundant or unimportant data. There are two types of challenges in data extraction. The first challenge is that due to the complexity of data generated, deciding which data to keep and which to discard gradually depends on the context in which the data was initially generated. For instance, security surveillance camera recording with similar frames may be discarded. However, similar data generated by a heart-rate sensor is not to be discarded. The lack of a common platform to integrate the wide variety of data is the next major challenge.

- *Data Collation*: The data collected from a single source is insufficient for making a good prediction. Therefore, data from more than one data source are often combined to make a better decision. For example, a forest fire detection application must collect data from different sensors like pressure, humidity, temperature sensor, etc. to summarize the health information of the customer.

Similarly, a weather monitoring application must collect the data from a variety of sensors to get information about daily humidity, temperature, pressure, precipitation, etc. This type of data convergence is often very important while processing the big data so as to achieve a better decision.

■ *Data Structuring*: This deals with various methods for organizing the data in a particular schema. Once all the data is aggregated, it is very important to present and store data in a structured format for further use. The structure of the data is important to execute the queries effectively. One of the major considerations in big data analytics is to provide real-time decisions and therefore aggregation of data can be done rapidly.

■ *Data Visualization*: The next step after structuring the data is visualization. The data should be presented using different charts and other visual representation that will help the user to understand the underlying data. In the case of data about water consumption rates, it should be presented with average temperatures to show the relationship between them.

■ *Data Interpretation*: The final step in big data analysis is data interpretation. The data should be interpreted correctly to gain useful information from data being processed. This step uses various visualization techniques to help users understand and interpret the analyzed results.

## 8.2.3 Privacy and Security Challenges in Big Data Cyber-Physical System

CPSs are more complex than they appear; ensuring security and privacy of big data cyber-physical systems was very difficult until CPSs were thoroughly understood. The fruitfulness of cyber-physical systems depends on the computer and communication technologies. The vulnerabilities and privacy threats in CPSs increased with the increase in complexity and connectivity of critical infrastructure (Altawy and Youssef 2016). Even though big data CPS can be used effectively for making a decision about the future, the exploding amount of data has increased the potential privacy breach. For example, Amazon and Google can study our shopping favorites and browsing behavior. Social networking sites such as Facebook store all the information about personal and social relationships. Popular video-sharing websites such as YouTube recommend us videos based on past search details.

With all of the above benefits, collecting, storing, and reusing personal information for the purpose of gaining commercial profits have put a threat to privacy and security. Organizations should have restricted access to use the information without individual knowledge to gain some benefit from the usage of our data. Several techniques and mechanisms were introduced to maintain a balance between security and the availability of big data.

Big data in cyber-physical systems face various challenges in terms of privacy (Jain et al. 2016) and security (Rushanan et al. 2014; Camara et al. 2015;

Sadeghi et al. 2015) in addition to the problem in big data storage and analysis. There is a meaningful difference between security and privacy. Security focuses on preventing unauthorized access to the data and resources, whereas privacy means the individual's privilege to control personal information when it is shared over the Internet. The information shared over the Internet cannot be used by any third-party without the knowledge of the owner (HIPAA 1999).

The difference between security and privacy can be easily understood by the situation where the customer purchases a product from ABC Company and gives their details for product shipment. In this case, to ensure privacy, the company should not share the personal information provided by the customer to another person without the knowledge of the customer. Also, security can be achieved by implementing various techniques like encryption, decryption, and firewalls to prevent vulnerabilities in the network.

Many security and privacy issues emerged with big data cyber-physical systems that are not likely to be solved by conventional security solutions. One of the serious issues in big data processing is the identification of personal information while transmitting data over the Internet. Protecting personally identifiable information (PII) is increasingly difficult because the data are shared too quickly.

## 8.3 Frameworks for Security and Privacy in Big Data Cyber-Physical Systems

Any data set in the cyber-physical system used for an analytic purpose may have sensitive information (Ghinita et al. 2011) about the individuals. Publishing such data without any modification may violate the individuals' privacy. The current practice in data publishing relies on policies and guidelines, so as to deal with different types of data that can be published and on agreements on the usage of published data. Traditional methods for ensuring security and privacy are not sufficient to handle this huge amount of data.

The various *privacy-preserving data publishing* (PPDP) techniques discussed here are categorized into (1) anonymization-based approaches, (2) perturbation-based approaches, (3) key-based approaches, (4) hybrid approaches, (5) user preference–based approaches, and (6) privacy preservation in data aggregation.

### 8.3.1 Privacy Preservation Based on Anonymization

Data anonymization (Sweeney 2002b) is the most common, successful, and earliest approach used for ensuring privacy in big data. In PPDP, it mainly focuses on anonymization techniques for publishing useful data while preserving privacy. According to anonymization, the original data will be modified in such a way that it does not reveal the sensitive information about the individual.

The original data set is described by different types of attributes that are classified into four broad categories as follows:

1. Identifier (ID): These are attributes that uniquely identify any particular individual. For instance, attributes such as name, SSN, and Aadhaar number are called identifier attributes.
2. Quasi-Identifier (QID): These are attributes that, when combined with other external data, can be used to identify the individuals. For instance, attributes like age, gender, and ZIP Code are QID attributes.
3. Sensitive Attribute (SA): Attributes like disease and salary that are sensitive and hence protected from disclosure are SAs.
4. Non-sensitive Attribute (NSA): NSA represents attributes other than identifier, quasi-identifier, and sensitive attributes.

Before publishing data to others, the table must be anonymized, that is, identifiers are removed and quasi-identifiers are transformed. As a result, an individual's identity is protected and sensitive attribute disclosure is prevented; $k$-anonymity, $l$-diversity, and $t$-closeness and their variations are some of the anonymization-based techniques used to ensure privacy.

### 8.3.1.1 k-Anonymity, l-Diversity and t-Closeness

$k$-Anonymity (Pan et al. 2012) is the simplest form of data anonymization method used to prevent identity disclosure (Raghunathan et al. 2003). The idea of $k$-anonymity is to transform the values of quasi-identifiers in the original data set, so that every tuple in the anonymized table is indistinguishable from at least $k$-1 other tuples. The anonymized table is called a $k$-anonymous table. The techniques (Salini et al. n.d.) used to implement $k$-anonymity are discussed as follows:

■ *Generalization*: This replaces the quasi-identifiers values into more general values, i.e., values that are forming higher levels in the concept hierarchy. Typical generalization schemes are full-domain generalization, subtree generalization, multidimensional generalization, etc.
■ *Suppression*: This replaces the QID values that lead to identity disclosure with an asterisk "*" or some constant values like 0. The various suppression schemes are record suppression, value suppression, cell suppression, etc.
■ *Anatomization*: This does not modify the quasi-identifier or the sensitive attribute, but de-associates the relationship between the two. Anatomization-based method partitions the given data into two to ensure privacy.
■ *Permutation*: This divides the records into groups and distributes sensitive values within each group so as to separate the quasi-identifier from a numerical sensitive attribute.

Table 8.1 shows the sample data set containing five attributes. *Name* is the key or identifier attribute and *Age, Gender, State* are quasi-identifier attributes; *Disease* is the sensitive attribute.

**Table 8.1   Sample Data Set**

| Name | Age | Gender | State | Disease |
|------|-----|--------|-------|---------|
| Joy | 31 | F | Tuticorin | Diabetics |
| Nalini | 34 | F | Dindugul | Cancer |
| Shylu | 39 | F | Tuticorin | Tuberculosis |
| Abijith | 36 | M | Kerala | Heart disease |
| Jonna | 34 | F | Dindugul | No illness |
| Balu | 33 | M | Kerala | Tuberculosis |
| Ratheesh | 25 | M | Dindugul | Diabetics |
| Kannan | 39 | M | Kerala | No illness |
| Jony | 27 | M | Dindugul | Tuberculosis |
| Britto | 23 | M | Dindugul | Flu |

The identifier attribute *Name* would be removed before $k$-anonymization. Suppression would replace all the QID attribute values that do not have similar values to "*." Table 8.2 shows the data transformed using suppression.

In generalization, the quasi-identifiers are transformed into more general values. For example, the *Age* attribute in Table 8.2 is replaced by the generalized values and shown in Table 8.3.

**Table 8.2   2-Anonymous Data after Suppression**

| Age | Gender | State | Disease |
|-----|--------|-------|---------|
| * | F | Tuticorin | Diabetics |
| * | F | Dindugul | Cancer |
| * | F | Tuticorin | Tuberculosis |
| * | M | Kerala | Heart disease |
| * | F | Dindugul | No illness |
| * | M | Kerala | Tuberculosis |
| * | M | Dindugul | Diabetes |
| * | M | Kerala | No illness |
| * | M | Dindugul | Tuberculosis |
| * | M | Dindugul | Flu |

**Table 8.3   2-Anonymous Data after Generalization**

| Age | Gender | State | Disease |
|-----|--------|-------|---------|
| 30–40 | F | Tuticorin | Diabetics |
| 30–40 | F | Dindugul | Cancer |
| 30–40 | F | Tuticorin | Tuberculosis |
| 30–40 | M | Kerala | Heart disease |
| 30–40 | F | Dindugul | No illness |
| 30–40 | M | Kerala | Tuberculosis |
| ≤30 | M | Dindugul | Diabetics |
| 30–40 | M | Kerala | No illness |
| ≤30 | M | Dindugul | Tuberculosis |
| ≤30 | M | Dindugul | Flu |

Table 8.3 is 2-anonymous (Aggarwal et al. 2005), which means at least two rows in the table have exact values for the QID attributes *Age*, *Gender*, and *State*. Adversaries use the values of the QID attributes known to them to disclose the sensitive values. In *k*-anonymization, the QID attributes have similar values for at least *k* records in an equivalence class. *K*-anonymous data are prone to various attacks such as unsorted matching attack, temporal attack, and complementary release attack (Gupta and Shukla 2016).

The *l*-diversity model (Meyerson and Williams 2004; Li et al. 2007; Sun et al. 2011; Tian and Zhang 2011) is an extended version of the *k*-anonymity model, which is used to protect the data from attribute disclosure; *l*-diversity is a group-based anonymization, where privacy is achieved by both generalization and suppression techniques so that any given record is equivalent to at least *k*-1 other records in the table. It requires at least *l* "well-represented" values for each sensitive attribute in each equivalence class to protect sensitive attributes. The term "well represented" means that at least *l* distinct values for the sensitive attribute should be in the equivalence class. The *l*-diversity technique is prone to attacks such as skewness and similarity attack, as it is inadequate to prevent attribute disclosure.

*Skewness Attack*: In this type of attack, each equivalence class has an equal number of positive records and negative records. It satisfies distinct 2-diversity, entropy 2-diversity requirements. This will create a serious privacy risk, because anyone in the class would be considered to have a 50% possibility of being positive, as compared with the 1% of the overall population.

*Similarity Attack*: When the sensitive attributes in an equivalence class are diverse but semantically similar, an adversary can learn more information about the sensitive values. Table 8.4 represents the original disease table and Table 8.5 shows an anonymized version satisfying distinct and entropy 3-diversity.

*Salary* and *Disease* are the sensitive attributes and *Zip Code* and *Age* are QID attributes. If an intruder knows Bob's age, then he/she knows that Bob's salary is in the range [4K–5K] and can infer that Bob's salary is relatively low. This attack not only applies to numeric attributes like "Salary," but also to categorical attributes like "Disease." Because all three diseases in the first equivalence class are heart-related, therefore, one can easily conclude that Bob has some heart-related problems by knowing the details about Bob's QID values.

Because *l*-diversity introduced the faster pruning algorithm, the performance of *l*-diversity is slightly better than *k*-anonymity (Samarati and Sweeney 1998; Bredereck et al. 2011).

An equivalence class is said to be *t*-close (Li et al. 2007) if the distribution of sensitive attributes in the equivalence class and the distribution of the sensitive attributes in the whole table is limited by the threshold *t*. A table is said to be *t*-close if all equivalence classes in the table satisfy *t*-closeness. It is an extended model of *l*-diversity achieved by effectively distributing the different data values for the attributes.

*t*-Closeness uses the concept of anonymization to hide the sensitive attributes of records. This technique does not affect the quasi-identifiers. In this method, an accurate anonymized table can be obtained by hiding only limited sensitive attributes of the records instead of suppressing whole records. Removing a value decreases diversity and may smooth a distribution and bring it closer to the overall distribution.

**Table 8.4   Original Disease Table**

| ZIP Code | Age | Salary | Disease |
|----------|-----|--------|---------|
| 45678 | 33 | 5K | Stroke |
| 45612 | 39 | 5K | Coronary disease |
| 45678 | 32 | 4K | Blood Pressure |
| 45905 | 50 | 6K | Pneumonia |
| 45909 | 54 | 12K | Viral infection |
| 45906 | 56 | 5K | Tuberculosis |
| 45605 | 25 | 6K | Asthma |
| 45674 | 26 | 8K | Asthma |
| 45607 | 28 | 2K | Ulcer |

**Table 8.5    3-Diversity Anonymized Table**

| Sl.No | ZIP Code | Age | Salary | Disease |
|---|---|---|---|---|
| 1 | 456** | 3* | 5K | Stroke |
| 2 | 456** | 3* | 5K | Coronary disease |
| 3 | 456** | 3* | 4K | Blood Pressure |
| 4 | 4590* | ≥50 | 6K | Pneumonia |
| 5 | 4590* | ≥50 | 12K | Viral Infection |
| 6 | 4590* | ≥50 | 5K | Tuberculosis |
| 7 | 456* | 2* | 6K | Asthma |
| 8 | 456* | 2* | 8K | Asthma |
| 9 | 456* | 2* | 2K | Ulcer |

**Bob**

| ZIP | Age |
|---|---|
| 45678 | 33 |

These techniques are combined with generalization and suppression to achieve better data quality.

*t*-closeness effectively solves the problem of attribute disclosure. The probability of re-identification increases with increase in size and variety of data. This leads to the significant downside of data anonymization while applied to big data.

## 8.3.1.2 Modified Approaches for Anonymization

One of the challenges of anonymization-based data privacy approaches is preserving the data utility while ensuring privacy. Data utility is measured in terms of information loss. A simple framework for producing high-utility anonymized data sets is utility-based anonymization (Xu et al. 2006a, 2006b). It is a hybrid greedy approach that combines the features of both local recoding and global recoding. Local recoding performs better than global recoding in terms of information loss. Hybrid recoding techniques exploit the advantages of both global and local recoding. Global recoding partitions the table into non-overlapping sub-tables, and then it employs local recoding to go on to divide each sub-table into smaller ones if possible. When we compare this hybrid recoding using utility metrics like normalized certainty penalty (NCP) and query answerability, it performs better in terms of utility view and gives high-quality results in terms of anonymization.

An enhanced model of *k*-anonymity is the (α,*k*)-anonymity (Sweeney 2002a; Salini et al. n.d.) algorithm for preserving privacy in data publishing of cyber-physical systems. In (α,*k*)-anonymity algorithm (Wong et al. 2006), α is a fraction and *k* is an integer. This method not only protects individual identification but also protects sensitive relationships by hiding multiple sensitive values using the simple *k*-anonymity model. This can be achieved by combining all sensitive values into a single class, and then the *k*-anonymity algorithm is applied, so that the inference confidence of each individual sensitive value is less than or equal to the inference confidence of combined sensitive value, α, i.e., the frequency of sensitive value no more than α.

More recent clustering techniques are used in combination with *k*-anonymization to form equivalence classes. One such approach is the clustering algorithm

(Aggarwal et al. 2006; Nayahi and Kavitha 2015) for ensuring data privacy protesting against both identity disclosure and attribute disclosure. The algorithm first determines the domain of the sensitive attribute ($S$) in the data set. Based on that, it will construct $K$ clusters each with $S$ diverse sensitive values for the sensitive attribute. The number of different sensitive values in each cluster would be equal to the number of sensitive values in the given data set. Then the resultant clusters are anonymized by replacing the actual values with the centroid ($G$) of the cluster. Based on this algorithm, each cluster formed has maximum possible diverse sensitive values, making it robust to similarity attacks. The method achieves a high degree of privacy with very minimal information loss, and the privacy degree is equivalent to a number of distinct values in the sensitive attribute.

## 8.3.2 Perturbation for Privacy

Perturbation is a simple and efficient method to preserve statistical information like mean and correlations while publishing the data and ensuring privacy. The general idea behind this approach is to replace the original data with some synthetic data values, so that the statistical information computed from the perturbed data is the same as that of the statistical information computed from the original data. One limitation of perturbation when compared to anonymization is that the published records in perturbation are synthetic in nature, so that it does not consider real-world entities represented by the original data. Therefore, individual records in perturbation are no longer helpful for the intended recipients and only the statistical properties are preserved. The publisher can publish only the statistical information rather than the perturbed data.

The commonly used methods in perturbation are additive noise, data swapping, and synthetic data generation. Additive noise is a widely used method for statistical disclosure control (Li et al. 2007); it is used to hide sensitive numerical data (e.g., salary). The general idea is to replace the sensitive value $s$ with $s + r$, where $r$ is any random value. Data swapping is another method of perturbation used to preserve data by exchanging values of sensitive attributes among individual records and also maintaining the frequency of swaps as a minimum. Unlike the additive noise method, this method is used to protect both numerical attributes (Reiss 1984) and categorical attributes (Adam and Worthmann 1989). The synthetic data generation method is used to ensure the record owner's privacy and retain the useful statistical information as in the case of statistical disclosure method (Reiss et al. 1982). The idea here is to build the statistical model and generate sample points from the model. These sampled points are used for data publication rather than the original data set.

### 8.3.2.1 Differential Privacy

Differential privacy (Gosain and Chugh 2014; Microsoft 2015) is a perturbation-based concept used for ensuring the privacy of information in the cyber-physical system. Differential privacy is a technique that allows the user to get useful information from a large volume of data without violating the privacy of the individual.

**Figure 8.4 Differential privacy.**

This is achieved by doing some distortion to the results provided by the database. The amount of distortion is either increased or decreased, based on privacy risk. The higher level of distortion leads to a high level of protection.

In this technique, individuals can't have direct access to the entire data. The intermediate layer between the users and data will supply all the required information to the users. This intermediate interface is a piece of software called a privacy guard. The privacy guard will accept the query from the user and get the answer from the database based on its evaluation. The resultant data can be distorted in some way based on the privacy requirements and sent to the user. Figure 8.4 shows the workflow diagram of differential privacy.

The distortion level is chosen based on the privacy risk. The distortion level is small if the risk of privacy is low and large if the privacy risk is high.

According to differential privacy, based on the type of data queried by the database analyst, the resultant data is distorted or modified using some mathematical calculations. In this, it just alters the resultant data without modifying the original data set.

## 8.3.3 Key-Based Approaches for Privacy and Security

Cryptographic techniques are used to protect the data from intruders. The public and private key pairs are generated and the data is encrypted at the source by the sender and decrypted at the destination by the receiver. This ensures that the data is secured against unauthorized access. Once decrypted, the data is plain and can be viewed by any authorized person. Privacy is securing the sensitive data not only from unauthorized access but also from authorized people. Cryptographic techniques are modified to ensure security as well as privacy.

### 8.3.3.1 Homomorphic Encryption

One of the techniques that supports both security and privacy is homomorphic encryption (Naehrig et al. 2011; Ogburn et al. 2013; Potey et al. 2016; Rahul et al. 2017). Homomorphic encryption allows the users to do basic computations on ciphertext, and generate an encrypted result as an output. When this encrypted result is decrypted, it will give similar results as applied to the plaintext. The main purpose of using homomorphic encryption is to perform computation on encrypted data without knowing the original data. It involves four functions, namely, key

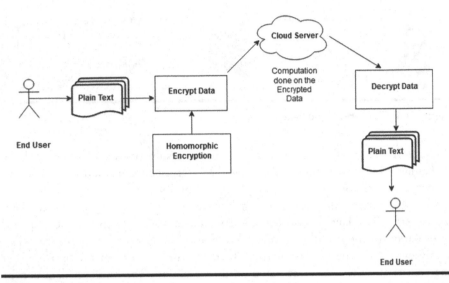

**Figure 8.5  Fully homomorphic encryption.**

generation, encryption, evaluation and decryption. Homomorphic encryption is of two types, namely, partially homomorphic encryption and fully homomorphic encryption. Partially homomorphic encryption schemes including RSA, ElGamal and Paillier allows only a restricted number of addition and multiplication on encrypted data, which have only limited practical applications.

Fully homomorphic encryption supports any arbitrary computations on encrypted data and has more practical use. Figure 8.5 shows the architecture of fully homomorphic encryption (Hayward and Chiang 2015), which is one of the many methods used to ensure data privacy in the cloud where the encrypted data are processed, and it returns encrypted results as an answer.

However, fully homomorphic encryption has a low processing speed and hence by parallelizing (Sowmya and Nagaratna 2016) the fully homomorphic encryption, the processing speed can be increased. The processing dispatcher will take an iterative set of operations on fully homomorphic encrypted data and split them across a number of parallel processing engines. Each processing engine behaves like a private cloud and does its task independently. Finally, the results of individual processing engines are combined to get the final encrypted result.

## 8.3.4 Hybrid Approaches for Preserving Privacy in Cyber-Physical Systems

The characteristics of big data make it difficult for any one of the existing approaches to deal with ensuring security and privacy. Recently, there are hybrid techniques exploiting the advantages of one or more existing approaches.

### 8.3.4.1 k-Anonymity with Privacy Key

An advanced method (Gupta and Shukla 2016) for ensuring big data privacy is using anonymity technique with privacy key. This method is based on $k$-anonymity and also incorporates features of differential privacy. The data set is anonymized using a $k$-anonymity algorithm and then a privacy key is added before publishing the data. The unique privacy key is generated using a privacy key generation algorithm and this private key is known only by the original user. While extracting, the user must have that key to extract the data from the database. Even the original user requires the key to access the data. Figure 8.6 explains the workflow of this hybrid anonymization technique.

Using this method, one can achieve two levels of security of data, one based on $k$-anonymity and the second based on privacy key. Thus, it supports both data privacy and security. The idea of anonymization with the key is a significant milestone in the development of privacy-preserving algorithms for data storage in the cloud.

### 8.3.4.2 Notice and Consent

The data owner is solely responsible for deciding which attributes are sensitive, and hence user preferences can be a significant attribute while ensuring privacy. The level of privacy or the confidentiality level of the attribute can be best decided by the data owner. The privacy preservation based on user preferences is Notice and Consent. This method considers the individual privileges to secure the data; i.e., it lets an individual decide on the privacy parameters. Based on this approach, whenever any individual is trying to access some application on the Internet, the owners are notified regarding the privacy concerns. It informs the user about the privilege to control their data. Existing frameworks on Notice and Consent may not be suitable for continuously flowing data, but they would be sufficient enough to protect privacy in the cyber-physical system when integrated with other approaches.

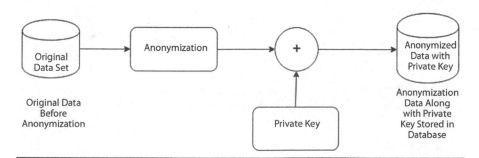

**Figure 8.6** *k*-Anonymity with private key.

### 8.3.4.3 Identity-Based Anonymization

Digital identity management is one of the challenging tasks in cloud computing infrastructure. To provide efficient access control to the cloud storage system, a user must be authenticated based on their identity and past interaction histories. At the same time, the confidentiality of the user and their browsing histories must be maintained. This is feasible by minimizing the possibility of identity disclosure. Identity-based anonymization (Govinda and Ravitheja 2012; Sedayao et al. 2014) is introduced to anonymize the data in a better way so as to ensure the confidentiality of the user using the cyber-physical system. The anonymization architecture has a security enclave to do de-identification (and re-identification when necessary). The anonymization process is initiated by encrypting any personally identifying information like IP address, username, and user ID in log files using AES symmetric key encryption. After anonymization of the data, the log files are moved to the Hadoop Distributed File System (HDFS)–based storage. This data is now available for the analysts who study the usage of a cyber-physical system. When the analysts need to re-identify log data, the log files must be moved back to the security enclave and the sensitive fields decrypted with the same symmetric key. Figure 8.7 explains the entire process in a systematic way.

### 8.3.4.4 k-Anonymization with MapReduce

*k*-anonymity using the MapReduce programming paradigm (Dean and Ghemawat 2008; Lämmel 2008) is a recent technique developed on the Hadoop distributed framework. Hadoop is an open-source implementation for reliable, scalable, distributed computing and data storage. It allows one to easily write and run applications that process vast amounts of data. MapReduce is the programming paradigm that involves two processes, namely Mapper and Reducer. Map task is an

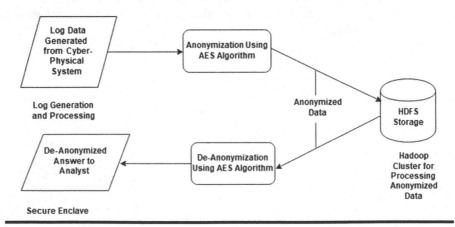

**Figure 8.7 Identity-based anonymization.**

initial ingestion and transformation step, which takes input data and converts it into a set of data, where individual elements are broken down into tuples (key/value pairs). Reduce task is an aggregation or summarization step, where it takes the output from a map task as an input and combines those data tuples into a smaller set of tuples.

This method works based on the basis of data partitioning. Initially, a data set is stored on a Hadoop Distributed File System (HDFS) (Shvachko et al. 2010) in the format of key/value pairs. Each key/value pair is represented as tuples in the data set. The key denotes the combination of quasi-identifiers and the value is the content of the tuple. It parallelizes the anonymization approach by splitting the data set into two equal-sized data sets and broadcasting it to all mappers. Each MapReduce program performs a partitioning process; i.e., it partitions the key space to determine which reducer instance will process which intermediate key and value. All keys that are the same go to the same reducer. The reducer output sets are then used to produce the final result. This algorithm is scalable and capable of anonymizing big data in order to support privacy-preserving data mining (Wu et al. 2014)

### 8.3.4.5 Privacy-Preserving e-Health System

Privacy-preserving e-health system (Yang et al. 2018) is a fusion of the Internet of Things, big data, and cloud storage. This system is constructed to realize secure IoT communication and confidential medical big data storage. The medical IoT network is used to monitor a patient's physiological data, which are gathered to form the electronic health record (EHR). The medical big data may contain a large volume of EHR stored on the cloud platform. The privacy should be ensured while sharing EHRs among patients and other users.

Attribute-based encryption (ABE) (Goyal et al. 2006; Bethencourt et al. 2007; Ostrovsky et al. 2007; Waters 2011) is an ideal solution to realize fine-grained access control that can be adopted in the medical big data system. A non-interactive and authenticated key distribution security model is designed for the medical IoT network. The patients distribute an IoT group key to other nodes in an authenticated way without any interaction. The IoT messages are encrypted using the IoT group key and transmitted to the patient, who can batch-authenticate the messages. The attribute secret key assigned to system users is used to encrypt the patients' EHRs with defined access policy. The users with the proper attribute secret key are authorized to access to patients' encrypted EHRs. Security models based on ABE are more efficient for securely sharing the patient's records in the e-health environment when compared to other approaches.

### 8.3.4.6 Scalable Privacy-Preserving Big Data Aggregation (Sca-PBDA)

Data aggregation is a significant process involved in the Internet of Things because it involves the summarization of data collected from physical devices. The process of data aggregation would reveal sensitive information to the intruders, and hence privacy has to be ensured during the process of data aggregation.

The amount of data generated by the sensors and other devices in wireless sensor networks (WSNs) is growing day by day. The process of aggregating, storing, transmitting, and analyzing the big sensor data becomes a significant challenge for researchers. The algorithm named Scalable Privacy-preservation Big Data Aggregation (Sca-PBDA) (Wu et al. 2016) introduces a scalable method for preserving the privacy of large-scale WSNs, and at the same time it supports energy-efficient parallel aggregation of sensor data. This method supports intra- and inter-cluster aggregation of sensor data to reduce energy consumption. It involves three steps:

- Clustering of sensor nodes used for data collection.
- Each sensor node is configured based on the privacy requirements of the sink node.
- Then the sink node will combine the aggregated results.

Sca-PBDA gives better results in terms of resource consumption and sensor data privacy when aggregating big sensor data from large-scale WSNs.

### 8.3.4.7 Fault Tolerance–Based Privacy-Preserving Big Data Aggregation in Smart Grid

A smart grid involves the installation of numerous smart meters that are used to collect real-time data from users, and these data are used to optimize energy utilization. The energy consumption data is used to support various basic services like load balancing and optimal dispatch strategy. Though the big energy data collected from consumers helps to provide better services to the consumers, it raises privacy issues.

This privacy-preserving data aggregation methods used in smart grid are fault tolerant and based on secret sharing scheme (Guan and Si 2017). The smart grid is divided into four parts, namely, the control center (CC), the key initialization center (KIC), the data aggregation device (DA), and the users. Initially, the consumers are divided into various groups based on their geographical locations, and every consumer is equipped with a smart meter (SM) to collect real-time data about their house appliances every 15 minutes. The key initialization center is responsible for initializing all the keys for the smart center and control center. The data aggregation device will integrate all the data sent by SMs, and then it calculates the sum of encrypted text based on a homomorphic algorithm. The aggregated sum of encrypted text can be uploaded to the control center along with the fault-tolerance function. Then the control center will decrypt the resultant encrypted text using decrypt key to obtain the summary of real-time data in a smart grid. This method is fault tolerant and better than other techniques in terms of privacy and security.

## 8.4  Comparison and Discussion

This section compares the different techniques discussed in the earlier section in terms of different metrics. Various metrics are available to measure the effectiveness of different privacy-preserving approaches. The existing metrics (Mendes and Vilela 2017) are classified into three categories, namely, privacy, data utility, and complexity. Metrics that are used to evaluate the privacy level measure the degree to which the data is secured against disclosure; data utility measures the loss of information, loss of originality or applicability of the data; and complexity metrics measure the efficiency and scalability.

### *8.4.1 Privacy Metrics*

The privacy-preserving algorithms show a trade-off between privacy and data utility. The privacy-level metrics measure the robustness to attacks. They are also classified into data privacy metrics and result privacy metrics. The data privacy metrics evaluate the feasibility of inferring the original sensitive data from the transformed data. The resulting privacy metrics deal with evaluating the disclosure feasible after the publication of outputs of analysis. The various metrics for measuring the data privacy are confidentiality level, average conditional entropy, and variance.

#### *8.4.1.1 Confidence Level*

The metric confidence level is used to measure the data privacy of additive the noise-based privacy-preservation method. This metric measures the confidence with which the original values may be reconstructed from the transformed data set. Suppose, if an original value of data set estimated to lie between the interval $[x1:x2]$ with C% confidence, then the amount of privacy at C% confidence is $(x2 - x1)$.

#### *8.4.1.2 Average Conditional Entropy*

This is based on the concept of information entropy. Given two random variables $X$ and $Z$, the average conditional privacy of $X$ given $Z$ is $H(X|Z) = 2^{h(X/Z)}$, where $h(X/Z)$ is conditional differential entropy of $X$ and is given by:

$$h(X \mid Z) = - \int_{\Omega X, Z}^{Z} f_{X,Z}(x, z) \log_2 f_{X|Z=z}(x) dx dz \qquad (8.1)$$

#### *8.4.1.3 Variance*

This data privacy metric is used in perturbation-based privacy-preservation methods. It measures the difference between the original data and perturbed data and tells about how closely one can recover the original data from the perturbed data.

Consider $x$ is the value in the original data and $z$ is the value in the distorted data; the variance can be calculated by using Equation (8.2).

$$\text{Variance} = \text{Var}\ (x{-}z)/\text{Var}\ (x) \tag{8.2}$$

The anonymization approaches such as $k$-anonymity, $l$-diversity, and $t$-closeness have a certain level of control over the privacy level based on the parameters $k$, $l$, and $t$ respectively.

## 8.4.2 Data Utility Metrics

Data utility metrics is also called information loss metrics and are used to quantify the loss of utility. Privacy-preserving techniques achieve privacy by affecting the quality of data being published. Data utility metrics are used to compare the difference between original data and the transformed data. The various parameters used for measuring the data quality are accuracy, completeness, and consistency. Accuracy deals with the estimation of the closeness between the transformed data and the original data, completeness refers to the loss of individual data in the transformed data, and consistency refers to the loss of correlation in the transformed data. The various commonly used metrics for measuring the loss of utility are discernibility metric, average equivalence class size, and the Kullback–Leibler divergence metric.

### 8.4.2.1 Discernibility Metric

The discernibility metric (DM) is a measure used to evaluate utility loss by measuring the size of the equivalence classes. That is, it measures how many records are identical to the given record due to generalization. The higher the value, more information that is lost. Therefore, a lower value of DM is preferred, because lower values refer to equivalence classes of small sizes and hence the utility loss would be less. In the case of $k$-anonymity, at least $k$-1 records are identical to any given record. The generalization and suppression methods are increased with the increase in $k$ value. This will automatically increase the discernibility, so it leads to a high loss of utility. The discernibility metric is given by the formula:

$$\text{DM} = \sum\nolimits_{E \in D*} (|\,E\,|2) \tag{8.3}$$

where $E$ is the equivalence class created by the algorithm and $D*$ is the privacy-preserved data set. The higher value of DM leads to high utility loss.

### 8.4.2.2 Average Equivalence Class Size

Another measure to quantify the loss of utility is average equivalence class size (CAvg). When the number of an equivalence class is high, the CAvg value becomes low. Usually, lower values for CAvg are preferred to minimize the utility loss. It is calculated using the formula:

$$\text{CAvg} = (|D|/\text{NOB})/k \tag{8.4}$$

where $|D|$ is the size of the data set, NOB is the number of equivalence classes formed by the algorithm, and $k$ is the anonymization parameter.

### 8.4.2.3 Kullback–Leibler Divergence Metric

The Kullback–Leibler divergence metric (KL) is used to measure the difference between the distribution before and after the process of anonymization. The distortion is calculated based on the KL value. Lower values of KL denote lower distortions in the distributions. When the KL value is zero, it means the two distributions are exactly the same. It is calculated based on the formula:

$$KL = p(x) \log [p(x)/q(x)] \tag{8.5}$$

where $p(x)$ and $q(x)$ are two distributions for which the divergence is to be calculated. The KL divergence value is also computed using the entropy or mutual information as follows:

$$KL = H(Y) - H(Y/X) \tag{8.6}$$

where $X$ is the set of quasi-identifier attributes, $Y$ is the sensitive attribute and $H(Y)$ is the entropy of the sensitive attribute $Y$, and $H(Y|X)$ is the conditional entropy of $Y$ conditioned on the quasi-identifier attributes $X$.

## 8.4.3 Complexity Level

The complexity of privacy-preserving techniques is concerned with the efficiency and scalability of the algorithm used for analysis. Efficiency can be measured in terms of time and space. Time efficiency refers to how fast the algorithm works and is measured by the CPU time and computational cost. Space efficiency refers to how much extra memory the algorithm requires for execution. Another measure called communication cost used in distributed cloud computing environment. It is computed based on the time, a number of messages exchanged, and the consumption of bandwidth.

Scalability refers to how well the privacy-preservation algorithm performs under the increasing size of data. In a distributed cloud computing environment, increase in the amount of data severely increases the amount of communication. Existing methods of privacy preservation are not sufficient to support this increasing nature of data in the cyber-physical system. Therefore, an efficient mechanism should be identified to solve the scalable nature of big data cyber-physical systems.

A comparison in Table 8.6 shows the comparison between various privacy-preserving approaches with its merits and demerits.

**Table 8.6  Comparison of Privacy-Preserving Techniques**

| Sl. No. | Frameworks | Method Used | Merits | Demerits |
|---|---|---|---|---|
| 1 | Sweeney (2002b) | *k*-anonymity | • Simple and easy to understand.<br>• Protect identity disclosure. | • Does not protect sensitive relationships in a data set.<br>• Cannot prevent attribute disclosure.<br>• Susceptible to homogeneity and background knowledge attacks. |
| 2 | Machanavajjhala et al. (2006) | *l*-diversity | • Handle attribute disclosure. | • Prone to attacks such as skewness and similarity attack.<br>• Insufficient to prevent attribute disclosure. |
| 3 | Li et al. (2007) | *t*-closeness | • Handle attribute disclosure. | • The probability of re-identification increases when size and variety of data increases. |
| 4 | Nayahi and Kavitha (2015) | (G, S) Clustering for privacy preservation | • Overcome the possibility of similarity attack.<br>• A high degree of privacy with very minimal information loss.<br>• Run over horizontally partitioned data in a distributed environment. | • Computational complexity possible with generalization. |
| 5 | Gosain and Chugh (2014) | Notice and consent | • More secure than anonymity. | • Not powerful and user-friendly.<br>• Impose the burden on individuals. |

*(Continued)*

**Table 8.6 (Continued)  Comparison of Privacy-Preserving Techniques**

| Sl. No. | Frameworks | Method Used | Merits | Demerits |
|---|---|---|---|---|
| 6 | Jain et al. (2016) | Differential privacy | • Robustness against powerful adversaries. <br> • Applicable to a wide range of data analysis problems. | • Better privacy can be achieved with high distraction. |
| 7 | Naehrig et al. (2011) | Partially homomorphic encryption | • Computation is performed on the encrypted text. <br> • More secure. | • Not more practical. <br> • A lot of computational overhead. |
| 8 | Hayward and Chiang (2015) | Fully homomorphic encryption | • Computation is performed on the encrypted text. <br> • More secure. | • Fully homomorphic encryption runs slow. <br> • A lot of computational overhead. |
| 9 | Wong et al. (2006) | $(\alpha, k)$-Anonymity | • Protects both identity and sensitivity relationships in data. | • NP-Hard Problem. <br> • The execution time and distortion ratio depends on the value $\alpha$. |
| 10 | Gupta and Shukla (2016) | $k$-anonymity with a privacy key | • Achieves two level of security. | • More space is required to store the privacy key. |
| 11 | Sedayao et al. (2014) | Identity-based encryption | • Ensures the confidentiality of the user. | • Less efficient in terms of privacy and utility. |
| 12 | Xu et al. (2006a) | Utility-based anonymization | • Produces high utility anonymized data set. <br> • Less information loss. | • Quality of analysis is compromised with an increased level of utility. |

*(Continued)*

**Table 8.6 (*Continued*)   Comparison of Privacy-Preserving Techniques**

| Sl. No. | Frameworks | Method Used | Merits | Demerits |
|---------|-----------|-------------|--------|----------|
| 13 | LeFevre et al. (2006) | Mondrian multidimensional k-Anonymity | • Performs better in terms of runtime and less information loss. | • NP hard. |
| 14 | Dean and Ghemawat (2008) | k-anonymization with MapReduce | • Supports parallelization of privacy-preserving algorithms.<br>• Improves scalability. | • A lot of computational overhead. |
| 15 | Goyal et al. (2006) | Attribute-based anonymization | • Ensures fine-grained sharing of encrypted data over the Internet. | • Complexity is increased. |

Any single method is not sufficient to protect privacy in the area of big data cyber-physical system; also, it will reduce the utility of data. The reduction of data utility is usually represented by *information loss*: Higher information loss means lower utility of the anonymized data. Therefore, we must have the different mechanism to reduce the information loss, to increase the data utility. The comparison in Table 8.7 shows the difference between different privacy-preserving approaches based on privacy, security, utility, scalability, and computational complexity.

**Table 8.7  Comparison of Privacy-Preserving Techniques Based on Privacy, Security, Utility, and Scalability**

| Sl. No. | Framework | Method Used | Security | Privacy | Utility | Scalability |
|---|---|---|---|---|---|---|
| 1 | Sweeney (2002b) | *k*-anonymity | | ✓ | | |
| 2 | Machanavajjhala et al. (2006) | *l*-Diversity | | ✓ | | |
| 3 | Li et al. (2007) | *t*-closeness | | ✓ | | |
| 4 | Nayahi and Kavitha (2015) | (G, S) Clustering for privacy preservation | | ✓ | | ✓ |
| 5 | Gosain and Chugh (2014) | Notice and consent | | ✓ | | ✓ |
| 6 | Jain et al. (2016) | Differential privacy | | ✓. | | |
| 7 | Naehrig et al. (2011) | Partially homomorphic Encryption | ✓ | ✓ | | |
| 8 | Hayward and Chiang (2015) | Fully homomorphic encryption | ✓ | ✓ | | |
| 9 | Wong et al. (2006) | (α, k)-Anonymity | | ✓ | | |
| 10 | Gupta and Shukla (2016) | *k*-anonymity with privacy key | ✓ | ✓ | | |
| 11 | Sedayao et al. (2014) | Identity-based encryption | ✓ | ✓ | | |

*(Continued)*

**Table 8.7 (*Continued*)    Comparison of Privacy-Preserving Techniques Based on Privacy, Security, Utility, and Scalability**

| Sl. No. | Framework | Method Used | Security | Privacy | Utility | Scalability |
|---|---|---|---|---|---|---|
| 12 | Xu et al. (2006a) | Utility-based anonymization | | ✓ | ✓ | |
| 13 | LeFevre et al. (2006) | Mondrian multidimensional k-anonymity | | ✓ | ✓ | |
| 14 | Dean and Ghemawat (2008) | k-anonymization with MapReduce | | ✓ | | ✓ |
| 15 | Goyal et al. (2006) | Attribute-based anonymization | | ✓ | | |

# 8.5 Limitations and Challenges in Big Data Security and Privacy

Except for a few approaches, all existing approaches discussed in Section 8.3 are applicable to traditional data. Data privacy and security are stringent need for data processing in CPS. The traditional approaches have to be enhanced to handle the large volume of unstructured and continuously flowing data in CPS. Scalability is an important challenge to be addressed by big data CPS. Approaches have to be redefined to extract data from heterogeneous data sources and have to be converted either to structured or semi-structured format. k-anonymization and cryptographic techniques would not be scalable for big data, whereas perturbation techniques would be scalable. Moreover, both of these techniques involve computational overhead. Though perturbation techniques are scalable, it would compromise the data utility to achieve privacy. Hence, parallel computation of these algorithms would be beneficial to deal with scalability issues, and it achieves a better trade-off between privacy and utility.

# 8.6 Conclusion

Cyber-physical systems are inevitable in the evolving era of smart devices, smart cities, smart homes, etc. These systems are widely used in a range of domains, including medicine, aerospace, automobiles, smart grids, etc. Data is an integral part of these systems and ensuring the security and privacy of these data is essential.

Various state-of-the-art literature on privacy preservation of data is studied and the merits and demerits of each of these techniques are identified. Anonymization is one of the earliest approaches, and it is more promising to preserve the identity of individuals against attacks. *l*-diversity, *t*-closeness, and other similar variations help to protect the data from identity disclosure as well as attribute disclosure. In order to preserve the data utility while achieving privacy, clustering techniques are integrated with anonymization. Perturbation-based methods can ensure a high degree of security and privacy by compromising the data utility. Homomorphic encryption techniques are used to perform operations on the encrypted data without disclosing the original data. Recently there are hybrid approaches integrating anonymization with encryption techniques, and these techniques are suitable for handling the data of large volume. MapReduce frameworks are also helpful to execute the privacy-preserving algorithms in a parallel manner and hence improve scalability. In general, anonymization would be always worthwhile to prevent the data against identity disclosure but these algorithms should be modified to run in a parallel manner. Differential privacy does not release the entire data and reveals only the answers of queries in a secured manner. This technique can be used to preserve the outputs of standing queries and ad hoc queries in data streams. Hybrid approaches including the merits of all the techniques would be beneficial when scalability, privacy, security, and data utility are taken care of.

# References

Adam, Nabil R., and John C. Worthmann. 1989. "Security-Control Methods for Statistical Databases: A Comparative Study." *ACM Comput. Surv.*, 21 (4): 515–556. doi:10.1145/76894.76895.

Aggarwal, Gagan, Tomás Feder, Krishnaram Kenthapadi, Samir Khuller, Rina Panigrahy, Dilys Thomas, and An Zhu. 2006. "Achieving Anonymity via Clustering." In *Proceedings of the Twenty-Fifth ACM SIGMOD-SIGACT-SIGART Symposium on Principles of Database Systems, PODS'06*, pp. 153–162. New York: ACM. doi:10.1145/1142351.1142374.

Aggarwal, Gagan, Tomás Feder, Krishnaram Kenthapadi, Rajeev Motwani, Rina Panigrahy, Dilys Thomas, and An Zhu. 2005. "Anonymizing tables." In *Proceedings of the 10th International Conference on Database Theory, ICDT'05*, pp. 246–258. Berlin, Germany: Springer-Verlag. doi:10.1007/978-3-540-30570-5_17.

Alguliyev, Rasim, Yadigar Imamverdiyev, and Lyudmila Sukhostat. 2018. "Cyber-Physical Systems and Their Security Issues." *Comput. Ind.*, 100: 212–223. doi:10.1016/j.compind.2018.04.017.

Altawy, Riham, and Amr M. Youssef. 2016. "Security Tradeoffs in Cyber Physical Systems: A Case Study Survey on Implantable Medical Devices." *IEEE Access*, 4: 959–979. doi:10.1109/ACCESS.2016.2521727.

Bethencourt, John, Amit Sahai, and Brent Waters. 2007. "Ciphertext-Policy Attribute-Based Encryption." In *2007 IEEE Symposium on Security and Privacy (SP'07)*, pp. 321–334. doi:10.1109/SP.2007.11.

Bredereck, Robert, André Nichterlein, Rolf Niedermeier, and Geevarghese Philip. 2011. "The Effect of Homogeneity on the Complexity of *k*-Anonymity." In Olaf Owe, Martin Steffen, and Jan Arne Telle (Eds.), *Fundamentals of Computation Theory*, pp. 53–64. Berlin, Germany: Springer.

Camara, Carmen, Pedro Peris-Lopez, and Juan E. Tapiador. 2015. "Security and Privacy Issues in Implantable Medical Devices: A Comprehensive Survey." *J. Biomed. Inform.*, 55: 272–89. doi:10.1016/j.jbi.2015.04.007.

Cardenas, Alvaro A., Saurabh Amin, and Shankar Sastry. 2008. "Secure Control: Towards Survivable Cyber-Physical Systems." In *2008 The 28th International Conference on Distributed Computing Systems Workshops*, pp. 495–500. doi:10.1109/ICDCS. Workshops.2008.40.

Chaudhari, Neetu, and Satyajee Srivastava. 2016. "Big Data Security Issues and Challenges." In *2016 International Conference on Computing, Communication and Automation (ICCCA)*, pp. 60–64. doi:10.1109/CCAA.2016.7813690.

Dean, Jeffrey, and Sanjay Ghemawat. 2008. "MapReduce: Simplified Data Processing on Large Clusters." *Commun. ACM*, 51 (1): 107–113. doi:10.1145/1327452.1327492.

Ghinita, Gabriel, Panos Kalnis, and Yufei Tao. 2011. "Anonymous Publication of Sensitive Transactional Data." *IEEE Trans. Knowl. Data Eng.*, 23 (2): 161–174. doi:10.1109/ TKDE.2010.101.

Giraldo, Jairo, Esha Sarkar, Alvaro A. Cardenas, Michail Maniatakos, and Murat Kantarcioglu. 2017. "Security and Privacy in Cyber-Physical Systems: A Survey of Surveys." *IEEE Des. Test*, 34 (4): 7–17. doi:10.1109/MDAT.2017.2709310.

Gosain, Anjana, and Nikita Chugh. 2014. "Privacy Preservation in Big Data." *Int. J. Comput. Appl.*, 100 (17): 44–47. doi:10.5120/17619-8322.

Govinda, Kannayaram, and Perla Ravitheja. 2012. "Identity Anonymization and Secure Data Storage Using Group Signature in Private Cloud." In *Proceedings of the International Conference on Advances in Computing, Communications and Informatics, ICACCI'12*, pp. 129–132. New York: ACM. doi:10.1145/2345396.2345418.

Goyal, Vipul, Omkant Pandey, Amit Sahai, and Brent Waters. 2006. "Attribute-Based Encryption for Fine-Grained Access Control of Encrypted Data." In *Proceedings of the 13th ACM Conference on Computer and Communications Security, CCS'06*, pp. 89–98. New York: ACM. doi:10.1145/1180405.1180418.

Guan, Zhitao, and Guanlin Si. 2017. "Achieving Privacy-Preserving Big Data Aggregation with Fault Tolerance in Smart Grid." *Digital Commun. Netw.*, 3 (4): 242–249. doi:10.1016/j.dcan.2017.08.005.

Gupta, Amit Kumar, and Neeraj Shukla. 2016. "Privacy Preservation in Big Data Using *k*-Anonymity Algorithm with Privacy Key." *Int. J. Comput. Appl.*, 153: 25–30.

Hayward, Ryan, and Chia-Chu Chiang. 2015. "Parallelizing Fully Homomorphic Encryption for a Cloud Environment." *J. Appl. Res. Technol.*, 13 (2): 245–252.

HIPAA. 1999. "HIPAA Health Insurance Portability and Accountability Act of 1999." http://www.hhs.gov/ocr/privacy/hipaa/administrative/privacyrule.

Jain, Priyank, Manasi Gyanchandani, and Nilay Khare. 2016. "Big Data Privacy: A Technological Perspective and Review." *J. Big Data*, 3 (1): 25. doi:10.1186/s40537-016-0059-y.

Karthikeyan, P., Jayavel Amudhavel, A. Abraham, Dananjayan Sathian, R. S. Raghav, and P. Dhavachelvan. 2015. "A Comprehensive Survey on Variants and Its Extensions of Big Data in Cloud Environment." In *Proceedings of the 2015 International Conference on Advanced Research in Computer Science Engineering & Technology (ICARCSET 2015), ICARCSET'15*, pp. 32:1–32:5. New York: ACM. doi:10.1145/2743065.2743097.

Kenyon, Tony. 2018. "5—Transportation Cyber-Physical Systems Security and Privacy." In Lipika Deka and Mashrur Chowdhury (Eds.), *Transportation Cyber-Physical Systems Chowdhury*, pp. 115–151. Elsevier. doi:10.1016/B978-0-12-814295-0.00005-8.

Lämmel, Ralf. 2008. "Google's MapReduce Programming Model—Revisited." *Sci. Comput. Program.*, 70 (1): 1–30. doi:10.1016/j.scico.2007.07.001.

LeFevre, Kristen, David J. DeWitt, and Raghu Ramakrishnan. 2006. "Mondrian Multidimensional *k*-Anonymity." In *22nd International Conference on Data Engineering (ICDE'06)*, 25. doi:10.1109/ICDE.2006.101.

Li, Ninghui, Tiancheng Li, and Suresh Venkatasubramanian. 2007. "*t*-Closeness: Privacy Beyond *k*-Anonymity and *l*-Diversity." In *2007 IEEE 23rd International Conference on Data Engineering*, pp. 106–115. doi:10.1109/ICDE.2007.367856.

Machanavajjhala, Ashwin, Johannes Gehrke, Daniel Kifer, and Muthuramakrishnan Venkitasubramaniam. 2006. "\ell-Diversity: Privacy Beyond\kappa-Anonymity." In *22nd International Conference on Data Engineering (ICDE'06)*, p. 24.

Matturdi, Bardi, Xianwei Zhou, Shuai Li, and Fuhong Lin. 2014. "Big Data Security and Privacy: A Review." *China Commun.*, 11 (14): 135–145. doi:10.1109/CC.2014.7085614.

Mendes, Ricardo, and Joao P. Vilela. 2017. "Privacy-Preserving Data Mining: Methods, Metrics, and Applications." *IEEE Access*, 5: 10562–10582. doi:10.1109/ACCESS.2017.2706947.

Meyerson, Adam, and Ryan Williams. 2004. "On the Complexity of Optimal *k*-Anonymity." In *Proceedings of the Twenty-Third ACM SIGMOD-SIGACT-SIGART Symposium on Principles of Database Systems, PODS'04*, pp. 223–228. New York: ACM. doi:10.1145/1055558.1055591.

Microsoft. 2015. "Microsoft Differential Privacy for Everyone [Online]." http://download. microsoft.com/.../Differential_Privacy_for_Everyone.Pdf.

Mitchell, Robert, and Ing-Ray Chen. 2014. "A Survey of Intrusion Detection Techniques for Cyber-Physical Systems." *ACM Comput. Surv.*, 46 (4): 55:1–55:29. doi:10.1145/2542049.

Mourtzis, Dimitris, and Ekaterini Vlachou. 2016. "Cloud-Based Cyber-Physical Systems and Quality of Services." *TQM J.*, 28 (5): 704–733. doi:10.1108/TQM-10-2015-0133.

Mueller, Frank. 2006. "Challenges for Cyber-Physical Systems: Security, Timing Analysis and Soft Error Protection." In *High-Confidence Software Platforms for Cyber-Physical Systems (HCSP-CPS) Workshop*.

Naehrig, Michael, Kristin Lauter, and Vinod Vaikuntanathan. 2011. "Can Homomorphic Encryption Be Practical?" In *Proceedings of the 3rd ACM Workshop on Cloud Computing Security Workshop, CCSW'11*, pp. 113–124. New York: ACM. doi:10.1145/2046660.2046682.

Nayahi, J. Jesu Vedha, and V. Kavitha. 2015. "An Efficient Clustering for Anonymizing Data and Protecting Sensitive Labels." *Int. J. Uncertainty Fuzziness Knowl.-Based Syst.*, 23 (5): 685–714. doi:10.1142/S0218488515500300.

Neela, T. Jothi, and N. Saravanan. 2013. "Privacy Preserving Approaches in Cloud: A Survey." *Indian J. Sci. Technol.*, 6 (5): 4531–4535.

Ogburn, Monique, Claude Turner, and Pushkar Dahal. 2013. "Homomorphic Encryption." *Procedia Comput. Sci.*, 20: 502–509. doi:10.1016/j.procs.2013.09.310.

Ostrovsky, Rafail, Amit Sahai, and Brent Waters. 2007. "Attribute-Based Encryption with Non-monotonic Access Structures." In *Proceedings of the 14th ACM Conference on Computer and Communications Security, CCS'07*, pp. 195–203. New York: ACM. doi:10.1145/1315245.1315270.

Pan, Yun, Xiao Ling Zhu, and Ting Gui Chen. 2012. "Research on Privacy Preserving on *k*-Anonymity." *J. Softw.*, 7 (7): 1649–1656. doi:10.4304/jsw.7.7.1649-1656.

Potey, Manish M., Chandrashekhar A. Dhote, and Deepak H. Sharma. 2016. "Homomorphic Encryption for Security of Cloud Data." *Procedia Comput. Sci.*, 79: 175–181. doi:10.1016/j.procs.2016.03.023.

Raghunathan, Trivellore E., Jerome P. Reiter, and Donald B. Rubin. 2003. "Multiple Imputation for Statistical Disclosure Limitation." *J. Off. Stat.*, 19 (1): 1–16. doi:10.1007/s13398-014-0173-7.2.

Rahul, Mohd, Hesham A. Alhumyani, Mohd Muntjir, and Minashi Kambojl. 2017. "An Improved Homomorphic Encryption for Secure Cloud Data Storage." *Int. J. Adv. Comput. Sci. Appl.*, 8 (12): 441–446. doi:10.14569/IJACSA.2017.081258.

Reddy, Yenumula B. 2014. "Cloud-Based Cyber Physical Systems: Design Challenges and Security Needs." In *2014 10th International Conference on Mobile Ad-Hoc and Sensor Networks*, pp. 315–322. doi:10.1109/MSN.2014.50.

Reiss, Steven P. 1984. "Practical Data-Swapping: The First Steps." *ACM Trans. Database Syst.*, 9 (1): 20–37. doi:10.1145/348.349.

Reiss, Steven P., Mark J. Post, and Tore Dalenius. 1982. "Non-reversible Privacy Transformations." In *Proceedings of the 1st ACM SIGACT-SIGMOD Symposium on Principles of Database Systems, PODS'82*, pp. 139–146. New York: ACM. doi:10.1145/588111.588134.

Rushanan, Michael, Aviel D. Rubin, Denis Foo Kune, and Colleen M. Swanson. 2014. "SoK: Security and Privacy in Implantable Medical Devices and Body Area Networks." In *Proceedings—IEEE Symposium on Security and Privacy*, pp. 524–539. doi:10.1109/SP.2014.40.

Sadeghi, Ahmad-Reza, Christian Wachsmann, and Michael Waidner. 2015. "Security and Privacy Challenges in Industrial Internet of Things." In *2015 52nd ACM/EDAC/IEEE Design Automation Conference (DAC)*, pp. 1–6. doi:10.1145/2744769.2747942.

Salini, Soman, Sreetha V. Kumar, and R. Neevan. n.d. "An Improved and Efficient Data Privacy in Big Data with K-Anonymity and Alpha Dissociation."

Samarati, Pierangela, and Latanya Sweeney. 1998. "Protecting Privacy When Disclosing Information: K-Anonymity and Its Enforcement Through Generalization and Suppresion." In *Proceedings of the IEEE Symposium on Research in Security and Privacy*, pp. 384–393. doi:10.1145/1150402.1150499.

Sanislav, Teodora, George Mois, Silviu Folea, Liviu Miclea, Giulio Gambardella, and Paolo Prinetto. 2014. "A Cloud-Based Cyber-Physical System for Environmental Monitoring." In *2014 3rd Mediterranean Conference on Embedded Computing (MECO)*, pp. 6–9. doi:10.1109/MECO.2014.6862654.

Sedayao, Jeff, Rahul Bhardwaj, and Nakul Gorade. 2014. "Making Big Data, Privacy, and Anonymization Work Together in the Enterprise: Experiences and Issues." In *2014 IEEE International Congress on Big Data*, pp. 601–607. doi:10.1109/BigData.Congress.2014.92.

Shvachko, Konstantin, Hairong Kuang, Sanjay Radia, and Robert Chansler. 2010. "The Hadoop Distributed File System." In *2010 IEEE 26th Symposium on Mass Storage Systems and Technologies (MSST)*, pp. 1–10. doi:10.1109/MSST.2010.5496972.

Song, Houbing, Glenn A. Fink, and Sabina Jeschke (Eds.). 2017. *Security and Privacy in Cyber-Physical Systems: Foundations, Principles, and Applications*. Chichester, UK: John Wiley & Sons.

Sowmya, Y., and M. Nagaratna. 2016. "Parallelizing *k*-Anonymity Algorithm for Privacy Preserving Knowledge Discovery from Big Data." *Int. J. Appl. Eng. Res.*, 11 (2): 1314–1321.

Sun, Xiaoxun, Min Li, and Hua Wang. 2011. "A Family of Enhanced (L,α)-Diversity Models for Privacy Preserving Data Publishing." *Future Gener. Comput. Syst.*, 27 (3): 348–356. doi:10.1016/j.future.2010.07.007.

Sweeney, Latanya. 2002a. "Achieving *k*-Anonymity Privacy Protection Using Generalization and Suppression." *Int. J. Uncertainty Fuzziness Knowl.-Based Syst.*, 10 (5): 571–588. doi:10.1142/S021848850200165X.

Sweeney, Latanya. 2002b. "*k*-Anonymity: A Model For Protecting Privacy." *Int. J. Uncertainty Fuzziness Knowl.-Based Syst.*, 10 (5): 557–570. doi:10.1142/S0218488502001648.

Tian, Hongwei, and Weining Zhang. 2011. "Extending ℓ-Diversity to Generalize Sensitive Data." *Data Knowl. Eng.*, 70 (1): 101–126. doi:10.1016/j.datak.2010.09.001.

Waters, Brent. 2011. "Ciphertext-Policy Attribute-Based Encryption: An Expressive, Efficient, and Provably Secure Realization." In Dario Catalano, Nelly Fazio, Rosario Gennaro et al. (Eds.), *Public Key Cryptography—PKC 2011*, pp. 53–70. Berlin, Germany: Springer.

Wong, Raymond Chi-Wing, Jiuyong Li, Ada Wai-Chee Fu, and Ke Wang. 2006. "(α,k)-Anonymity: An Enhanced *k*-Anonymity Model for Privacy Preserving Data Publishing." In *Proceedings of the 12th ACM SIGKDD International Conference on Knowledge Discovery and Data Mining, KDD'06*, pp. 754–759. New York: ACM. doi:10.1145/1150402.1150499.

Wu, Dapeng, Boran Yang, and Ruyan Wang. 2016. "Scalable Privacy-Preserving Big Data Aggregation Mechanism." *Digital Commun. Netw.*, 2 (3): 122–129. doi:10.1016/j.dcan.2016.07.001.

Wu, Xindong, Xingquan Zhu, Gong-Qing Wu, and Wei Ding. 2014. "Data Mining with Big Data." *IEEE Trans. Knowl. Data Eng.*, 26 (1): 97–107. doi:10.1109/TKDE.2013.109.

Xu, Jian, Wei Wang, Jian Pei, Xiaoyuan Wang, Baile Shi, and Ada Wai-Chee Fu. 2006a. "Utility-Based Anonymization for Privacy Preservation with Less Information Loss." *SIGKDD Explor. Newsl.*, 8 (2): 21–30. doi:10.1145/1233321.1233324.

Xu, Jian, Wei Wang, Jian Pei, Xiaoyuan Wang, Baile Shi, and Ada Wai-Chee Fu. 2006b. "Utility-Based Anonymization Using Local Recoding." In *Proceedings of the 12th ACM SIGKDD International Conference on Knowledge Discovery and Data Mining, KDD'06*, pp. 785–790. New York: ACM. doi:10.1145/1150402.1150504.

Yang, Yang, Xianghan Zheng, Wenzhong Guo, Ximeng Liu, and Victor Chang. 2018. "Privacy-Preserving Fusion of IoT and Big Data for e-Health." *Future Gener. Comput. Syst.*, 86: 1437–1455. doi:10.1016/j.future.2018.01.003.

# Chapter 9

# Big Data Technologies–Supported Generic Visualization System in an Enterprise Cyber-Physical Environment

Ferda Özdemir Sönmez and Banu Günel

## Contents

## 9.1 Introduction

Since the 1940s, the variety and use-cases of information and communication devices have grown high. Initial computers converted to PCs with network connections. PCs changed to laptops with wireless adapters. Mobile phones and other GPS devices take the place of laptops from time to time. PDAs and smartphones dominate everything both in private and business life. And, lastly, the functionalities that wearable devices offer increases day by day. The computational efforts changed over time from mainframe computing, desktop computing, ubiquitous computing, and Cyber-Physical Systems (CPSs). CPSs are engineered systems, which combine computing, communication, and data storage capabilities with the aim of coordinating, monitoring, and controlling environmental entities.

There are many projects done so far which focus on providing new computational ways for various cyber-physical environments. The National Institute of Standards and Technology, NIST, had a project named Smart Space (Fillinger et al. 2009). As a part of this project, they developed a Meeting Room Recognition application (Stanford et al. 2003) which forms a prototype for future meeting rooms, and command control centers. Future Computing Environments Group in Georgia Institute of Technology developed the Classroom 2000 project (Abowd et al. 1996) as a prototype to monitor the effects of the impact of ubiquitous computing in education. Interactive Workspaces Project (Johanson et al. 2004) by Stanford University investigated new technologies to form a multi-person, multi-device, collaborative working place.

It is a known fact that the cost of not giving enough importance to information security may be very high. To prevent these losses, each organization should prioritize information security management. The primary sources of security management in an enterprise are log files, alerts produced by security systems and devices, and network traffic data. In this work, an enterprise network with its information security–related infrastructure elements will be the focus. Security visualization outlines may have diverse purposes, for example, condensing the information, reenacting past occurrences, permitting design disclosure, the location of malignant exercises, inconsistencies, misconfigurations, and anomalies. Security visualization may give various perspectives of similar information all the while or it might picture diverse information in a similar view. The aim of this study is to create a design which targets to use the data produced by environmental entities and allows the creation of visualizations. Existing security visualization systems are mostly rigidly attached to the log file type. The proposed system is based on generic parsers for each file format type (JSON, CSV, and TXT). The proposed system also enables feedback from the users, such as systems analysts, operation center users, heads of departments, or senior managers.

There are many data sources and many visualization tools. In order to use a visualization tool for a specific data source a considerable amount of effort is required. A generic visualization system which may visualize data in multiple forms with little preparation, based on metadata and selection of visualization type on the fly would be very beneficial. Some of the proposed features are distinctive for a security visualization solution. A security visualization knowledge base is aimed to be formed through the use of user feedbacks in the proposed system. These feedbacks will form an enterprise security visualization knowledge base and may also be used for automatic processing for various purposes. This knowledge base is expected to accelerate the learning and to help the creation of more successful visualizations in the future.

During this study, a set of requirements is prepared to specify the proposed system. The main source of the defined requirements is the results of a requirements analysis survey. The reason for making such an inquiry was to enable a user-oriented specification set for this domain. The size of the enterprise, a set of business processes, and infrastructure of the organization would directly affect the visualization tasks. The target is to provide a set of security visualization requirements for all type of organizations. It does not point out a specific organization. However, the restricted scope depends on the outputs of the survey and common design features from the literature search results. Thirty participants with various levels of solid information security knowledge and experience attended this survey. The raw results of the survey are not included due to space limitations. However, for each requirement or group of requirements, the rationality of the requirement is described briefly. This rationality may include reference to the parts of security visualization survey results or the known issues from literature. The authors also elicited these requirements based on applicability, consistency with other needs, and compatibility with the overall structure of the projected enterprise visualization solution.

Being a cyber-physical system, the human entities of the selected case may be technical such as system operators and security analysts, and nontechnical such as nontechnical managers or report writers. The environmental entities for the proposed system are the enterprise infrastructure elements, such as hardware or software firewalls, honeynets, intrusion detection systems, and operating systems. The security-related outputs of these environmental entities are aimed to be monitored by the use of visualizations prepared by human entities.

One of the main concerns related to developing a visualization system for an enterprise is the existence of a variety of data formats and data types which are originated from various devices. Standardization is required to use the outputs of these devices. There have been some attempts to standardize the log files which are Extended Log File Format by W3C (Hallam-Baker and Behlendorf 1996), Common Event Format by ArcSight (Arcsight 2009), Syslog by IETF (Gerhards 2009), IDMEF by IETF (Debar et al. 2007), and SDEE/CIDEE by Cisco (Cisco 2009). While some of these standardization attempts are outdated, the others are in their early stages. The lack of standardization for these files is a serious problem for the security visualization efforts (Marty 2009; Chuvakin et al. 2012). Another important concern is the scalability of these systems. As time passes the number of data accumulated can grow very high and may require specific storage and computational requirements. Creating a visualization depends on many factors. There are a variety of issues, issues related to data preparation and issues related to selecting correct display type. In this paper, the concerns of the security visualization domain are not the main subject. The focus will be mostly the features of the proposed system, but while describing and discussing these features the concerns of the domain will also be mentioned to some extent. In order to gain more knowledge on the security visualization domain please refer to Sonmez and Gunel's extended review (2018).

Similar to all CPSs, the proposed system has some other (not domain-specific) concerns, such as security, privacy, fault tolerance, safety, and reliability. Some of these concerns will be satisfied with the properties of the proposed system. Others will be fulfilled by the features of underlying infrastructure elements. The modular structure and service-based design of the proposed system allow it for further progress and adding new features. It has advantages regarding genericness and scalability. The safety of infrastructure elements is left to the enterprise policies. Physical protection mechanisms may be included whenever possible to increase the safety and physical security of enterprise network infrastructure. Other security concerns and reliability concerns, such as timing and ordering of events in a distributed system and fault tolerance, are fulfilled by depending on industry standard technologies and architectures. These issues are depicted more in the design description.

The aim of this study can be briefly summarized as to form the design of a generic security visualization system which is capable of visualizing data coming from multiple sources in an enterprise and which forms a knowledge base as the result of these visualization tasks. Thus, the enterprise is defined as a CPS first. The provided design can be applied for any enterprise with varying types and sizes as long as it possesses the necessary infrastructure suitable for its data. The security-related data sources supported with the provided design are limited to structured and uncompressed CSV, TXT, and JSON files. Other data sources can be included using third-party parser tools. The visualization systems in the scope are the JavaScript-based visualization systems which are presented via the HTML pages.

The rest of the chapter is constructed as follows. Section 9.2 includes the description of an enterprise network as a ubiquitous environment. Section 9.3 has the functional requirements for the proposed system. Section 9.4 has the initial design features, and Section 9.5 has the improvements made by integrating the initial design with big data technologies. Section 9.6 includes the results achieved. Sections 9.7 and 9.8 are devoted to the discussion and concluding remarks respectively.

## 9.2 Description of an Enterprise Network as a Ubiquitous Environment

Enterprise security is defined by Sherwood as protecting business goals and assets for an enterprise (Sherwood et al. 2005). The security requirements will have huge differences among various enterprises due to the dependence of enterprise security on many factors such as criticality of business models, number and types of internal and external users, size of the data stored, types of business and/or infrastructure protection software, hardware used, and network architecture. The approaches to information security management also vary from organization to organization.

The current trends of the enterprises help the growth of potential risks such as moving to e-business, increased mobility, fast and flexible change management, and cloud computing. While the majority of the threats originate from the Internet for an organization, there may also be malicious actions that originate from insiders. Vendors of systems and devices offer specific precautions. These precautions have a variable set of behaviors and produce log files, which are far from being processed using standard procedures in different organizations.

The owners of the IT investments in an enterprise may assume that everything looks perfect after building up the IT infrastructure by spending a non-ignorable amount of money. However, they will eventually understand that the vendors of the hardware and software elements of the IT infrastructure usually undertake the security issue. The best security solution is as good as how it is administered. Enterprise network and security management may include many devices and tools layered in various layers of the network. Enterprise infrastructure and devices are pictured well and in detail in Shin (2017). In order to have a base to examine enterprise-level security further, a possible set of hardware and software solutions is illustrated in Figure 9.1, which forms a sample IT infrastructure model for an enterprise.

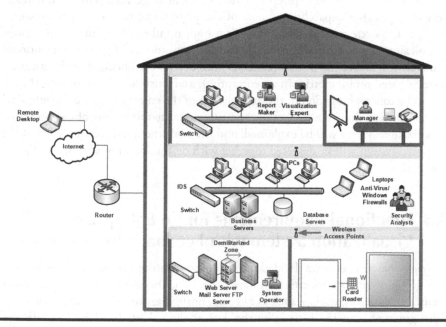

**Figure 9.1   Typical IT infrastructure of an organization.**

In a cyber-physical environment, close coupling of cyber and physical devices is required. In the proposed system, the data produced by physical and computerized systems are aimed to be processed by cyber systems. To put it concretely, data to be treated may be from the intrusion detection system, card reader devices, or firewall logs. Although in the picture mostly familiar structures are shown, the limit is the networking and processing capability of the enterprise; thus, other devices which have other cyber capabilities, such as active RFID tags to protect business-critical assets, sensors to monitor the heat, humidity in the system operation room, and biometric access control systems may as well be in the picture.

A generic visualization system which is capable of storing and visualizing data coming from multiple sources will result in many benefits in such an enterprise. Due to the increase of data in information technologies, visualization has become a popular technique for analyzing, communicating, and decision making for big data. Using visualization in the security domain is a relatively new research area. The first published work was in 2004. The major reason for the emergence of security visualization is the necessity of analyzing security-related vast data on time. Security visualizations enable human assessment of large-size log files efficiently, which results in rapid and improved decision making. Marty (2009) described the benefits of using visualization in the security domain, as "it answers questions, it poses new questions, it allows exploration and discovery, it supports decisions, it communicates information, it increases efficiency, and it inspires."

Characteristics of cyber-physical systems include large-scale wired and wireless networking, cyber capability in most of the physical components, the networking speed in extreme scales, existence of a high number of systems with various complexity varying from simple to too complex, existence of some unconventional systems with cyber-physical capabilities (for example embedded authentication systems based on biometrics in an enterprise), and existence of nontechnical people in the control loop. So far, the explanation of the enterprise as a cyber-physical system is made. In the next section, the functional requirements of the proposed visualization system will be explained, and the nonfunctional requirements of the proposed system will be depicted together with design features, and in the discussion part.

## 9.3 Functional Requirements for an Enterprise Visualization System with Feedback from Users

Security analyses tasks are divided into three consecutive groups of activities. The first group of activities focuses on data collection; the second group of activities focuses on data preparation, such as filtering, normalization, and sampling; and the third group of activities focuses on the data analysis. Visualization is an effective way of data analysis. As mentioned before, prior to the system design, requirements are identified through the use of examination of enterprise needs, literature search, and inquiry results. During the preparation of these requirements, data preparation tasks such as cleansing, conversion, formatting, and normalization are excluded. These are earlier undertakings which require manual reasoning, and hence they did not fit as a part of an automated structure.

The requirements captured can be grouped into two categories. The first group is predominantly specific to the proposed design, and the second group is either already implemented by other studies in different ways or are known design issues. A few of the requirements are included for the sake of completeness of the design. In the text below, for each requirement, a number is provided in parentheses for traceability during design and validation.

A display type library is the first feature detected (Req. 1). It is designed to be in a form to store information related to display types aiming toward proper use and selection of various display types. Most of the enterprise users are not experts of display types or visualization technologies. The level of visualization knowledge is not questioned explicitly during the survey. Nevertheless, a set of display type thumbnail views were included in the survey content. Some participants asked simple questions regarding these display types, which shows that although they have expertise in the security domain, they need more support on display types. Each display type is powerful to exhibit some data classes. For example, scatter plots are more effectual to display large datasets. The reason is each point allocates small space, and this allows visualization of extensive data in a small space in scatter plots. Treemaps are more suitable to display hierarchical data. Departmental data, data coming from hierarchical network devices, may be more appropriate to be visualized with this kind of display. Circular display types allow visualization of data including what, how, when, and where forms of information, such as events and alerts occurred in specific devices/hosts. A dictionary-like platform including such information is required.

Ability to read data in multiple various formats (Req. 2) such as JSON, TXT, and PCAP files are required due to the high variety of security data sources. During the survey, the usefulness of examining 12 independent data sources was asked. The results show that all 12 data sources were nearly equally crucial to the enterprises. It is not desired that security analysts should give importance to one or two data sources and leave the others unanalyzed. Each of these sources has specific formats. Formats also change based on the brand of the security systems. Being ready for such a diverse set of security data sources is difficult for most of the enterprises. The system should facilitate the addition of a new type of security data sources for examination and visualization purposes (Req. 3).

The survey results show that users are not familiar enough about their security data resources for their organizations, which may affect the data preparation tasks. Survey results show that a platform regarding sharing such information among users may be very beneficial. In this platform, the information such as file locations, file access information, excepted file sizes, the frequencies of renewal for looped files, data formats, responsibilities, tasks of staff regarding analyses of these log files and experiences can be shared (Req. 4). These feedbacks from the users can also be used for the automatic creation of visualizations (as a future work). During the design phase, how these feedbacks from users can be used to form a closed control loop in an enterprise which is essential for a cyber-physical environment will also be discussed.

A generic metric definition system is also an inherent requirement (Req. 5). Not all security visualization solutions enable the creation of data queries during runtime. Most of them run on predefined metrics. On the other hand, there are many attributes for each mentioned data source. These attributes are meaningful in specific ways. For example, some attributes between intervals, count of some qualities, min or max of some traits, and the set of some characteristics might be meaningful for various purposes. Allowing the user to define the queries for these attributes easily during runtime would result in more user-centric metrics, rather than the predefined ones. Some of the general purpose visualization tools have excellent properties of forming generic user queries, and some enable selection of display types on the fly using very sophisticated user interactions such as drag and drop. However, these tools lack security perspective and do not help to form an enterprise security knowledge base while visualizing the data.

Threats definition system (Req. 6) is another requirement detected by the survey. During the requirements analysis survey, the participants were asked to group analyses that they make using security data sources to monitor or detect a set of threats. As a continuation of this question, they were also asked to associate the attributes such as "number of events in a time duration," "types of alerts," and "list of source IPs" to the threat sets for each analysis group. Using right associations of visualizations to threats and other visualization purposes will be the key to success for the analysts. If people know what to do or what to look for, they are more likely to succeed. However, the survey showed that people have issues related to making these decisions. Although all of the survey participants were familiar with well-known threats, some other risks are not very well known. Admitting that they knew the threat mechanism, nearly the majority of them had difficulty associating one threat with a particular data source or with a particular data source attribute, resulting in many illogical associations of data sources to the threats or data attributes to the threats.

Similar to the feedback from the users related to security data sources recognized before, storing associations of threats to the visualizations (Req. 7) and associations of threats to data sources (Req. 8) in the knowledge base is useful. The problem is to find correct associations. When asked theoretically as in the survey, people are having difficulty in accurately giving answers to these questions. Showing something concrete, as the created visualizations, may allow getting more effective, essential feedback.

Examining data through the use of visualizations to seek possible threats is one purpose itself. Purpose definition system (Req. 9) is also included in requirements to enable the definition of other intents. Even some infrastructure elements may have different intents when installing differently. For example, a web server may serve to intranet users or Internet users; a firewall may protect the overall organization or a department; a honey network may be used for protection or educational purposes. As the size of the enterprise increases, there may be multiple installations of the same hardware and software elements which have different purposes. Knowing the goals of these infrastructure elements may result in a better analysis of the generated data. Similar to the threats, purposes can be associated with data sources (Req. 10) or visualizations (Req. 11). This requirement arose due to the investigation of data sources. Besides associating threats and purposes to the visualizations, users are free to give other feedback to the viewings (Req. 12). These feedbacks can also be automatically processed (as a future work).

Use of the various type of displays with different complexity levels (Req. 13) is another requirement. The literature study showed that existing studies used various display types. Some display types are more mainstream contrasted with others. Survey results point that although users have an interest in more complicated charts, such as 3-D charts, results indicate that simple charts are easier to understand and have higher usability for the majority of the participants. However, some of the more complex display types, such as parallel coordinates, are more proper for some specific cases. Association of display types with the generic data definitions on the fly is required (Req. 14). Selection of the display type during runtime is a feature which exists mostly in some of the dashboard designs. For the sake of completeness, it was necessary to combine this feature with the generic data file and generic data metric definitions in the proposed design.

Assigning difficulty levels to the display types is also necessary (Req. 15). Not every display type has a similar difficulty level. This requirement arose due to the same rationality as in display type descriptions. The majority of the security experts are not visualization experts, so the association of difficulty levels to the display types may allow better selection of displays during the visualization tasks for inexperienced users.

Easy access to external visualization tools is required (Req. 16). During the survey, the participants were asked about their familiarity with a long list of security visualization solutions. Unfortunately, the participants were familiar with only a few of these visualization tools. These are the visualization tools which are used in conjunction with other tools such as scanner tools. Due to this reason, encapsulating access information for such products, such as links to websites to download or use (for online tools), may be beneficial to increase familiarity with these tools for the enterprise users.

There are requirements which are captured through the literature search. These were also questioned during the survey. Having the ability to depict relatively large data (Req. 17), the ability to represent data from more than one security log simultaneously (Req. 18), the ability to save detected patterns (Req. 19), the ability to work with real-time data (Req. 20), and the ability to depict most types of attacks (Req. 21) are some significant requirements among them. A few of the requirements from the literature search are mainly related to display-type technology selection. Displaying incident time (Req. 22), having thick boundaries to separate different classes of information (Req. 23), ability to use the visualization without mouse (Req. 24), being interactive (Req. 25), being searchable (Req. 26), being zoomable (Req. 27), and being scalable in terms of the amount of data displayed (Req. 28) are in this group.

Besides the listed ones, some design decisions are made based on technical background, industrial development standards, and the properties of existing display libraries, which provide a high level of diversity in form and difficulty. Due to the variety and richness of JavaScript-based display-type technologies, using JavaScript-based display-type technology was decided. Java development language is selected due to background knowledge and its high compatibility with the mentioned display-type libraries, and a web-based design is to enable easy access from any computing device for an enterprise.

## 9.4 Initial Design Features

There are many concerns related to the design and development of cyber-physical models. While deterministic models allow the creation of more robust models, it is not possible or feasible to create a deterministic model consisting of many parts in a distributed manner at all times (Lee 2015). Thus, the authors decided to implement a prototype which may be used for the materialization of some design problems through trial and error.

In top-down design, the modeler starts with the domain at large and starts with the design of the upper-level application modules and divides the top level structure into smaller pieces, generally in the form of classes. This results with classes with either complex or coercion relations. In the bottom-up approach, the design of the necessary data structures is completed first. Later, first simple, then more complex functions are added to form an integrated structure to process the anticipated design structure and to fulfill the requirements. A bottom-up approach is taken during the prototype design. The data entities are extracted from the requirements, and the data model shown in Figure 9.2 is created as a first step. The detailed structure and Java code of these data entities will be shared on Git-Hub under the name of "Data Entities for Generic Security Visualization Solution with Knowledge Base." As a next step, the functions and data structures are mapped to modules. For each module, the front-end and back-end classes are implemented later.

The overall functionality, including relations to data structures, is described in the state diagram shown in Figure 9.3. The solid lines correspond to service calls

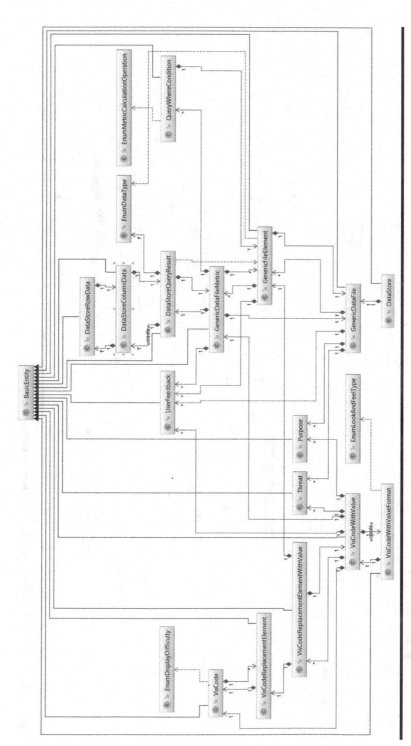

**Figure 9.2  Class diagram of the proposed structure.**

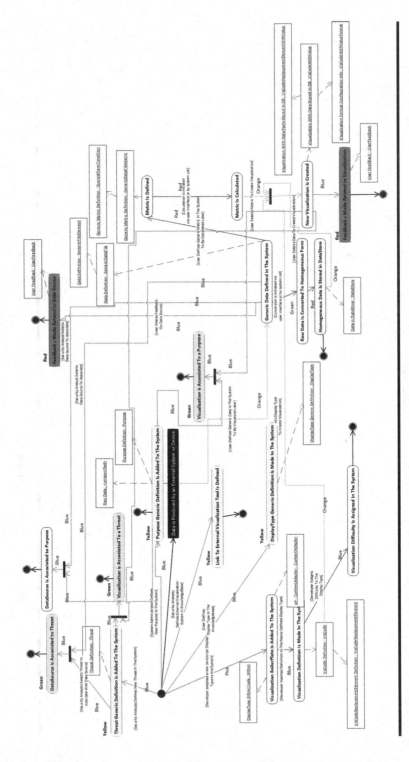

**Figure 9.3   States and activities of the proposed system.**

which transform the data into another state. The dashed lines are used to show the relations of the data structures from the proposed design. These boilerplates are based on external JavaScript-based visualization libraries.

The brief explanation of how each part of the data structure is associated with the features of the proposed system is as follows. *BasicEntity* is the abstract class which is on top of the hierarchy for all entity classes. Instances of *GenericDataFile* and *GenericFileElement* classes are used to define various types of security data sources which possibly exist in an enterprise network. The list of *GenericFileElements* is used in order to parse the element values from the data. Two separate parsers are required: one for JSON formatted files and one for the TXT formatted files. For PCAP files, third-party PCAP to TXT converters might be integrated while a PCAP parser is implemented. Regardless of the file format and number of attributes, the element values are stored in a data structure named *DataStore* in tuple formats. In order to define and execute User Requested Queries on Generic Data Store, "SQL"-like queries are defined involving selected fields, grouped by fields and query conditions. These queries are converted to hibernate queries by using associations of *GenericDataFileElements* to DataStore elements which select from generic DataStore. Each query is given a name and called a *GenericDataFileMetric*. Each query result is stored in a structure named *DataStoreQueryResult*. Selection of destination IPs and the ports initiated from a specific Source IP, selection of the number of source IPs grouped by honeynets, min port number accessed between a time interval are sample queries that are defined in the prototype system using test data. The prototype allows association of query results, which are called data metrics, to display types. A query result may be associated with more than one display type, or one display type can be associated with more than one query result. It is important to make correct associations of DataStoreQueryResult structure data types to display type data fields. For example, in the bar chart visualization script data array, d1 can be any numeric type such as Integer, Double, or Long, and data array d2 can be any categorical or numeric types.

As the DataStoreQueryResult instances encapsulate a list of result data types, the correctness of user-defined association can be made by the DisplayType specific *ContentAdapter*. Associations of these metrics to display types are stored in the system, *VisCode*. When the display is actually associated with data, an instance of *VisCodeWithValue* is created. The visualization in the proposed system is designed to be in dashboard style. This will allow visualization of selected metrics of the data coming from multiple sources simultaneously encapsulating subviews in the same screen. The classes named *Threat, Purpose, ExternalVisualizationSystem,* and *UserFeedback* are not directly related to visualization generation but part of enterprise visualization knowledge base. The states marked with colors are associated with phases with data entrance to the visualization knowledge base: Yellow-marked states are relevantly static—in big data terminology, having low velocity, low volume, and medium variety. Green-marked states indicate points that will grow when the associations are made to threats or other purposes, having low volume, low velocity, and medium variety. The red-marked states point out user feedbacks with topics which are expected to be the most dynamic part; that is, growing fast in time and having high volume, high velocity, and high variety.

### 9.4.1 Extensible Display Type Library and Dashboard Design

In order to become knowledgeable on display types, some JavaScript-based visualization libraries have been explored. The libraries which are examined so far are Flotr (Humble Software 2018), FlotCharts (Laursen 2018), Data-Driven Documents (D3) (Bostock 2018), and Sparklines (Splunk 2013). Flotr allows drawing simple static charts such as bar chart, line chart, pie chart; FlotCharts allows more interaction such as zooming in and out; D3 allows custom visualization of data; and Sparklines allows better integration of text and data by using inline charts and visualizing more data by encapsulating Sparklines in a table. The JavaScript-based visualization libraries are not limited to this list: there are other alternatives.

The existence of a high number of available display-type libraries with different properties and difficulty levels resulted in the decision of using these third-party scripts in the proposed design. In the proposed design, for each display type, an XHTML file, which embeds the necessary display container and the corresponding reference to the JavaScript library, and the required JavaScript source have to be prepared. In order to fill the XHTML page correctly with data, a Java class which is called as *ContentAdapter* has to be implemented for each display. The use of Java interfaces standardizes this part of the implementation. XHTML page structure, a sample hierarchy of ContentAdapter for bar chart visualization based on Flotr JavaScript library, and a sequence diagram showing the visualization generation and exhibition via dashboard is shown in Figures 9.4 through 9.6 respectively.

This design allows the use of JavaScript display libraries with various designs. The standardized structure allows easy integration of new libraries. A significant portion of the requirements, such as zoomable design due to interactive display type library, incident time display due to proper chart design selection, are fulfilled due to this extensible display library design structure.

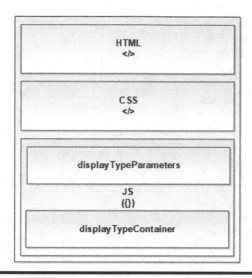

**Figure 9.4    XHTML content for a JavaScript-based display.**

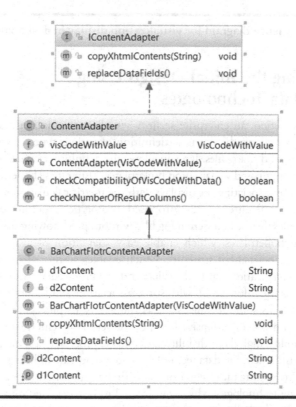

**Figure 9.5    ContentAdapter structure for Flotr JavaScript library-based bar chart visualization.**

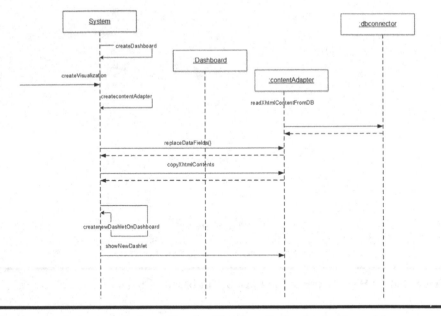

**Figure 9.6  Sequence diagram for visualization display in dashboard form.**

## 9.5 Evolving the Initial Design Using Big Data Technologies

In the previous section, an enterprise security visualization solution design was introduced. Basically, it allows the dynamic definition of various types of data files, which happen to correspond to log files. It again allows the definition of metrics on these data files. It has JavaScript based visualization boilerplates. Using these boilerplates, visualizations are created by associating the displays with the metric calculation results.

Distinguishable features of the proposed prototype structure are its genericness, which enables the use of different log files without pre-knowing the data structure; its visualization boilerplates, which enable easy adaptation of available JavaScript-based displays; its permanent structure, which enables storing of metric values, and corresponding visualizations created earlier; and the knowledge base formed through the use of some static information with some associations, and user feedback.

However, it lacks some features such as the ability to process real-time data, the ability to store and process extensive log files for very large enterprises, encapsulating horizontal scalability, high data reliability, and parallel processing abilities. Thus, another step is taken in the study: big data technologies are examined, and the prototype design is improved by integrating the design features with the appropriate big data technologies. First, the big data technologies which are selected to be integrated into the design will be introduced shortly. Following this introduction, the intersections of the enterprise security visualization solution and the related design decisions will be described.

## 9.5.1 Big Data Technologies

Apache Hadoop is selected cause it is the most popular big data ecosystem at the present day. This ecosystem envelopes big data technologies for various purposes. Technologies related to storing of big data in a distributed manner, related to analyzing big data, running queries and algorithms in a distributed manner, related to streaming big data from multiple sources possibly distributed in multiple machines, and related to collecting big data in real-time from multiple nodes are part of Hadoop. Some of the technologies which are demonstrated as a part of Hadoop also have stand-alone designs which allow them to work independently. The methodology used during the evolving of the initial design with big data technologies is shown in Figure 9.7. First, Hortonworks Hadoop Sandbox (Hortonworks 2018) is installed on a virtual machine. All of the technologies available in the sandbox are reviewed. All along this revision, besides responsibilities, available application program interfaces (APIs), compatibility with Java language, and structures (structured, semi-structured, non-structured) are examined and tested to some extent. In order to integrate the existing structure with the big data technologies, the data structures which are part of the first design are reviewed. During this work, for the selected technology, the integration is planned either as code or configuration structure. As a result of this effort, two new designs, which are called second design and third design, for security visualization solution are prepared. In this part, the technologies which are part of the evolved security visualization solution will be briefly reviewed. In the next section, how these technologies are integrated into the new design structures will be explained.

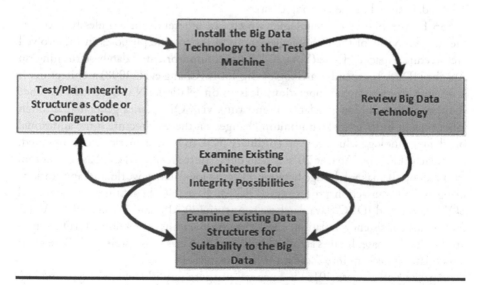

**Figure 9.7   Methodology to find integration points with big data technology.**

Some of the big data technologies are used inherently due to their roles in the Hadoop (Zikopoulos and Eaton 2011) ecosystem. Hadoop Distributed File System (HDFS) (Cohen and Acharya 2013) is a file system for large volumes of data. It consists of name nodes and data nodes, providing a distributed file system encapsulating multiple nodes in possibly multiple Hadoop clusters. It is optimized for handling large files, but it can also handle small files. It stores files by breaking them into blocks. These blocks are distributed among several computers. In order to handle failures, it stores multiple copies of any block. Name node keeps track of where each copy of each block is stored.

The role of Apache Zookeeper (Haloi 2015) is to maintain the Hadoop configuration information and act as a name lookup service. It is the prime building block providing distributiveness of the Hadoop environment. Apache Hadoop Yarn (Vavilapalli et al. 2013) is responsible for separating the resource management and processing tasks. The Yarn is also responsible for arranging that the processes using some data from one HDFS block runs on the same node with that HDFS block. Even if it not explicitly initiated by any application, it is there managing the Hadoop cluster's resources.

Apache Ambari (Wadkar and Siddalingaiah 2014) is an open source management platform which provides management, securing, and monitoring of Hadoop clusters. Ambari is not directly part of the proposed solution; however, when the flat files and some part of the database are moved to Hadoop, Ambari will be a valuable tool to manage the Hadoop and will act as a file manager and a database client. Moreover, using Ambari will provide users new interfaces to run custom analyses based on other distributed big data processing methods such as running Apache Map-Reduce (Dean and Ghemawat 2008) tasks, Apache Spark MLLib (Meng et al. 2016) machine learning libraries, or Apache Pig Latin (Olston et al. 2008) scripts on the data stored in the Hadoop clusters.

So far the big data technologies which are inherently or seamlessly used are mentioned. Some of the technologies are used for specific purposes in the evolved architecture. Apache HBase (Vora 2011) is a column-oriented database running on Apache HDFS. It depends on Google's Big Table (Chang et al. 2008) architecture. It is designed for low-latency operations. It is a non-relational NoSQL database. Since the rest of the solution is relational and runs via SQL queries generated through hibernate, in order to make minimum changes on the architecture some additional big data technology which converts ordinary SQL to HBase queries is encapsulated.

Apache Phoenix (Apache 2018a) is an open source relational database driver tool for Hadoop. It makes a bridge between HBase's low latency world and applications using online transaction processings (OLTP). It enables benefiting with standard SQL queries and JDBC (Java database connectivity) API, and it also enables ACID (atomicity, consistency, isolation, durability) transactions over a non-ACID compliant HBase database. It takes standard SQL and converts it to a series of HBase scans which later transform into standard JDBC ResultSet.

Apache Kafka (Garg 2013) is a general purpose publish/subscribe type messaging system–based data streaming tool. Streaming technology allows processing new data as it is generated. Kafka servers store all incoming messages from publishers for some period and publish them into a data structure called topic. Kafka consumers subscribe to one or more topics. In the proposed design Kafka will act as a data collector for the visualization system.

The proposed system has its intrinsic features. However, there are also highly enhanced commercial or open source big data visualization systems available in the ecosystem. Some of these big data visualization systems are developed as a part of the Hadoop ecosystem, such as Apache Zeppelin (Apache 2018c). Apache Zeppelin is a web-based notebook which enables data exploration and supports technologies like Apache Spark (Apache 2018b), SQL, or Phyton. Some of the external visualization tools are developed independently from Hadoop but provide ways to be integrated with big data stored in Hadoop clusters, such as Tableau (2018) and Qlik (2018). There are also common-purpose visualization systems based on technologies such as D3.js (2018) or R Foundation (2018), which are also commonly used for the visualization of big data. In the evolved version of the proposed visualization system, some new ways will be offered to associate visualizations created with these enhanced visualization systems.

Apache Sqoop (Jain 2013) is a data transfer tool which is used to transfer data from structured resources such as relational databases to Hadoop. Use of this tool is anticipated as a part of the third design. Similar to Apache Sqoop, Apache Spark (Kane 2017; Zaharia et al. 2016) is part of the third design structure. Apache Spark is a technology which allows processing of massive amounts of data in various ways. It has features including data streaming, machine learning, and graph analysis. Spark streaming receives a stream of data, divides it into batches, and processes to generate a final stream of data.

## 9.5.2 Big Data Technologies–Related Design Decisions for Generic Enterprise Security Visualization Solution

There is a variety of big data technologies. Conceding that the target is to adopt a web-based application to big data technologies, the application can evolve into many different structures embracing various integration items. There is not one right structure for a big data–related visualization design. In this design study, the main principle is to make the change while sticking to the original service methods and data structure substantially.

The first version of the security visualization prototype is a Java web application using Spring (Johnson et al. 2009) and Hibernate (Konda 2014) frameworks, and a relational database structure running on MySql database for data storage and processing purposes. The input data was expected to be flat files stored in the operating system disk space in different formats such as TXT, CSV, or JSON. Users make a request for various phases of the visualization generation from the web interface. The service layer fulfills these requests. During the evolving of the first design, besides the technical work, existing data structures are examined mainly for their suitability to be saved and to be processed by big data technologies. Expected size, processing requirements, level of being relational regarding existing relations to other data structures, and level of usability by third-party analysis tools were the focus points during this examination. As a result Table 9.1 is formed. This table does not include the data enumeration structures and the abstract data structures.

The CRUD (create, read, update, delete)–like operations initiated from the user interfaces are used to create and modify the data structures related to file structure definition, metric definition, and visualization boilerplate definition. ORM (object-relational mapping) (Myerson 2002) technologies are selected due to their ability to provide increased performance and scalability while protecting from SQL injections. This part of the data mainly is the metadata required for the visualization. In all three designs, this metadata stays in the relational database.

In the first design, flat data files are static files which are defined and accessed through their operating system path and file name. In the second design, these files are collected through the use of Apache Kafka file connectors and stored in the Hadoop HDFS structure. This time, the service layer reaches these files via the HDFS URL value assigned for each. The third design includes streaming via Apache Spark technology. By this way, the data from flat files collected by Apache Kafka are directed to Apache Spark for streaming. In order to associate the file structure with the segmented streamed data, Apache Spark SQL is used. Through this, a fundamental part of data processing tasks is moved to Hadoop in the third design. This design will also result in real-time processing of the input data.

DataStore is a structure used to store data in various forms in a homogenous structure after parsing of input data. This DataStore is used together with the GenericDataFileElements while producing and executing user-defined queries (metrics). In the first version, this DataStore was part of the relational database. Due to its expected size and usability from external analysis tools, in the second design, this data structure is moved to Apache HBase non-relational database. Since Apache HBase cannot be directly reached from Java-SQL environment, Apache Phoenix, which is called SQL skin for HBase, is used as a layer. This layer allows conversion of normal JDBC calls to HBase scans.

In the first design, users make feedbacks using web user interfaces. The user interfaces are additionally used to share the previous feedbacks. Each feedback comprises of a topic, which is chosen from a predefined set and a text-based feedback content. The reason for including a structured feedback mechanism in the visualization system is to permit the automatic processing of these feedbacks later on. Automatic processing of the feedbacks may cause a timely response to events and increment the speed of information sharing. Users of the visualization system can enter remarks/assessment results/report entries/commands based on the data analysis results immediately. Each topic may relate to a particular purpose, such as the automatic communication of data in various ways such as e-mail or reports (topic:WEEKLY_REPORT, content:check IDS alerts in detail), or the automatic command execution, such as commands for firewalls and for active directory (topic:ACTIVE_DIRECTORY_COMMAND, content:NET USER loginname newpassword /DOMAIN).

**Table 9.1  Examination of Data Structures for Suitability to Big Data Technologies**

| Data Structure | Expected Size | Being Relational in Nature | Processing Difficulty | Useful for Third-Party Tools |
|---|---|---|---|---|
| Flat Files (Text File, JSON File) | Small to very large | N/A | Easy to Difficult based on file size and format | Yes |
| Generic Data File Definition (Generic_Data_File, Generic_Data_File_Element) | Small | Has complex relations with other data structures | Easy | No |
| Generic Data File Metric (Generic_Data_File_Metric, Query_Where_Condition) | Small | Has complex relations within each other and with other data structures | Easy | No |
| Metrics Calculation Result (Data_Store_Query_Result, Data_Store_Column_Data, Data_Store_Row_Data) | Small to medium | Has complex relations within each other and with other data structures | Easy to Medium based on visualization boilerplate | No |
| Visualization BoilerPlate Definition (Vis_Code, Vis_Code_Replacement_ Element) | Small to medium | Has complex relations within each other and with other data structures | Various difficulty based on chosen JavaScript library | No |

*(Continued)*

**Table 9.1 (Continued)   Examination of Data Structures for Suitability to Big Data Technologies**

| Data Structure | Expected Size | Being Relational in Nature | Processing Difficulty | Useful for Third-Party Tools |
|---|---|---|---|---|
| Visualization (Vis_Code_ With_Value, Vis_Code_ Replacement_Element_ With_Value, Vis_Code_ With_Value_Format) | Medium to large | Has complex relations within each other and with other data structures | Easy to Medium based on visualization dashboard features | No |
| KnowledgeBase-Direct Feedback (UserFeedBack) | Medium to very large | Has simple relations with other data structures | Medium for Automated Processing | Yes |
| KnowledgeBase-Static (Threat, Purpose, DisplayType, ExternalVisualizationSystem) | Low to medium | Has simple relations with other data structures | Easy | Yes |
| KnowledgeBase-Associations (Vis_Code_With_Value_ Threats, Vis_Code_With_ Value_Threats, Generic_Data_File_Threats, Generic_Data_File_ Purposes) | Small to medium | Has simple relations with other data structures | Medium for automated processing | Yes |
| DataStore | Small to very large | Has one-way relation to generic data file | Medium to Difficult based on chosen metric | Yes |

The knowledge base data is created with CRUD activities through the user interface and supposed to be viewed from there, combined with other metadata information. The amount of feedback as a part of the knowledge base is likewise expected to grow quickly in time. Thus, to permit automated processing of these feedbacks in the future, in the third design, user feedbacks are decided to be replicated in a denormalized form in HBase using Apache Sqoop. Sqoop can be configured to replicate the views having joins of multiple tables on the relational database to the HBase as the new tuples of data entered in the knowledge base.

The last change made to the first design is encapsulating a visualization boilerplate for external URL-based visualization systems. In the first design, a visualization boilerplate was required to use a JavaScript-based visualization. For each boilerplate, there is a content adapter class and an XHTML-based display type code definition which include the JavaScript code for that specific display type, as mentioned before. There are various external visualization systems associated with visualization of big data. Most of these visualization systems are accessible via web URLs. In the second version of the enterprise visualization system, in order to benefit from these external visualization systems, an XHTML page which merely includes embedded external content and a content adapter, which is not responsible for any data processing task, are prescribed. Utilizing a boilerplate, which allows integrating URL-based visualizations created by external visualization systems, will cause the following benefits. Figures 9.8 through 9.12 show all three versions of the enterprise security

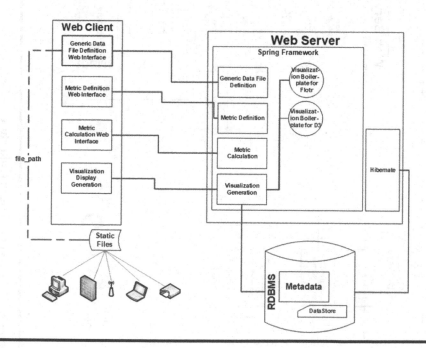

**Figure 9.8   The first design architecture.**

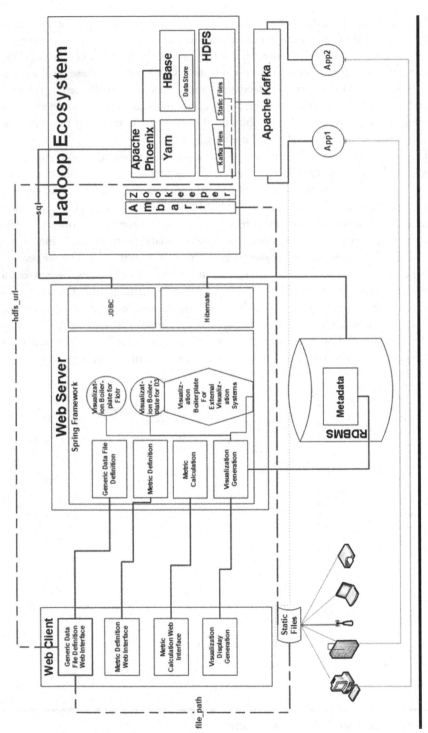

**Figure 9.9  The second design architecture.**

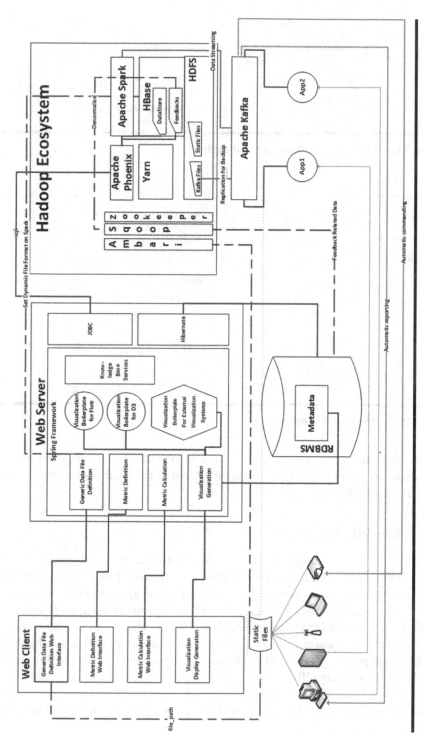

**Figure 9.10 The third design architecture.**

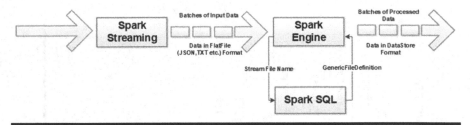

**Figure 9.11    Streaming details for third design.**

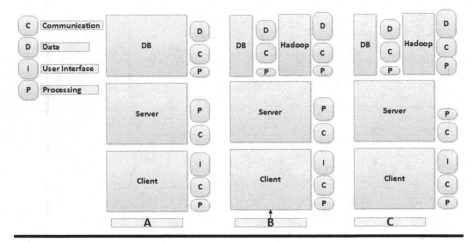

**Figure 9.12    Evolution summary.**

visualization design structure and Spark streaming details. As the design improved, the amount of processing and data storage increased in the big data environment, as illustrated in Figure 9.12.

## 9.5.3  Security Concerns

As the ubiquity of the designs rises, the privacy and security requirements of the system will increase eventually. A multilayered Internet of Things, IoT, system has the following layers: perception layer, network layer, and application layer. In some architectures, a service layer which is responsible for service management and service discovery, is added to the other three layers (Lin et al. 2017). Each layer has its security concerns. Security concerns of an IoT application which has a web-based interface were previously depicted in Özdemir Sönmez's IoT case study running in IBM BlueMix platform (2016). The security concerns of the proposed system have similarities to that list. However, in the proposed system these concerns are fulfilled in different ways, mainly depending on the security features of industry-standard platform choices.

The security of the environment running the proposed design is the first concern. The proposed system is designed to be used in an enterprise environment which would have a firewall, an intrusion prevention system, and network security controls; these protections may be supported by server-level hardening mechanisms. The second concern for the cloud IoT system is the genuineness of the cloud platform, which does not apply to this design. The third concern is the authentication of the web application, which will be fulfilled using a standard way to implement user authentication, Spring Security (Mularien 2010). The fourth concern is the authentication of the devices to the application. In a standard Kafka installation, any user can write any messages to any topic; however, in a more advanced setup, Kafka provides authentication of Kafka clients via Secure Sockets Layer (SSL) or Simple Authentication and Security Layer (SASL). The fifth concern is the security of the authentication data stored in the cloud, which does not apply to the current solution. However, this time the security of data stored in the relational database and the Hadoop is the concern for which the system will rely on the protections of the underlying technologies. Each Hadoop component has its authentication, authorization, encryption of data at rest, and encryption of data in transit (Sharma and Navdeti 2014). Similarly, contemporary database systems have advanced protections systems (Basharat et al. 2012). The sixth concern is a secure gateway between various platforms in the distributed system. Although not included in the proposed design, Hadoop has multiple gateway structures. The seventh concern is secure messaging between devices and the application. In the proposed design, the devices/applications are not expected to send direct messages to the application, but Kafka file connectors are in charge of reading device/application log files. The eighth concern is preventing the data leakage between devices, and the ninth concern is to prevent the data leakage between devices and the application. The devices are not expected to communicate as a part of the proposed solution; as mentioned earlier, all issues related to the communication of devices and application depend on Kafka security.

## 9.6 Results

In order to test critical implementations and the proposed data structure, a web-based prototype is developed as a part of this study. The authors evaluated the coverage of the requirements based on this prototype. Table 9.2 is a traceability matrix which shows whether the requirements are met, not met, or require modification by the decided design features. The modifications are thought of as future work, which may include implementation of new modules, new adaptors, and parsers for new file formats.

One of the critical concerns is the provided level of genericness of the system. The system evaluated with sample data, and the suitable examples, is provided in Table 9.3. Although there are known issues with the parsers, such as allowing

**Table 9.2 Traceability of the Requirements to Detailed Design Features**

| # | Short Description | Generic Data File Def. | Metric Def. | Metric Calc. | Visuali-zation Gen. | Standard-ized Display Boiler-plate | Dash-board Design | Know-ledge base | Use of HDFS and HBase | Generic Display Boiler Plate | Kafka Data Collection | Use of Other Structures | Avg. Review Results 1–5 Scale |
|---|---|---|---|---|---|---|---|---|---|---|---|---|---|
| 1 | Display type library | | | | | | | (+) | | | | | 3.75 |
| 2 | Read data in various formats | (+) | | | | | | | | | | | 4.50 |
| 3 | Addition of new data sources easily | (+) | | | | | | | | | | | 4.50 |
| 4 | Feedbacks for data files | (+) | | | | | | (+) | | | | | 2.25 |
| 5 | Non-predefined metrics | | (+) | (+) | | | | | | | | | 4.50 |
| 6 | Threat definition | | | | | | | (+) | | | | | 3.75 |
| 7 | Associating threats to visualizations | | | | (+) | | | (+) | | | | | 3.75 |

*(Continued)*

**Table 9.2 (Continued) Traceability of the Requirements to Detailed Design Features**

| # | Short Description | Generic Data File Def. | Metric Def. | Metric Calc. | Visuali-zation Gen. | Standard-ized Display Boiler-plate | Dash-board Design | Know-ledge base | Use of HDFS and HBase | Generic Display Boiler Plate | Kafka Data Collection | Use of Other Structures | Avg. Review Results 1–5 Scale |
|---|---|---|---|---|---|---|---|---|---|---|---|---|---|
| 8 | Associating threats to data sources | (+) | | | | | | (+) | | | | | 3.75 |
| 9 | Purpose definition | | | | | | | (+) | | | | | 3.75 |
| 10 | Associating purposes for data sources | (+) | | | | | | (+) | | | | | 3.75 |
| 11 | Associating purposes to visualizations | | | | (+) | | | (+) | | | | | 3.75 |
| 12 | Feedback to generating visualizations | | | | (+) | | | (+) | | | | | 2.25 |
| 13 | Use of display types with various complexity | | | | | | | | | | | | 4.75 |

*(Continued)*

**Table 9.2 (Continued)   Traceability of the Requirements to Detailed Design Features**

| # | Short Description | Generic Data File Def. | Metric Def. | Metric Calc. | Visuali-zation Gen. | Standard-ized Display Boiler-plate | Dash-board Design | Know-ledge base | Use of HDFS and HBase | Generic Display Boiler Plate | Kafka Data Collection | Use of Other Structures | Avg. Review Results 1–5 Scale |
|---|---|---|---|---|---|---|---|---|---|---|---|---|---|
| 14 | Display type on the fly | | | | | | | | | | | | 4.50 |
| 15 | Visualization display difficulty | | | | (+) | | | (+) | | | | | 3.75 |
| 16 | Access to external visualization | | | | | (+) | | | | (!) | | | 3.25 |
| 17 | Depict large data | | | | | | | | (+) | | (+) | (+) | 4.50 |
| 18 | Simultaneous display | | | | | | (+) | | | | | | 4.50 |
| 19 | Save detected pattern | | | | (*) | | (+) | | | | | | 3.50 |
| 20 | Work with real-time data | | | | | | | | | | (+) | (+) | 4.00 |
| 21 | Depict most type of attacks | (+) | (+) | | | | | | | | | | 4.50 |

*(Continued)*

**Table 9.2 (Continued)  Traceability of the Requirements to Detailed Design Features**

| # | Short Description | Generic Data File Def. | Metric Def. | Metric Calc. | Visualization Gen. | Standardized Display Boilerplate | Dashboard Design | Knowledge base | Use of HDFS and HBase | Generic Display Boiler Plate | Kafka Data Collection | Use of Other Structures | Avg. Review Results 1–5 Scale |
|---|---|---|---|---|---|---|---|---|---|---|---|---|---|
| 22 | Displaying incident time | | | | (#) | | | | | | | | 4.00 |
| 23 | Thick boundaries between classes | | | | (#) | | | | | | | | 4.00 |
| 24 | Visualization without mouse | | | | (#) | | | | | | | | 4.00 |
| 25 | Being interactive | | | | (#) | | | | | | | | 4.00 |
| 26 | Being searchable | | | | (#) | | | | | | | | 3.00 |
| 27 | Being zoomable | | | | (#) | | | | | | | | 4.00 |
| 28 | Being scalable ITO data display | | | | (#) | | | | | | | | 4.50 |

(+) This requirement is met through this design feature.
(#) This requirement can be met through the integration of proper visualization libraries.
(*) This requirement requires updates to this design feature.
(1) This requirement is allowed using primitive ways.

**Table 9.3   Sample Data Sources and Their Representation in the Proposed Design**

| |
|---|
| **Modern Honey Network Alert Log:** Type=JSON, , Separator = "-", Elements: oid, destination_ip, protocol, hp_feed_id_oid, timestamp_date, source_ip, source_port, destination_port, identifier, honeypot |
| Line 409755: { "_id" : { "$oid" : "58c1d06f58e5cf04aff99ea3" }, "destination_ip" : "200.200.200.201", "protocol" : "pcap", "hpfeed_id" : { "$oid" : "58c1d06e58e5cf04aff99ea0" }, "timestamp" : { "$date" : "2017-03-10T00:00:14.147+0200" }, "source_ip" : "221.229.162.121", "source_port" : 4405, "destination_port" : 22, "identifier" : "fea0bde0-5d6d-11e6-9709-000c297e338e", "honeypot" : "p0f" } |
| **Web Access Log:** Type =TXT, Separator = "-", Elements: IP, Dummy1, DateAndTime, Dummy2, MethodAndURL, SystemInfo, |
| 117.201.11.139 - - 02/Jan/2017:02:35:43 -0800 "GET /wp-login.php HTTP/1.1" 404 295 "-" "Mozilla/5.0 (Windows NT 6.1; WOW64; rv:40.0) Gecko/20100101 Firefox/40.1" 117.201.11.139 - - 02/Jan/2017:02:35:49 -0800 "GET /wp-login.php HTTP/1.1" 404 295 "-" "Mozilla/5.0 (Windows NT 6.1; WOW64; rv:40.0) Gecko/20100101 Firef8ox/40.1" |
| **Hardware Firewall Log:** Type =TXT, Separator = "\|", Elements: Count, fw1src, fw1service , fw1proto, fw1action, fw1tcpflags |
| \| 2 \| 192.168.184.5 \| 80 \| tcp \| accept \| NULL \|<br>\| 2 \| 172.16.224.16 \| 80 \| tcp \| accept \| NULL \|<br>\| 1 \| 172.16.100.38 \| 80 \| tcp \| accept \| NULL \| |

single separators and encapsulating nested collections as a part of row data, these missing points can be improved without any side effect to the overall structure. The proposed design also does not handle the compressed file inputs as is.

A significant concern to be tested is the storing of generic data in a nongeneric format to be queried. No issue has been detected related to this concern. However, there are limitations to the current design which may be extended easily. The first limitation is the 94 generic query where conditions identified in the current design, which may be extended. The second limitation is the maximum number of fields for each data type (present limit is 10) in a log file.

The proposed system uses the JavaScript-based displays as is. Thus, any advanced JavaScript-based display type having advanced display properties, such as interactivity, zooming, or having proper display designs such as thick boundaries and displaying incident time, can be integrated to the proposed design based on two conditions. The first condition is the display code may be represented as shown in Figure 9.4. The second condition is that developing display type–specific ContentAdapter Java code must be both probable and feasible.

Another significant concern was to display the generated visualizations in dashboard form. For this purpose, the PrimeFaces (Çalışkan and Varaksin 2013) dashboard control is used in the prototype. This component has built-in dashboard features such as drag and drop, resize, and reorder of dashboard parts.

Attempts to integrate the initial design with big data technologies resulted in small changes in the original design. However, these changes resulted in an extensive list of benefits. For example, moving the input files to HDFS resulted with a more scalable structure for storage of large files and low latency due to Hadoop's vast data file processing capabilities. Moving the input files to shared storage, they become reachable from other data analysis and data visualization tools. Moving the data, DataStore, and feedbacks to the HBase increased the scalability and performance of data processing and data storage; besides, this allows execution of other big data analyses available in Hadoop. The original web interface, controller, and service structure were based on building dynamic queries on DataStore based on SQL capabilities, including the mathematical SQL functions. Since the original analysis methods are protected, the same queries are arranged to be run on Hadoop HBase non-relational database. At this step, Apache Phoenix is the main catalyzer of the overall process. As mentioned in the previous section, it allows running SQL queries on HBase by converting SQL to HBase scans, resulting in low-latency queries. It provides ACID properties which allow OLTP over data, and it returns a standard JDBC structure which can further be converted to hibernate objects so that metric calculations results can be stored and can be associated to the selected display types in the relational database. In the first design, static files stored in the operating system disks are used for visualization purposes. In the improved design, Apache Kafka is integrated with the design to enable automatic data collection from multiple points for visualization purposes. Integrating Apache Kafka with the original design resulted in near real-time examination of raw data files, a painless collection of visualization data from multiple points/ nodes, and the standardization. As a next step, Apache Spark improved the near real-time design to be real time.

It is necessary to depict the critical points when processing large data, which depends on several factors. The benefits of Hadoop for processing and storing big data is already mentioned. Other issues include the capability of visualization script to display large data, and the processing algorithms. The authors found out some of the mentioned display libraries are already known for their proper performance with big data. Some of these, for example, D3, provide online performance test tools. There are three main algorithms, one parsing and storing task algorithm, a querying task algorithm, and a content adapter task algorithm. In general, these have the following algorithmic complexities in Big O (Abu Naser 1999) notation: respectively, $O(n)$, $O(\log(n))$, and $O(n)$, where n is the number of rows in the dataset. However, if the complexity of the content adapter algorithm for a specific display increases the latest complexity may change.

Feedbacks are included in the system for increasing learning, for information sharing. Moving the feedbacks to the HBase allows further examination of these feedbacks. Making queries to ask the number of feedbacks for each topic is, for each topic and feedback item pair, a straightforward way to process the feedbacks. These feedbacks may help users during the data preparation and visualization generation tasks. User feedback may also trigger certain conditions and events in cyberspace when appropriately processed. These feedbacks may form closed control loops as shown in Figure 9.13. Identification of feedback topic types and other ways of feedback processing is left as future work.

Benefits of generic visualization boilerplate include displaying of same raw data with external visualization systems and displaying the visualizations created by external visualization systems without leaving the enterprise visualization system. Simultaneous view of data by external systems and the proposed system in the same dashboard will enable comparison and may result in showing different aspects of the same data.

As a part of the validation effort, a series of semistructured expert judgment interviews are done. The last column of Table 9.2 corresponds to average reviewer scores on a scale of 1–5. The validation efforts and results are further discussed in the next section.

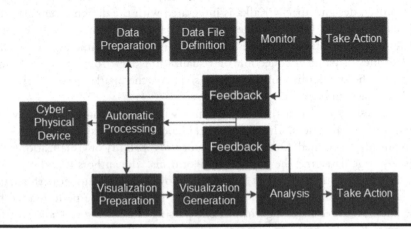

**Figure 9.13 Feedback loops.**

## 9.7 Discussion (Including Validation Efforts)

Difficulties of validating cyber-physical systems and conceptual systems are known concepts. The issues that can be or cannot be verified vary based on the validation subject. The proposed design is partly conceptual. For the conceptual parts, tests are made to check the interoperability for most of the parts. For the implemented elements, the prototype is used to test the functionalities and coverage. Hence, some of the issues are verified conceptually, and some parts are verified through the run of the prototype and by inspection. During the evolution of the design, two presentations are made to the academics, and feedbacks from these meetings resulted in the improvement of the design in stages.

The soundness of the requirement set and soundness of the design were two factors that were examined in phases during this study. If we were to talk about the requirements, the requirements are tested against survey results and the literature findings. As a result of these checks, the rationality for each item is provided based on the combined results.

Qualitative methods are commonly used for the cases when it is difficult or expensive to run the experiments, such as in distributed systems, systems with big data, when there are conceptual design issues, when the sample datasets are not adequate to show all aspects of the design, and when the number of evaluators is low (Seaman 1999). The soundness of the design and selected technologies are checked through the use of a series of semistructured interviews with the experts. These interviews also included questions to check the ability of the design to fulfill the initial requirement set. The participants include one faculty member who has long-term experience and position in the information systems field with particular focuses on software architecture, computer networks, and the Internet of Things; a second faculty member with long-term experience and position in information systems with particular focus on technologies, design patterns, and software testing; one senior manager who has more than 30 years of experience in the IT and more than 20 years of experience in information security and risk management supported with certificates (CISA, CISM, CGEIT, CRISC-ISACA Chapter Founder); one current chief researcher in a public research center, who is a former Microsoft engineer and has more than 20 years of experience in software development and software project management, and more than 4 years in information security.

During these interviews, IEEE 1471-2000 (Hilliard 2000) standard is used, which is a recommended practice for Architectural Description of Software-Intensive Systems. This standard requires that the system should be introduced systematically by means of a system definition, including environment description, mission and stakeholders' identification, and architecture descriptions. These architecture descriptions include a series of architectural views and model definitions, each having viewpoints and concerns. In order to demonstrate the proposed system, a presentation based on IEEE 1471-2000 was made prior to each interview. This presentation included conceptual graphs, top-level architectural views, class diagrams, detailed views for critical parts, data structures, definitions, scenarios, user interface screenshots, demonstrative information for the available display type libraries and their integration to the proposed system, and code parts to describe various interoperability or algorithmic details. While other standard information is presented in the slides, the rationality of each presented item is explained verbally to the reviewer. Adequateness of the development infrastructure is a significant issue that is questioned for cyberphysical systems because wrong development infrastructure will result in unscalable solutions with low performances or bad security. Numerical scores given related to each requirement item are shown in Table 9.2. Other information is in Table 9.4.

One more issue which is verified for some cyber-physical systems is the compactness of the overall design. Compactness of the proposed system can be specified based on the compactness of the model-view-controller with service layer architecture, compactness of the Java Spring framework, and the compactness of Hadoop sandbox. Although the reviewers made no explicit evaluation for compactness, the authors claim that depending on the industry standard technologies, the architectures will add on to the level of compactness of the proposed design.

Reliability of CPSs is another significant issue. Reliability of a CPS will depend on the reliability of the system structure and the reliability of the underlying network. Reliability of the system is contingent on the reliability of the selected technologies and underlying structure. Reliability of the algorithms is tested for correctness using sample data. No other experiments are done to test the reliability of the overall system due to conceptual parts.

The main focuses were the interoperability of various systems and technologies, and the scalability and the reliability of the overall system. The main limitation of this study is that the proposed design is partly conceptual. Thus, it does not provide quantitative outputs to compare the overall performance of the final design. Certification is a way which helps judgment of a design's adequateness, safety, and reliability in a specified environment. CPSs may be subjected to legal assurance and the certification in real life, commonly before the production phase, which does not apply for this specific case. How this CPS can be certified and what type of certificate is more proper is left as future work. Another future work detected is improving the structure of free text feedbacks from the users. This improvement may allow further ways of automatic processing other than automatic reporting of these feedbacks. The reviewers also mentioned the shortcomings of the feedback system, and suggested to create a taxonomy of probable feedback topics, and improve the free text feedback structure for proper automatic processing in the future.

In general, although a systematic introduction of the study objective and design was made based on a recommended standard, and the review material included a considerable amount of details, there had been times that reviewers had difficulties to understand and give answers based on available documentation, which affected the review scores.

**Table 9.4     Expert Judgment Semi-structured Interview Results**

| *Reviewer 1: Faculty Member* | *Review Duration: Two and half hour* |
|---|---|
| Notes: In its current structure of feedbacks, future automatic processing of the feedbacks will be limited to making topic-based queries. The free text format may be extended to enable automatic command generation type of processing. Some aspects of the design, such as working in real time or near real time cannot be observed during this type of validation. Although the underlying structure allows the real-time execution of big data, more experiments or other quantitative validation methods may be included to improve the validation. A knowledge base as in the presented form may enhance the learning process in an enterprise. The ways of this enhancement can be investigated separately in another study. | |
| *Reviewer 2: Faculty Member* | *Review Duration: One hour* |
| Notes: Preparing a taxonomy for feedback topics in order to improve their future processing is suggested. Including assignment of display type difficulties in the system is good. However, more study can be made to clarify how these difficulty levels will be used in user selections. Access to external visualization systems may be improved by feeding these systems by data rather than simply linking them. More time is required to investigate this system for more detailed validation; there may be misunderstood issues in limited present duration. | |
| *Reviewer 3: ISACA Chapter Founder* | *Review Duration: Two hours* |
| The stakeholders of this system may include outsiders (non-corporate users, users from contracted companies). Processing compressed files are suggested. Including non-functional requirements explicitly in the requirements list is recommended. Depending on file names on Spark Streaming to match the data with file definitions may have handicaps. A secondary integration item may be included. Some of the validation questions are vague. | |
| *Reviewer 4: Former Microsoft Engineer, Security Researcher* | *Review Duration: Two hours* |
| Predefined metrics such as those synthesized from common vulnerabilities and exposures (CVE) databases can be incorporated. Threat definition structures may be improved. With real-time visualization, scalability might be an issue for high volumes of data. | |

Principally the review results related to design decisions and the scope do not include major issues, and are parallel to the author's evaluation results, as shown in Table 9.2. The technology selections are also found appropriate by the users. Some lacking issues, such as working with compressed files, are suggested to be included by a reviewer. Another improvement proposed is relying on multiple integration items for Spark. We believe that these are minor issues which can be injected into the system smoothly as future work. Some of the reviewer recommendations, such as adding more structured threat definitions, adding more structured feedbacks, working on a taxonomy of feedback topics, and including some predefined metrics besides allowing user-defined metric definitions, can be structured on top of the proposed design.

## 9.8 Conclusion

In this study, an enterprise security visualization system which targets generic processing of log data files, non-predefined metrics, and a knowledge base design is presented. The first contribution is the identification of its unique scope using a requirement analysis survey. The second contribution is its generic and standardized design, which allows adaption and extension of new files and new display types. The third contribution is the methodology and design adopting a web-based application to Hadoop big data technologies, which may also be exemplary for the integration of legacy applications with big data technologies. We believe that the final version of the design results in a more scalable system regarding raw data storage place and corresponding processed data stores, and higher performance due to low latency. The big data technologies integrations included easily implemented changes which resulted in extensive benefits. The data collection mechanism included in the second and third design shifts the initial design to a better spot and provides better usability. Finally, enabling integration with external visualization systems increases the overall quality and usability of the proposed system by making simple changes in the original design.

The final design has an easily implemented structure with enhanced qualities mainly in terms of performance, scalability, interoperability, and security, due to the design structure and the underlying technologies. This design also has good abstraction and multiplicity features due to generic and/or standardized definitions. The authors believe that including a knowledge base in the enterprise security visualization system may help learning and may more easily allow the creation of better visualizations.

The proposed design may likewise be expanded effectively with future work. A mechanism for automated processing of the user feedbacks given for the visualizations can be formed. Taxonomy of these feedbacks can be done as a part of the visualization knowledge base. It is constantly conceivable to add new kinds of parsers or adapters to support other file types. The current system is primarily intended for JavaScript-based visualizations. It might be tried with other visualization libraries which are applicable to web-based applications. Another significant limitation is the allowed file types for the generic parsers. The current data structure did not allow to move all the data parts to NO-SQL database. A more detailed data denormalization study may allow further using the NO-SQL database.

# References

Abowd, Gregory D., Chris Atkeson, Ami Feinstein, Yusuf Goolamabbas, Cindy Hmelo, Scott Register, Nitin Nick Sawhney, and Mikiya Tani. 1996. *Classroom 2000: Enhancing Classroom Interaction and Review.* Technical GIT 96-21, Atlanta, GA: Georgia Institute of Technology.

Abu Naser, Samy S. 1999. "Big O Notation for Measuring Expert Systems complexity." *Islamic University Journal* 57. Accessed August 17, 2018. http://www.scs.ryerson.ca/~mth110/Handouts/PD/bigO.pdf.

Apache. 2018a. *OLTP and Operational Analytics for Apache Hadoop,* June 12. https://phoenix.apache.org.

Apache. 2018b. *Unified Analytics Engine for Large-Scale Data Processing,* June 12. https://spark.apache.org.

Apache. 2018c. *Web-Based Notebook that Enables Data-Driven, Interactive Data Analytics and Collaborative Documents with SQL, Scala and More,* June 12. https://zeppelin.apache.org.

Arcsight. 2017. *Common Event Format Implementing ArcSight Common Event Format (CEF). Standard.* Berkshire, UK: Microfocus.

Basharat, Iqra, Farooque Azam, and Abdul Wahab Muzaffar. 2012. "Database Security and Encryption: A Survey Study." *International Journal of Computer Applications* 47 (12): 0975–888.

Bostock, Mike. 2018. *Data Driven Documents,* June 7. Accessed August 7, 2018. https://d3js.org/.

Çalışkan, Mert, and Oleg Varaksin. 2013. *PrimeFaces Cookbook.* Birmingham, UK: Packt Publishing Ltd.

Chang, Fay, Jeffrey Dean, Sanjay Ghemawat, Wilson C. Hsieh, Deborah A. Wallach, Mike Burrows, Tushar Chandra, Andrew Fikes, and Robert E. Gruber. 2008. "Bigtable: A Distributed Storage System for Structured Data." *ACM Transactions on Computer Systems* 26 (2): 4.

Chuvakin, Anton, Kevin Schmidt, and Christopher Phillips. 2012. *Logging and Log Management: The Authoritative Guide to Understanding the Concepts Surrounding Logging and Log Management.* Newnes, Australia: Elsevier.

Cisco. 2009. "Cisco Intrusion Detection Event Exchange (CIDEE) Specification." Accessed April 28, 2016. http://www.cisco.com/c/en/us/td/docs/security/ips/specs/CIDEE_Specification.html.

Cohen, Jason, and Subatra Acharya. 2013. "Towards a More Secure Apache Hadoop HDFS Infrastructure." *International Conference on Network and System Security.* Berlin, Germany: Springer, pp. 735–741.

D3.js. 2018. *D3,* June 12. https://d3js.org.

D'Amico, Anita, Kirsten Whitley, Daniel Tesone, Brianne O'Brien, and Emilie Roth. 2005. "Achieving Cyber Defense Situational Awareness: A Cognitive Task Analysis of Information Assurance Analysts." *Proceedings of the Human Factors and Ergonomics Society Annual Meeting* (SAGE) 49 (3): 229–233.

Dean, Jeffrey, and Sanjay Ghemawat. 2008. "MapReduce: Simplified Data Processing on Large Clusters." *Communications of the ACM* 51: 107–113.

Debar, H., D.A. Curry, and B.S. Feinstein. 2007. "The Intrusion Detection Message Exchange Format (IDMEF)." IETF.

Fillinger, Antoine, Imad Hamchi, Stéphane Degré, Lukas Diduch, Travis Rose, Jonathan Fiscus, and Vincent Stanford. 2009. "Middleware and Metrology for the Pervasive Future." *IEEE Pervasive Computing Mobile and Ubiquitous Systems* 8 (3): 74–83. Accessed July 31, 2018. https://www.nist.gov/information-technology-laboratory/iad/mig/nist-smart-space-project.

Fumy, Walter, and Joerg Sauerbrey (Eds.). 2013. *Enterprise Security: IT Security Solutions, Concepts, Practical Experiences, Technologies*. John Wiley & Sons.

Garg, Nishant. 2013. *Apache Kafka*. Birmingham, UK: Packt Publishing.

Gerhards, Rainer. 2009. "The Syslog Protocol." IETF. Accessed May 17, 2016. https://tools.ietf.org/html/rfc5424.

Hallam-Baker, Phillip M., and Brian Behlendorf. 1996. "Extended Log File Format." *WWW 3* (W3C).

Haloi, Saurav. 2015. *Apache Zookeeper Essentials*. Birmingham, UK: Packt Publishing.

Hilliard, Rich. 2000. "IEEE-Std-1471-2000 Recommended Practice for Architectural Description of Software-Intensive Systems." *IEEE Standards*, pp. 16–20.

Hortonworks. 2018. *Hortonworks Sandbox on a VM*. https://hortonworks.com/products/sandbox/.

Humble Software. 2018. *Flotr2 JavaScript Visualization Library*, August 7. Accessed August 7, 2018. http://www.humblesoftware.com/flotr2/.

Jain, Ankit. 2013. *Instant Apache Sqoop*. Birmingham, UK: Packt Publishing Ltd.

Johanson, Brad, Armando Fox, and Terry Winograd. 2004. *The Stanford Interactive Workspaces Project*. Technical, Stanford, CA: Stanford University.

Johnson, Rod, Jürgen Höller, Alef Arendsen, Thomas Risberg, and Colin Sampaleanu. 2009. *Professional Java Development with the Spring Framework*. John Wiley & Sons.

Kane, Frank. 2017. *Frank Kane's Taming Big Data with Apache Spark and Python*. Birmingham, UK: Packt Publishing Ltd.

Konda, Madhusudhan. 2014. *Just Hibernate: A Lightweight Introduction to the Hibernate Framework*. Sebastopol, CA: OReilly.

Laursen, Ole. 2018. *FlotCharts*, August 7. Accessed August 7, 2018. http:/www.flotcharts.org.

Lee, Edward A. 2015. "The Past, Present and Future of Cyber-Physical Systems: A Focus on Models." *Sensors* 15: 4837–4869.

Lin, Jie, Wei Yu, Nan Zhang, Xinyu Yang, Hanlin Zhang, and Zhao Wei. 2017. "A Survey on Internet of Things: Architecture, Enabling Technologies, Security and Privacy, and Applications." *IEEE Internet of Things Journal* 4 (5): 1125–1142.

Marty, Raffael. 2009. *Applied Security Visualization*. Boston, MA: Addison Wesley Professional.

Meng, Xiangrui, Joseph Bradley, Burak Yavuz, Evan Sparks, Shivaram Venkatamaran, Davies Liu, Jeremy Freeman et al. 2016. "MLlib: Machine Learning in Apache Spark." *Journal of Machine Learning Research* 17: 1–7.

Mularien, Peter. 2010. *Spring Security 3*. Packt Publishing Ltd.

Myerson, Judith M. 2002. *The Complete Book of Middleware*. Boca Raton, FL: Taylor & Francis Group.

Olston, Christopher, Benjamin Reed, Utkarsh Srivastava, Ravi Kumar, and Andrew Tomkins. 2008. "Pig Latin: A Not So-foreing Language for Data Processing." *Proceedings of the 2008 ACM SIGMOD International Conference on Management of Data.* New York: ACM, pp. 1099–1110.

Özdemir Sönmez, Ferda. 2016. "Case Study: Development of Automated Remote Security Prototype Based on IOT Technologies on IBM Bluemix Platform." *International Conference on Information Security.* Ankara, Turkey: ISCTURKEY, pp. 88–97.

Özdemir Sönmez, Ferda, and Banu Günel. 2018. "Security Visualization Extended Review Issues, Classifications, Validation Methods, Trends, Extensions." In *Security and Privacy Management, Techniques, and Protocols*, Yassine Maleh (Ed.). Hershey, PA: IGI Global, pp. 152–197.

Qlik. 2018. *Data Analytics for Modern Business Intelligence*, June 12. https://www.qlik.com.

R Foundation. 2018. *The R Project for Statistical Computing*, June 12. https://www.r-project.org.

Seaman, Carolyn B. 1999. "Qualitative Methods in Empirical Studies of Software Engineering." *IEEE Transactions on Software Engineering* 4: 557–572.

Sharma, Priya P., and Chandrakant P. Navdeti. 2014. "Securing Big Data Hadoop: A Review of Security Issues, Threats and Solution." *International Journal of Computer Science and Information Technologies* 5: 2126–2131.

Sherwood, John, Andrew Clark, and David Lynas. 2005. *Enterprise Security Architecture: A Business-Driven Approach.* Boca Raton, FL: CRC Press.

Shin, Bongsik. 2017. *A Practical Introduction to Enterprise Network and Security Management.* Boca Raton, FL: CRC Press, Taylor & Francis Group.

Splunk. 2013. *JQuery Sparklines*, July 15. Accessed August 7, 2018. https://omnipotent.net/jquery.sparkline.

Stanford, Vincent, John Garofolo, Olivier Galibert, Martial Michel, and Christophe Laprun. 2003. "The NIST Smart Space and Meeting Room Projects: Signals, Acquisition, Annotation and Metrics." *Proceedings of IEEE International Conference on Acoustics, Speech, and Signal Processing.* Hong Kong: IEEE, pp. 6–10.

Tableau. 2018. *Make Your Data Make an Impact*, June 12. https://www.tableau.com.

Vavilapalli, Vinod Kumar, Arun C. Murty, Chris Douglas, Sharad Agarwal, Mahadev Konar, Robert Evans, Thomas Graves et al. 2013. "Apache Hadoop YARN: Yet Another Resource Negotiator." *Proceedings of the 4th Annual Symposium on Cloud Computing.* New York: ACM, p. 5.

Vora, Mehul Nalin. 2011. "Hadoop-HBase for Large-Scale Data." *Proceedings of 2011 International Conference on Computer Science and Network Technology.* Harbin, China: IEEE, pp. 601–605.

Wadkar, Sameer, and Madhu Siddalingaiah. 2014. "Apache Ambari." In *Pro Apache Hadoop*, Sameer Wadkar, Madhu Siddalingaiah and Jason Venner (Eds.). Berkeley, CA: Apress, pp. 399–401.

Zaharia, Matei, Reynold S. Xin, Patrick Wendell, Tathagata Das, Michael Armbrust, Ankur Dave, Xiangrui Meng et al. 2016. "Apache Spark: A Unified Engine for Big Data Processing." *Communications of the ACM* 59: 56–65.

Zikopoulos, Paul, and Chris Eaton. 2011. *Understanding Big Data: Analytics for Enterprise Class Hadoop and Streaming Data.* Emeryville, CA: McGraw-Hill Osborne Media.

*Chapter 10*

# Searching for IoT Resources in Intelligent Transportation Cyberspace (T-CPS)— Requirements, Use-Cases and Security Aspects

Md. Muzakkir Hussain, Mohammad Saad Alam, M. M. Sufyan Beg and Rashid Ali

## Contents

# 10.1 Introduction

Due to rigorous research and development in state-of-the-art information and communication technologies (ICTs) and upsurge in the human population, Intelligent Transportation Systems (ITSs) became an integral part of contemporary human inhabitation (Ezell 2010). The ITS architecture is comprised of a set of advanced applications aimed at applying ICT amenities to provide quality of service (QoS) and quality of experience (QoE) guaranteed service for traffic management and transport (Singh and Gupta 2015). According to the conceptual framework of future ITS development planned by the U.S. Department of Transportation (DOT) and ITS-America, the relationship between ITS services was defined to ensure compatibility and interchangeability (Misra et al. 2014). Figure 10.1 depicts the fundamental components defined for a typical ITS architecture that supports 7 functions and 30 users' services provide to drivers and commuters (Ezell 2010). The dependence on transportation systems is indispensable, as is clear from the fact that nearly 49% of the global population devotes at least one hour commuting every day (Zhang et al. 2011; Bitam 2012; Pham et al. 2015). In fact, the competitiveness of a nation, its economic force, and its productivity rely heavily on how robustly its transportation infrastructures are installed (Zhang et al. 2011). However, the current landscape of vehicle penetration into transportation architectures comes up with numerous opportunities and challenges (Hussain et al. 2018). It may be in the form of traffic congestion, parking issues, carbon footprints, accidents, etc. Efficient transportation protocols and policies need to be employed to confront such issues. Odd-even policies adopted by China in the Beijing Olympics 2008 and by the Delhi government in 2016 are notable attempts to alleviate fleet congestion and air pollution in cities (Zhang et al. 2011).

These days, due to advancements in ITS data generation and storage technologies, the connected vehicles became the key driver of IoT (Bonomi et al. 2013). Such vehicles are leveraged with onboard sensor units and form connected networks with ITS subsystems such as roadside infrastructures, parking lots, smart vehicle charging

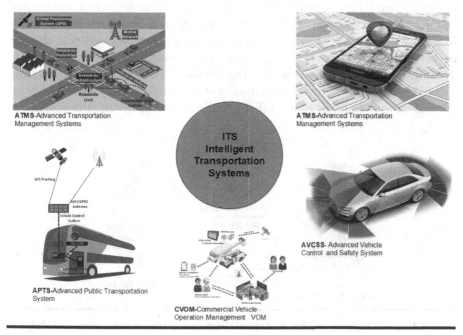

**Figure 10.1   Key components of a data-driven ITS.**

stations (Hussain et al. 2018a), advanced metering infrastructures (AMIs), etc. Also by enabling the IoT technology as a global standard and basis for ITS communications, new avenues will be created for maximizing the prospects for future innovations (Hussain et al. 2018b). The cyber-machineries in connected vehicular networks aim at providing real-time connectivity, automation and tracking of on-premise smart devices, deployed for analysis monitoring and control of the ITS infrastructures. The IoT evolves the legacy ITS architecture into a transportation cyber-physical system (T-CPS) (Figure 10.2), where the cyber half forms the basis of data-driven big data analytics, ultimately focused to manage the operation of physical subsystem (Hussain et al. 2019). The IoT-aided ITS turns into an integrated framework of real and virtual worlds where the attributes from the real (physical) as well as virtual (cyber) world are fed as input to the control/data centers (clouds), to generate simulation models for predicting future mechanisms and transitions (Hussain et. al. 2018c).

Since the T-CPS comprises billions of these intelligent communicating devices (vehicles, roadside infrastructures, sensors, etc.) that generate an enormous amount of data, performing analysis on this data is a significant task (Mashrur Chowdhury et al. 2017; Hussain et. al. 2018d). Using search techniques, the size and extent of data can be reduced and limited, so that an application can choose just the most important and valuable data items as per its necessities (Pattar et al., 2018). It is, however, a tedious task to effectively seek and select a proper device and/or its data among a large number of available devices for a specific application. Suppose that on a weekday morning, an electric vehicle (EV) driver needs to charge his car. In a crowded city, if he randomly goes to a nearby (maybe the nearest) charge station, he might have to wait a long time in the queue. One viable option for the driver is to use his GPS app (e.g., Google Maps) and find a noncongested-cum-shortest possible route. However, online maps of centralized search engines cannot provide dynamic information about the charging station. Even if the driver chooses the optimal route to reach the charge station, it is not guaranteed that he will obtain the highest priority token number. For specific EV charging case, the situation is

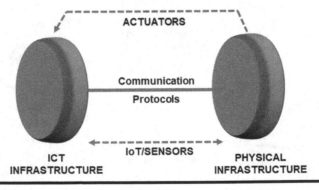

**Figure 10.2  A framework for transportation cyber-physical systems (T-CPS).**

even more complex because the power tariff is dynamic and may even vary from station to station. Thus, the earnest need is to have search (recommendation) strategies that capture the context and geo-distribution of every agent existing in the infrastructure.

Hence, efficient search techniques are essential to T-CPSs that are prone to challenges caused by a large number of IoT devices, dynamic availability, power- and resource-limited devices, real-time data streams in various types and formats, historical and behavioral monitoring, etc. (Bitam 2012; Stojmenovic and Wen 2014). The searching techniques are an integral part of "sensing as a service layer" in a T-CPS architecture. This architecture is built on top of the IoT-aided ITS infrastructure, where middleware solutions connect sensor devices to software systems and their related services.

In the past, several studies were conducted to examine the search techniques in IoT-aided systems. Based on the design space of different approaches, Römer et al. (2016) presented a bird's-eye view on search techniques in IoT. However, sensor measurement data has not been addressed in this survey. Zhang et al. (2015) compared search techniques in IoT with other spheres viz. ubiquitous computing, information retrieval and mobile computing. They identified architectural design, real-time, scalability, locality of the search, etc. as major research directions. However, they didn't perform the entire gamut of the investigation. Search techniques in the Web of Things (WoT) are analyzed by (2011) according to aggregation approaches (i.e., either pull-based or push-based). Similar efforts were made by Zhou et al. (2016), where they focused on search algorithms in the WoT domain, focusing on techniques applied, types of targeted search results and data representations. Keeping all these discussions and challenges in mind, in this chapter:

- We examined the evolution of century-old transportation system to a transportation cyber-physical system enabled by periodization of network service development (NSD) technologies such as wireless sensor networks (WSN), IoT, cloud computing and multi-access fog computing.
- We analyze the steps of resource searching in a T-CPS along with the requirement analysis and challenges in developing search engines for next-generation intelligent transportation systems.
- We propose a distributed edge based model for Transportation Cyber-Search (TCS) along with the key processing blocks and the workflows in TCS design.
- We highlighted the architecture and the key technology enablers of TCS design, the data acquisition technologies and the conceptual framework for big data analytics in TCS design.
- We examined the potential privacy and forensics challenges in T-CPS applications (taking TCS as an example) and examined the state-of-the-art solutions in T-CPS forensics.

The chapter is organized as follows: In Section 10.2 we explore the data infrastructure development solutions and NSD technologies, considering diverse T-CPS applications, their data workload characteristics, and corresponding requirements. In Section 10.3, we first lay out some requirements and challenges for developing a robust and comprehensive IoT search solution. Then we discussed the steps of resource searching in T-CPS, along with the key requirements and challenges. In Section 10.4 we propose our distributed search model "Transportation Cyber-Search (TCS)" and outlined the key processing steps of TCS search engines. Section 10.5 highlights the key technologies of TCS design. In Section 10.6 we analyzed the security challenges faced while going for TCS deployment, specifically TCS privacy and T-CPS forensics. Section 10.7 concludes the chapter.

## 10.2 Evolution of Network and Storage Technologies in Intelligent Transportation Systems

Due to rigorous research and investments in ICT, the network and system services are evolving at a very fast pace. ICT services are ubiquitous across every domain and their integration into the legacy systems such as transportation systems, power grids, vehicular networks, healthcare, etc. have transformed the existing architectures and functionalities (Hussain and Alam, 2018). Simply saying, the ICT tools and services became the bread and butter for all such emerging applications. In order to ensure services and functionalities (provided from millions of connected devices) available to billions of clients, network applications and distributed systems became two abutted pillars of the same trend. Similarly, the traditional web- and Internet-aided applications need to scale to billions of users' endpoints. Often hundreds of computers are clustered, and many clusters are geographically dispersed and connected so that users perceive them as single service (Luntovskyy and Spillner, 2017). The architectures must be trained for high performance, high reliability, high privacy and security, low cost, low effort and low energy consumption, among other factors (Gubbi et al. 2013). Services not offering all or a majority of these benefits will have dwindling chances to compete for clients hence will ultimately fail to be sustainable.

We explore the data infrastructure development solutions and NSD technologies, considering diverse ITS applications, their data workload characteristics, and corresponding requirements (Figure 10.3). An overview of infrastructures to support the requirements of data infrastructure capable of storing, processing, and distributing large volumes of data using different abstractions and runtime systems are presented. ITS application requirements are then mapped to a technical architecture for a data infrastructure. Different high-level infrastructures focusing on the different programming systems, abstraction, and infrastructures, and low-level infrastructure focusing on storage and compute management are summarized. In this section, we take the case of transportation systems and observe the

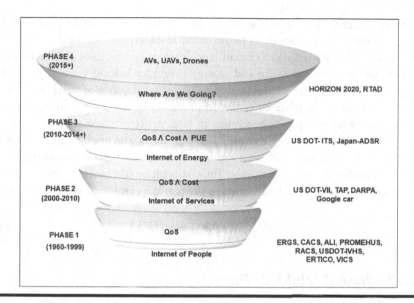

**Figure 10.3    Evolution in intelligent transportation technologies.**

brief history and periodization of ICT network services and distributed systems, and predict the complex landscape of distributed service systems in the future. Specifically, we identify four distinct phases of technological foundation toward achieving a breakthrough in this vision, as shown in Figure 10.3.

## 10.2.1  Phase 1 (1960–1999)

The first phase of network transformation can be traced back to the '70s (1970–2000), which marks the establishment of basic network services and early web applications. The key goal here was to offer more and more functionalities and improved QoS. Here, the QoS considerations were mostly confined to strict technical network characteristics, without taking end-to-end user experience (QoE) into account. Attempts were made to achieve increased bandwidth and reduced latency. For instance, in 1999, a 56 Kb/s modem connected to copper telephone networks was the norm for private users and replaced by faster digital subscriber line (DSL) connections with 768 Kb/s downstream bandwidth. It is noteworthy that this phase marks the first monopolization tendencies, i.e., the previous network protocols that were defined and implemented by multiple vendors are now standardized by a centralized third party called the World Wide Web Consortium (W3C) (Misra et al. 2014). Though it provides vendor-specific extensions, even today it still causes trouble and processing overhead.

Hence, the consumers could only rely on such numbers as upper bounds in a best-effort service market and could not easily translate these numbers into application benefits, for instance, video quality or file transfer performance. Besides, large computing centers began to be installed in enterprise markets as they were economically effective, thanks to broadband Internet connections that enabled the consolidation of many compute and storage resources behind a single data pipe. The system reliability was improved due to better availability of spare parts (hard drives, power units, switches etc.), the employment of redundant units wherever possible and emergency power generators in large centers, where they were feasible (Saleem et al. 2018; Porambage et al. 2018). Similarly, the application availability and scalability were increased with replicated setups in high-availability/failover and load-balancer setups, respectively. Thus, this phase has been about connecting people to the Internet, i.e., the Internet of People. We can formulate it as:

$$GOAL_{phase1}^{ITS} = Maximize(QoS) \tag{10.1}$$

### 10.2.2 Phase 2 (2000–2010)

Due to the advent of hardware consolidation and server virtualization in the next phase (2000–2010), cost optimization requirements were explicitly added to the QoS enhancement schemes. Service level agreements (SLAs) were defined that mandated a minimum cost achievement within strictly given QoS constraints (Mukherjee and Matam 2017). Similar to the first phase, the use of large computing centers (aka clouds) was still a culture, as the maintenance cost in the large computing centers is lesser than in smaller ones (Dastjerdi and Buyya 2016). The servers can be updated centrally with security patches, upgrades can be better tested before deploying and the maintenance actions are mostly the same at homogeneous servers (Porambage et al. 2018). In 2009, 16 Mbps/s ADSL connections were offered in many developed urban cities and even 55 Mbps/s VDSL2 connections were available in selected hotspots. Because of faster connections, peer-to-peer file-sharing applications such as Bittorrent, video conferencing, and crypto-currency became popular (Paper 2019; Zanella et al. 2014). However, the promise of many governments during this time to achieve 100% broadband coverage had (and still has) not been achieved anywhere. Among all, the second phase was characterized by the rollout of cloud-hosted applications. Figure 10.3 summarizes the status of ITS developments in the second phase. The formula to characterize this phase is:

$$GOAL_{phase2}^{ITS} = Maximize(QoS) \wedge Cost \leq constraints \tag{10.2}$$

### 10.2.3 Phase 3 (2010–2014+)

The third phase (after 2010) was earmarked by the trend of "green and sustainable" computing and increasing energy demand and prices. The enterprise mega data

**Table 10.1    Data Center Distribution Across the Globe**

| Name | Size (sq. ft) | Location |
|---|---|---|
| Range International Information Group | 6,300,000 | Langfang, China |
| Switch SuperNAP | 3,500,000 | Nevada, USA |
| DuPont Fabros Technology | 1,600,000 | Virginia, USA |
| Utah Data Centre | 1,500,000 | Utah, USA |
| Microsoft Data Centre | 1,200,000 | Iowa, USA |
| Lakeside Technology Centre | 1,100,000 | Chicago, USA |
| Tulip Data Centre | 1,000,000 | Bangalore, India |
| QTS Metro Data Centre | 990,000 | Atlanta, USA |
| Next Generation Data Europe | 750,000 | Wales, UK |
| NAP of the Americas | 750,000 | Miami, USA |

centers were built more often in colder regions of the globe (refer to data enter distribution in Table 10.1). More energy-efficient hardware was installed, and software was developed from an energy conservation perspective. In order to save electricity consumptions, overall idle periods and shrink carbon footprints, the processors were tuned to dynamic voltage and frequency settings. The metric power usage effectiveness (PUE) has gained prominence, and consumers are increasingly aware and demanding sustainable IT (Hussain et al., 2018). The use of mobile phones to host applications and even mobile services strengthens the awareness due to limited handset battery capacity. Smart grid (SG) installations are on the rise and lead to greater energy autonomy by introducing the integrated notion of *prosumers* (producer plus consumer) (Wu et al. 2012). The formula for the third phase is:

$$GOAL_{phase\ 3}^{ITS} = Maximize\left(PUE\right) \Lambda \left(QoS \geq QoS_{minimum}\right) \Lambda \left(Cost \leq Cost_{minimum}\right) \quad (10.3)$$

Besides, this era is characterized by the rollout of WSNs across all emerging infrastructures, viz. ITS, SG, Vehicular Adhoc Network (VANET) etc. WSN connectivity technologies led to the emergence of diverse abstractions: compute, storage and networking services which concentrate applications and services in shared clouds.

## 10.2.4  Phase 4 (2015+)

The fourth phase and the next development vector's inception was in 2016 and will cause a high impact on computing in the near future. Now, the focus is not just on networking services and distributed software applications, but to a truly user-focused Internet of Service (IoS) assisted by IoE environments, where "E" is comprised of things, vehicles, energy, etc. (Munir et al. 2017). For instance, the ITS services are executed via clouds, in the frame of the IoT with many connected vehicles (V2V), wearables (W2W), a diverse range of devices, T-CPSs and robots, mobile ad-hoc networks (MANETs) and ultimately fog, edge, and wearable computing (Shojafar et al. 2016). The upcoming architectures will provide services that are always on, always available, and deployed as pay-as-you-go utility. The relationship between all four phases can be defined as

$$Phase\,1 \subset Phase\,2 \subset Phase\,3 \subset Phase\,4 \tag{10.4}$$

Also, it can be observed from Figure 10.3 that

$$Phase\,1 \cup Phase\,2 \cup Phase\,3 = Phase\,4 \tag{10.5}$$

Hence, ITS applications will become even more critical to operating in upcoming scenarios. Connected vehicles will alleviate traffic congestion and increase traveler safety and environmental benefits. Autonomous vehicles will become available to the mass population to further improve mobility efficiency and safety. Smart and connected cities will emerge as a system of interconnected systems, including transportation, residencies, employment, entertainment, public services, and energy distribution. Federal bodies are funding several projects on recent ITS research and deployments focusing on connected (CVs) and automated vehicles (AVs). For instance, a 2015–2019 multimodal program plan was developed by the U.S. DOT's ITS Joint Program Office to promote AV research and deployments. Under this plan, the regulatory framework for AV operation on public roads was established in California, Florida, the District of Columbia, Nevada, and Michigan. Similarly, Japan introduced its ITS Spot program using dedicated short range communication (DSRC) radio with an aim to develop and verify the automated driving system (ADS) for safe operations on public roads by 2030. European nations are actively participating in projects like Horizon 2020, which defines the framework for safe and automated road transportation (Chowdhury et al. 2017). According to 2008 census data, nearly half of the world's populations were urban, and it is predicted to become 66% by 2050 (Chowdhury et al. 2017). With dense populations, future cities will need to confront severe transportation challenges characterized by managing safety and air pollution under conditions of excessive traffic congestion and inadequate infrastructures. All such attempts depend on developing sustainable future ITS systems characterized by key (but not the least) features summarized in Table 10.2.

**Table 10.2 Characteristics of Sustainable Future ITS Systems**

| | Features | Objective |
|---|---|---|
| 1 | QoS/QoS Guarantee | • To deploy streaming and real-time ITS applications with robust service interfaces<br>• To have low-latency, location-aware, energy-efficient use of heterogeneous hardware from large-scale computing centres to tiny nodes (aka Fog/Edge Nodes) |
| 2 | Secure Transportation | • The importance of utilizing emerging capabilities that demonstrate the potential to transform transportation at the same time that user and citizen privacy is protected |
| 3 | Autonomy | *To have ITS infrastructure which is:*<br><br>• Self-adaptive<br>• Self-organizing<br>• Self-configurable<br>• Self-described<br>• Self-optimized<br>• Self-protected<br>• Self-discoverable<br>• Self-matchmaking<br>• Self-powered |
| 4 | Interoperability and Backward Compatibility | • The evolution of standards and architectures to ensure that technological advancements are reflected, and the maintenance of backward compatibility and interoperability of different ITS components |
| 5 | Connectivity and Mobility | • Very big number of hardware nodes and their mobility, based on IPv6 connectivity |
| 6 | Smart City Integration | • The development of a workforce of transportation professionals trained to capture, manage and archive data collected from the smart city system |
| 7 | Data-Driven Public and Private Transportation | • Public agency acceptance of the integration of data analytics via public-private partnerships to provide public agency transportation professionals with the required skill sets to manage these systems |

## 10.3 Resource Searching in T-CPS—A Synoptic Overview

Before going to the searching techniques in T-CPS domain, we first lay out some requirements and challenges for developing a robust and comprehensive IoT search solution.

### 10.3.1 Searching the IoT: Requirements and Challenges

IoT data services are usually designed to be available to devices and users on request at any time and at any location. Quality, latency, trust, availability, reliability, and continuity are among the key parameters that impact efficient access and use of IoT data and services. However, the current data and service search, discovery, and access methods and solutions for the IoT are more suited for fewer (hundreds to millions) and static (or stored) data and service resources. With IoT resources (in the billions, with dynamic, always-on, and streaming sources), we are faced with a different ecosystem in terms of the number of resources, heterogeneity and complexity, and the amount of data. Efficient discovery, ranking, (automated) selection, access, integration, and interpretation and understanding of the data and services from various resources require coordinated efforts among networks, data/service provider resources, and core IoT components (such as discovery engines) to select and use the right resources at the right time. The distribution, scale, heterogeneity, multimodality, streaming data, and dynamicity of IoT environments imply that existing solutions for searching, accessing and using the information on the Internet and the web will not be applicable or will remain far from adequate for practical and large-scale dynamic IoT applications.

Next up in the evolution is the incorporation of finding relevant multimodal and multisource information that is collected and published from the physical world. This requires that bots respond to a user's information needs to be expressed in a conversational form or to machine-requested data based on context (for example, time, location, and type) rather than keywords typed by a user in a search box. The raw data is also often required to be integrated from different sources and further analyzed to extract information, events, and insights (rather than presenting raw observation and measurements). Figure 10.4 illustrates the data and query actors in a simplified IoT discovery scenario. The IoT domain has seen growing advancements in device and hardware manufacturing, networking and lower-layer protocols, and the integration of real-world objects into the fabric of the current Internet and the web. In the near future, IoTs will be an inherent extension of current networks. It will be possible for every connected thing to blend and interoperate using the underlying networking and communication technologies, data and service communications, and interactions with other things and resources on the Internet.

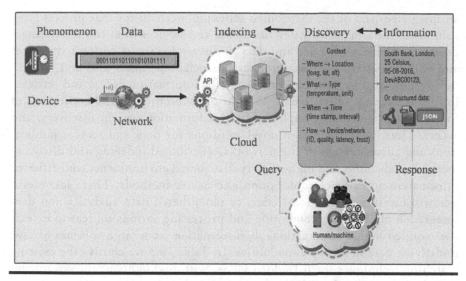

**Figure 10.4  Control flow for data discovery, Internet of Things (IoT).**

The data can be provided by various devices (wireless sensor networks) and by other smart objects (network-enabled devices), applications, and social media streams. The data can be published directly or relayed (and further analyzed) by gateways and other middleware components. Index and discovery services and their associated algorithms should be able to read parameters and attributes of the data providers (devices, smart and connected objects, and other sources) and create efficient, distributed, and scalable solutions to look up, discover, and access the data. The queries can be initiated by human users, applications, and other devices that might require obtaining the data based on different criteria and various options (such as location, time, interval, type, and value). However, the current information access and retrieval methods on the IoT are still at the same stage that the web and the Internet were when web search came on the scene in the late 1990s. Information retrieval on large-scale IoT systems is based on the assumption that the sources are known to the devices and consumers or that those opportunistic methods will send discovery and negotiation messages to find and interact with other relevant resources in their outreach.

The latter assumption can be interesting in scenarios such as autonomous vehicles and car-to-infrastructure communications and interactions; however, IoT systems will also require search and discovery of resources in large-scale distributed networks in scenarios such as environmental monitoring, smart city, and disaster response applications. In summary, IoT data requires efficient and scalable indexing and ranking mechanisms for distributed and dynamic IoT environments. Quality, trust, and availability analysis methods will be an integrated part of the distributed indexing mechanisms that process large volumes of the semantically described data and services from connected smart devices and objects. Service and data discovery for smart connected devices and objects will also involve automated associations, mash-ups, and integration to provide an extensible framework for information access and retrieval in IoT. We cannot separate data publication, subscription, and constraints of the resources and their complexity from information search, discovery, and access. Thus, we must design novel solutions for data and service publications and advertisement in the networks, distributed indexing and discovery methods in dynamic and high-velocity distributed environments, and efficient information and service subscription and access methods. These last methods will use in-network, local edge, or cloud-based data analytics and data integration methods for combining and processing various sources to extract the required higher-level, actionable information from large volumes of raw, underlying data in multiple modalities. In Table 10.3 we classify the existing searching techniques for IoT-added CPSs, with highlights of the advantages and challenges of each of them.

## 10.3.2 Resource Searching in T-CPS

Let us first define the steps for searching for resources in a T-CPS. Here, the search query will look for an IoT device, data or a result generated from both. For instance, an EV driver makes a search query to the VANET, to find the location of an optimal charging station for charging his vehicle or locating a free parking spot at a particular area at a given time. The flow for overall search process is shown in Figure 10.5. Sensors are embedded into the IoT objects that collect real-time data about the vehicular environment. In the real world, they detect events and then generate data about the detected events. Since the objects are networked at various levels (e.g., local or global), a middleware is employed to manage them. The objects register with the middleware through a subscription process; the APIs are provided by it for application development and to perform management operations. There may be different search principles used for searching ITS devices and their services. Some of the design principles and techniques are given in Table 10.3. However, for each search query generated, the following steps are commonly followed to obtain the required number of matching resources.

### 10.3.2.1 Query Generation

Users/Machines submit their query to middleware either by using an interface provided by the application or through the API. Once a query has been received by the search system, it can be processed (through techniques like transformation, filtering, normalization, etc.) and divided into sub-queries.

**Table 10.3  Existing Searching Techniques for IoT-Added Cyber-Physical Systems (CPS)**

| Approach | Features | Challenges |
|---|---|---|
| Content-Based (Data) | • Allows access to historical and real-time data.<br>• Statistical and prediction models can be employed.<br>• The reasonable accuracy of search results.<br>• Missing values, future outcomes, and error detection are relatively easy. | • **Dynamicity:** Prone to errors between two successive readings in a short interval of time.<br>• **Scalability:** Have to send a data request to all the nodes.<br>• **Heterogeneity:** Have to deal with data in various formats. |
| Context-Based (Data) | • Information about the deployment environment and IoT object's state are available.<br>• Do not need to deal with low-level sensor data.<br>• Provides efficient control to manage IoT objects through context-information.<br>• Generates high-level and meaningful results. | • **Management:** Requires dedicated context-aware server for management.<br>• **Data Acquisition:** Hard to retrieve context-information from IoT objects.<br>• **Generality:** Strongly integrated with specific application requirements. |
| Location-Based (Object) | • Supports indexing and ranking methods to reduce search space.<br>• Strongly coupled with query-routing techniques. | • **Identification & Naming:** Difficulty to uniquely identify IoT objects when they are collocated.<br>• **IoT-Object Safety and Security:** Prone to physical damage and operational condition when the location is revealed.<br>• **Mobility & Dynamicity:** Has to deal with mobile IoT objects, which frequently change their availability status. |
| Social-Structure-Based (Object) | • Leverage social networks formed by humans.<br>• User behavior and likeliness are well modeled. | • **Privacy:** Prone to identity thefts as personal and private information are collected. |

*(Continued)*

**Table 10.3 (*Continued*)   Existing Searching Techniques for IoT Added Cyber-Physical Systems (CPS)**

| Approach | Features | Challenges |
|---|---|---|
| Semantic and Ontology-Based (Object) | • Allows representation of complex real-world events.<br>• Results can be modeled in natural language.<br>• Crawlers and Indexer can be utilized effectively.<br>• Preprocessing and query subdivision are supported. | • **Interoperability:** Various ontologies across different domains have to be integrated.<br>• **Standardization:** No well-defined and accepted ontology for all applications. |
| Resource and Service Discovery | • Support for QoS parameters is well established.<br>• Complex relationship is managed. | • **Opportunistic Presence:** Devices and their services are not available all the time.<br>• **Specificity:** Limited to the application domain, due to lack of a general semantic model to represent all kind of IoT objects. |

## 10.3.2.2 Data Acquisition

Once the data is fed to middleware, the search algorithm is then executed. The data from IoT resources can be gathered by middleware in the following different ways:

### 10.3.2.2.1 Publish/Subscribe

In the IoT network, resources are loosely coupled with each other and the middleware. An explicit relationship between an IoT resource and middleware is established in this method, where the IoT resource acts as data publisher and the middleware as data subscriber.

### 10.3.2.2.2 Request/Response

In this method, IoT resources and middleware are tightly coupled with each other. A resource request for some kind of service is sent to the middleware, and the middleware responds to the request by providing the service. The data acquisition methodologies are explained in detail in Section 10.4.

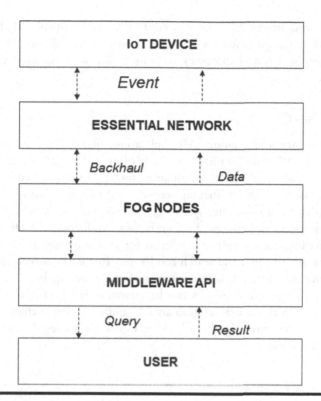

**Figure 10.5  Steps of resource searching in T-CPS.**

### 10.3.2.2.3 Fog Computing

The sub-queries are then presented to the fog nodes (e.g., vehicular fog nodes [Hussain et al., 2018], traffic light [Liu et al. 2018], etc.). The system then contacts the nodes (cloudlets) in the ITS network to retrieve a list of matching resources.

### 10.3.2.2.4 Result Aggregation

The sub-results/resources/data obtained from the fog nodes are further indexed and ranked based on some scoring method employed by the search application. Now, the ranked outputs are again mapped to the query request to find the required number of matching resources.

## 10.3.3 Requirements and Challenges

The search and discovery techniques for the T-CPS applications are faced with numerous challenges that reduce their performance quality. They need to support certain requirements to enhance their applicability and usability across different

IoT application domains. In order to identify the requirements and challenges of developing search engines for T-CPS applications, we first discuss the applicability of different search and discovery techniques across various use cases in IoT applications.

### 10.3.3.1 Use-Case 1

In aerial, aerospace and aviation ($A^3$) applications of T-CPS, a search algorithm can be used be to locate the present altitude of an airborne vehicle. The altitude data of this vehicle is collected through an altimeter sensor attached to the vehicle; for example, laser vegetation imaging sensor (LVIS) is an airborne altimeter sensor used to perform topography, hydrology, and vegetation studies [24]. However, this data is dynamic and changes frequently, as such during take-off and landing. Requirements for search engines developed for that purpose could be to obtain accurate data in real time, and search results may further be subjected to filtering based on different altitudes, where a range of altitudes can be displayed in different facets; however, these results can be viewed only by authorized personnel. Challenges faced in this search scenario are to acquire frequently changing data that is specific to $A^3$; the search algorithm here has to operate among different data communication protocols.

### 10.3.3.2 Use-Case 2

One of the crucial parts of secure transportation is anomaly detection viz. intrusion detection or theft detection. Real-time detection and control of unwanted vehicles or intruder can prevent harmful attacks. Such a scenario requires the deployment of sensors that monitor the entry of attackers or thieves. However, the number of such sensors utilized is enormous due to the size of the vehicular network and thus, possesses scalability challenges. Also, the search algorithm should be energy efficient and impervious to the noise gathered from the data source. Moreover, the system should also be able to identify the intent of the new vehicles joining the existing fleet.

### 10.3.3.3 Use-Case 3

The T-CPS is one of the most thriving application domains of IoT. Consider a vehicle routing problem where the driver in heavily congested traffic is assisted by his car's navigation system to find an optimal route, by contacting nearby cameras, traffic management system and peer vehicles in the same location to reach the destination in the stipulated time. For such applications, the search system should retrieve information on the volume of traffic on road networks of the commuting area in real time and the latency between subsequent queries about the route information should be kept to a minimum. However, the most critical requirement is that of the accuracy of the query result. To design such a search system, the connectivity of the vehicle with the T-CPS infrastructure is of major concern and also the myriad of communication standards used by the T-CPS pose a challenge for the effective design of a search solution. The search and discovery techniques for routing applications should also have an effective user interface design that provides updated status information about traffic congestion and vehicle mobility. Similarly, the smart traffic lights need to adapt themselves to the real-time traffic circumstances within a particular region. In this case, the reaction time for one or several smart traffic lights is too short, such that it is virtually impossible to traffic all the application execution to a distant cloud. Therefore, such traffic lights should be programmed in a way that they autonomously cooperate with each other and with all the locally available computing resources such as roadside units (RSUs) to coordinate their operations. Keeping such use cases in mind, in Figures 10.6 and 10.7 we respectively highlight the key data collection, processing and disseminating requirements and challenges of search algorithms implemented in ITS cyberspace.

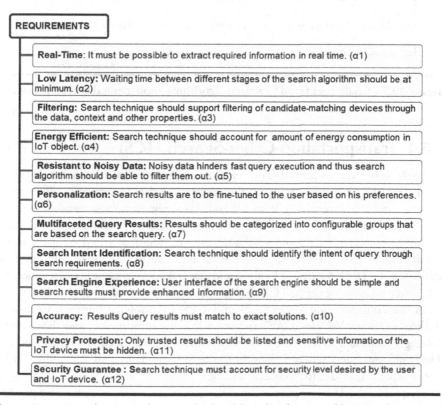

**REQUIREMENTS**

**Real-Time:** It must be possible to extract required information in real time. (α1)

**Low Latency:** Waiting time between different stages of the search algorithm should be at minimum. (α2)

**Filtering:** Search technique should support filtering of candidate-matching devices through the data, context and other properties. (α3)

**Energy Efficient:** Search technique should account for amount of energy consumption in IoT object. (α4)

**Resistant to Noisy Data:** Noisy data hinders fast query execution and thus search algorithm should be able to filter them out. (α5)

**Personalization:** Search results are to be fine-tuned to the user based on his preferences. (α6)

**Multifaceted Query Results:** Results should be categorized into configurable groups that are based on the search query. (α7)

**Search Intent Identification:** Search technique should identify the intent of query through search requirements. (α8)

**Search Engine Experience:** User interface of the search engine should be simple and search results must provide enhanced information. (α9)

**Accuracy:** Results Query results must match to exact solutions. (α10)

**Privacy Protection:** Only trusted results should be listed and sensitive information of the IoT device must be hidden. (α11)

**Security Guarantee :** Search technique must account for security level desired by the user and IoT device. (α12)

**Figure 10.6 Requirements for search algorithms implemented in ITS cyberspace.**

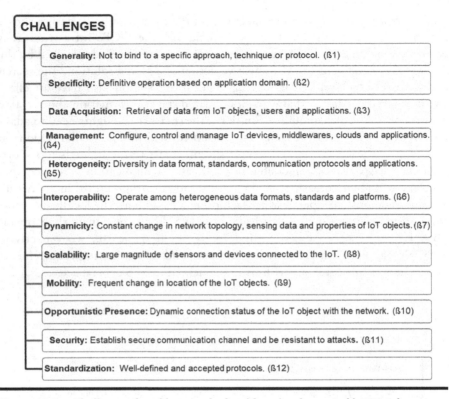

**Figure 10.7  Challenges faced by search algorithms implemented in ITS cyberspace.**

## 10.4 Transportation Cyber-Search (TCS)

In crude words, the traditional search engines face greater difficulty in meeting users' needs. For the ubiquitous transportation cyberspace that incorporates information, persons, and things, evolving Web 2.0 and 3.0 application modes and new search needs in an era of big data call for a new generation of revolutionary, milestone-making search engines that offer valuable, smarter solutions based on users' actual needs. Figure 10.8 depicts the workflow for big data processing in transportation cyber-search (TCS). Motivated by abovesaid discussions we present here the concept of TCS. Accordingly, we define TCS as a "search system that seeks to acquire knowledge from the Transportation Cyber-Space in T-CPS, by involving humans, Connected and Autonomous Vehicles (CAV) and other Internet information, and then provide intelligent solutions to meet the user's search intention with accurate understanding." Using efficient query retrieval algorithms, the multimodality and ambiguities in the user's input are eradicated. Since it leverages quick and accurate understanding of user's queries, the TCS not only achieves the goal of search accuracy but also provides an improved user experience with the simplified interaction with users.

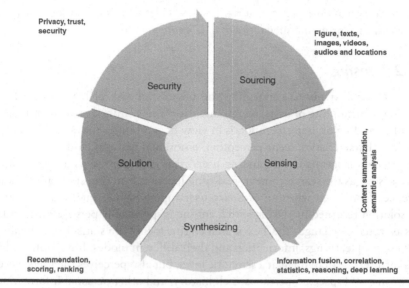

**Figure 10.8 Big data processing in transportation cyber-search (TCS).**

As discussed, the intelligent search solutions in a TCS system are a set of answers ranked on the basis of knowledge acquired from the T-CPS users' queries. The system employs different techniques viz. matching, reasoning, and crowdsourcing, etc., to preprocess the queries and gives the search result that captures both the user's context as well as intent. In the next section, we highlight the key processing steps of TCS search engines.

## 10.4.1 Sourcing

Sourcing from the cyberspace means gathering data from the ITS cyberspace with online vehicles and humans, things, information and their geospatial and temporal properties according to the given target or task. These data, which widely cover the various applications in cyberspace and may appear in different models, are the data foundation of big search. The data acquired from the cyberspace are those gathered from different channels and sources, as long as they are associated with the query request, including directly matched and semantically relevant data. Different from the traditional web search, the TCS engine not only obtains information from the World Wide Web, but also from the IoT, social transportation networks, smart parking lots, video surveillance systems, geographic information systems, intelligent traffic management systems, and other T-CPS applications. With these resources, TCS extracts valuable data of users' knowledge, relevant physical messages, and so on. For spatial and temporal data, TCS also records real-time attributes to keep them updated. In addition, TCS differs from the traditional search in data patterns.

With the combination of figures, texts, images, videos, audios and locations, the TCS has various forms and modalities of data expression.

## 10.4.2 Sensing

Sensing the context means confirming the target and task of big search with a semantic understanding of users' search intention and a unified expression model. The understanding of users' intention is in view of combining the queries' context, space-time characteristics, scene perception, emotional gestures, and other factors. Scene perception means understanding users' real intention by using their query scenarios, such as the search context, real-time location, emotion state, and unique preference. Thus, even when an ordinary user inputs his/her question, the intelligent solution returned by the big search engine is specifically personalized, and it differs as scenarios change. The TCS search engine focuses on statistic and dynamical data of people, things, information and their different modes. It not only understands the users' search intent at a semantic level, but also perceives the users' needs by their time and space properties, search history, and other personal information.

## 10.4.3 Synthesizing

This is the core of TCS that embarks to mining and discovering knowledge and wisdom from the data acquired from the T-CPS by employing state-of-the-art technologies viz. information fusion, correlation, statistics, reasoning, deep learning and even crowdsourcing in a unified representation model of knowledge and relationships. Here the knowledge is synthesized not only from a single channel but also from the integration of other cyberspaces and applications, covering a diverse range of representation models of data and knowledge. The knowledge is finally integrated after inference and refining, with comprehensiveness, accuracy, and abundance. Knowledge synthesis highlights the comprehensive analysis of possible relations among the multisource data or different schema data to mine or discover more novel and rich knowledge. For example, for the query "Autonomous Vehicles in Europe," the traditional search engines give only a collection of relevant web pages, but the TCS system search will obtain much more comprehensive knowledge of its basic information, controversial content, historical evolution, multimedia information, public discussion, development trend, and more. In order to produce such optimized results, the TCS needs a mechanism to process these relevant matching results, including association analysis, reasoning and computing, systematic overall organizing, and so on.

## 10.4.4 Solution

The TCS provides a ranked set of intelligent solutions to the users in an appropriate format, based on accurate understanding of search context and content and the knowledge discovered from T-CPS. The intelligence here refers to a comprehensive and wise answer that meets users' needs exactly. It may involve multifaceted elements of user needs and may be time critical. For example, when a user searches for information about vehicle servicing, the traditional search engines will only provide some relevant web pages, leaving the user to do specific plans by himself. Different from that, TCS will recommend particular reference renovation programs concerning the vehicular conditions that the driver can choose from. In this case, the intelligent solution is the one that focuses on the property and the user's historical preferences. Moreover, the TCS will ensure that the results have been organized with framework services or knowledge integration. Simultaneously, the engine will sort the results in an effective way, and eventually it will return consensus ranked results in the order of priority.

### 10.4.5 Trust and Security

It is the fundamental guarantee of big search—i.e., the entire process of big search, from users' intention and understanding, data acquiring, and knowledge synthesizing to intelligent solution returning—is credible and security-guaranteed. It also supports privacy protection and harmful information filtering. Credibility means that the data source for big search is reliable, trustworthy, and traceable; security means that the result of a big search will not be abused by unauthorized users; privacy protection is to ensure that users' privacy (such as personal information, location data, and others) will not be illegally leaked out during the search process. Nowadays, social networks such as Micro.blog and Facebook are currently flooded with rumors and hostile attacks because most users are anonymous online. This will have an adverse effect on social transportation. To solve such issues, we have to ensure the credibility of data and sources. With these requirements, TCS has emerged with a strong capability of correlation analysis and can associate and induct all the aspects of important persons and things to form new cognition or knowledge. However, the comprehensive information may contain personal privacy or even federal secrets, so the search results must be strictly judged and controlled. This is a prerequisite for the promotion of TCS. In general, the system must have the ability of accurate filtering of some specified content, such as microblogs, the flooded violence, pornography, reactionary, and other undesired information on the Internet. Only in this way can TCS be in accordance with national regulations and judicial constraints.

## 10.5 Key Technologies of TCS Design

The fundamental strategies that are to be considered when developing and implementing a search system for the IoT are:

### 10.5.1 Architecture

The TCS system for IoT applications is designed and implemented through either centralized or distributed architectural styles. In the former approach, a central middleware/server is responsible for the search system as opposed to a distributed approach. In decentralized architectures, the TCS system comprises of middleware that store indexes, data items of the IoT aided ITS resources are distributed geographically and the search algorithm is run locally on them. Results are then aggregated at the global level by combining local search results. Figure 10.9 depicts a distributed or decentralized architecture of TCS systems along with the basic information processing flow as described in Section 10.3.

## 10.5.2 Data Acquisition from ITS Cyber-Space (T-CPS)

It is a ubiquitous cyberspace-oriented, task-based information acquisition mode that refers to the process of collecting, acquiring, and exploring data and information needed by users with certain strategies and methods. The objective is to effectively organize, store, and manage the acquired data to lay the foundation for intelligent answering. Key domains of data acquisition are discussed in the following sections.

### 10.5.2.1 Internet Data Acquisition

Internet data acquisition means automatically collecting big data in a highly parallel manner and quickly gathering them in the system, according to the tasks deployed by users. It involves a web-page data acquisition mode, which mainly applies web

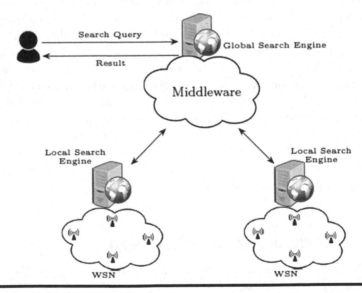

**Figure 10.9   Architecture of distributed search engine for the IoT-aided ITS (T-CPS).**

crawlers and the service data acquisition mode, which means adopting the way of service interface calling to obtain network service data. For example, if a user is concerned about the political situation in a remote area, he can submit to the system three keywords: "Unmanned Aerial Vehicle (UAV)." The system then shall automatically start the "web crawler" program to collect data from multiple sources.

### 10.5.2.2 Data Acquisition in IoT

IoT data acquisition includes the collection of data through radio frequency identification (RFID) and through wireless sensor network techniques. RFID data collection technique involves collecting basic information of humans, things, equipment, environment, and status on time or in real time through tag readers and tag receivers. Take an intelligent transportation system as an example. In recent years the United States, Japan, and some developed European countries have stepped up research on intelligent traffic monitoring systems based on wireless sensor networks. Sensors collect information on real-time traffic, real-time speed, road accidents, and road conditions, and then such information is transferred to the gateway nodes. After initial processing, the information will be uploaded to a data management platform to analyze the real-time traffic condition and provide support for traffic management and scheduling.

### 10.5.2.3 Traffic and Surveillance Data Acquisition

Traffic or surveillance data acquisition is a procedure for collecting and integrating video data from video surveillance systems and the Internet-based on users' demand. For instance, if users are concerned about the traffic conditions in the major national highways within a certain time span from October 1, 2014, they can submit a query about the "specific time" and "specific road section" to the system. The system will immediately start the "video acquisition program" to collect data from the traffic video surveillance system. The video request program can obtain the video data on transportation conditions with different places of the highways in this specific period of time from the video surveillance system, and return the result to users.

### 10.5.2.4 Data Acquisition from Social Transportation Networks

Data acquisition is the process of automatically collecting and rapidly integrating relevant data from various social network sites based on users' demand and their interactions. It mainly includes two aspects. One is that the server dynamically generates and returns web information resources through the database query interface. The other is that the registered users of the open web information have to log in before viewing the information that they are interested in. For example, if users are concerned about the discussions on the film *Assassination of Kim Jong-un* on the social networks, they can directly employ the application programming interface (API). If the discussions are presented in the form of topics, the user can obtain the hot or posts by calling API.

### 10.5.2.5 Data Acquisition from Miscellaneous Smart City Services

Since T-CPS is an important domain of smart city services, it is also conjoined to other application domains such as energy management (smart grid), healthcare, etc. For efficient operations, TCS also sources information and logs from these domains. For instance, data from the healthcare information systems and the healthcare-related big data on the Internet may be acquired according to the task deployed by the user in a highly parallel manner. If a user is concerned about the causes and treatment of liver cirrhosis, the system will start the healthcare data acquisition program and collect data from various sources, such as IBM Healthcare and Life Sciences Grid System, hospital information systems, clinical information systems, and others. Meanwhile, web crawlers will creep to take relevant content from medical news blogs, social network data, professional journals and other public medical information, online medical information systems such as electronic medical records (EMRs), medical data, and others. Then, all these data and the information will be collected and returned to the user. The whole set of operations (from 2 to 6) and the related techniques and tools are depicted in Figure 10.10.

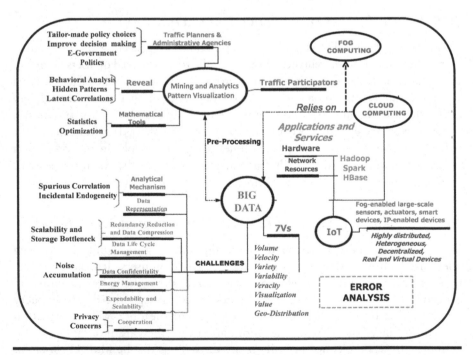

**Figure 10.10 Conceptual framework for TCS operation.**

## 10.5.3 *Knowledge Acquisition and Information Fusion*

Representation, acquisition and use of knowledge are the three main areas of research and application in artificial intelligence. In cyberspace, we need to synthesize knowledge from data through mining and referencing and enable search engines to make references, answer questions, and draw conclusions from masses of data to meet the challenge of extracting knowledge from terabytes of data. Knowledge fusion refers to the merging of information and knowledge from heterogeneous sources with varying conceptual, contextual, and typographical representations by a large number of tools, including versions of search and mathematical optimization, logic, methods based on probability and economics, and many others. Knowledge fusion in cyberspace involves two major tasks. The first is to locate and classify elements in structured or unstructured content into predefined categories, such as the names of persons, organizations, locations, and things; expressions of times, quantities, monetary values, and percentages; and to identify the relations between them. The second is to make inferences by applying a set of rules that make every implicit fact explicit or discover new knowledge from large data where the dimensionality, complexity, or volume of data is prohibitive for manual analysis. The facts may change over time, and temporal semantic heterogeneities need to be considered when reasoning. To synthesize knowledge from a wide variety of sources and provide good solutions to the queries, we may require the following key techniques:

- Knowledge representation of entities and their relations
- Relation-based acquisition
- Statistic-based acquisition
- Rule reasoning–based acquisition
- Crowdsourcing-based acquisition

## 10.5.4 *User Intent and Opinion Mining*

Understanding and representation of users' search intent mean accurately understanding the users' search intent at the semantic level based on keywords, voices, and gestures submitted and represented with a unified model supporting efficient query deduction. By converting the search input into a machine-readable representation language, deeply learning users' thoughts, and unifying query views, user queries can be converted into a machine-readable language model to help the machine understand the search intents. Its key technologies include the following elements:

- User intent understanding based on temporal and spatial characteristics
- User intent understanding based on statistical analysis
- User intent understanding based on body movements or haptic technologies
- User intent understanding based on sentiment analysis
- User intent understanding based on human-computer interaction

### 10.5.5 Search Space Structuring

Characteristics of the search space in an IoT-aided ITS network determine the type of search algorithm employed. It can be structured to effectively retrieve the query matching resources. Popular design strategies used to construct the search space are indexing, crawling, scoring, ranking, etc. [33]. In indexing, the data collected from the search space in the T-CPS network is stored and indexed at middleware for fast and efficient look-up. Indexers in the IoT domain can utilize features like context and content data. In crawling, the updated information about search space at middleware is maintained by the crawler. It visits every object in the IoT network and fetches its data and gives it to the indexer. The relevance of the resource matching a given query is then determined by the scoring and ranking algorithm employed. Quantitative scores are given to matching IoT resources based on their fitness to the context of the query and then sorted to retrieve the topmost results. Also, due to the dynamic and large size of the T-CPS network, where the IoT resources are computationally constrained, frequent communication between them and the search system can be reduced by constructing appropriate prediction and recommendation models.

### 10.5.6 Matching and Intelligent Answering

Intelligent answering is one of the most critical steps in TCS. It utilizes a unified user intent representation, matches users' intent with the integrated knowledge data, and finally outputs a ranked list of recommended answers. In the era of big search, with the emerging usage of graph structures to represent the information intention of the users, the traditional text-string matching methods need to be replaced by new intention-matching methods. The feasible improvements, such as the keyword query–based approaches and sub-graph search-based methods, can be adopted to match the search intention and reach the targeted objects in the corresponding search space. The key techniques of intelligent answering involved in this section are:

- Knowledge integration, management, and update based on the entity relation model
- Text-based matching
- Graph-based matching
- Matching based on audio-visual data
- Search result ranking and evaluation

## 10.6 Security Challenges of TCS Design

Privacy, safety and security are the key challenges for distributed TCS design because of the grave consequences of attacks and the broad attack surface. The T-CPS operates physical components that can damage themselves, people, or property when they are improperly used. Hence, T-CPSs and IoT integrated systems, because they deal with both the physical and cyber-world, must be designed and operated under a unified view of safety and security characteristics. While TCS inherits several features from the traditional search engines, most of the existing research is compartmentalized; no synergies have been explored.

The existing security and privacy measurements for existing searching techniques cannot be directly applied to TCS due to its characteristic features, such as mobility, heterogeneity, large-scale geo-distribution and autonomy. Figure 10.11 illustrates a spectrum of techniques to counter safety and security flaws both at design time and at runtime. The key security challenges in the context of TCS design are related to identity and authentication, access control mechanisms, network protocols, trust management, intrusion detection mechanisms (IDMs), virtualization, edge privacy and forensics.

The T-CPS prototypes are also sensitive to a wider range of attacks and design flaws that are within IT systems. A complete threat model requires both the environment, including the attacker, and the system under threat. Any threat, whether from an attacker or from a bug, may take advantage of a system failure to fully comply with a specification of its properties, whether implicit or explicit. For instance, disturbing the timing of real-time operations can cause the smart traffic management system (STMS) to fail. Insertion of false data as input to a control system

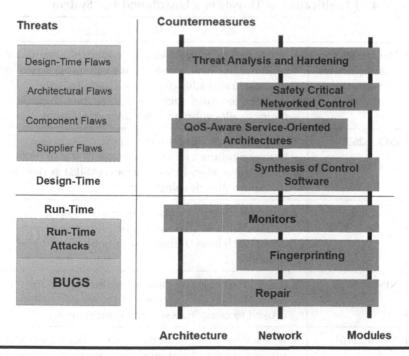

**Figure 10.11 Spectrum of techniques to counter safety and security flaws at design and runtime.**

can lead to wrong decisions and destructive actions in SG-powered vehicles (EVs). Battery-powered IoT nodes can have their batteries run down by denial-of-service (DoS) attacks, causing the system to be inoperative even after the DoS attack ends. All the traditional IT-centric attacks are still of great concern for T-CPSs, since they can be used as entry points to attack the power plants as well as more traditional targets such as billing information systems, etc. Interestingly, T-CPSs and IoT systems can be also exploited to launch DoS attacks against other targets, e.g., Mirai botnet attack (Stojmenovic and Wen, 2014). In a nutshell, the design of correct, reliable TCS is more challenging than for traditional search systems because many important nonfunctional properties of T-CPSs are non-compositional. Both real-time performance and power consumption have complex properties that depend on both the software architecture and computational platform. False data injection attacks, which insert false data as inputs to an otherwise uncompromised system, also constitute an attack class that has never been considered in traditional search systems. A variety of guidelines and standards has been developed for the design of reliable CPSs, such as in Table 10.4, can be adapted for T-CPS based solutions also. In this section, we consider the challenges and future prospects particular to TCS

**Table 10.4  Classification of Threats in a Distributed TCS System**

| | *Assets* | *Threats* |
|---|---|---|
| 1 | The MISRA C Guidelines for Automotive Software | • These coding guidelines are divided into mandatory, required (for which compliance can be accepted by a deviation) and advisory levels. Guidelines are directed to issues such as variable types, dynamic memory allocation and compiler differences. |
| 2 | ISO 26262 | • This standard concentrates on the relationship between product lifecycle and safety. It also defines automotive safety integrity levels (ASILs) as risk levels for automotive development. |
| 3 | DO-178C | • It defines design assurance levels (DALs): catastrophic, hazardous, major, minor, no safety effect. Each level defines a set of objectives to be satisfied. |
| 4 | NISTIR 7628 | • It describes risk assessment for smart grids and associated security requirements. A reference model is used to categorize security requirements. |
| 5 | ASTM F2761 | • It describes the required characteristics networked medical devices from multiple manufacturers used for the care of a single high-acuity patient. |

privacy and TCS forensics. We also holistically analyze the threats and forensics, challenges, and mechanisms inherent in distributed search engines, while highlighting potential synergies and avenues of collaboration.

## 10.6.1 Privacy in Transportation Cyber-Search

Since an IoT-aided ITS infrastructure comprises of multiple sensors, computer chips and devices etc., its deployment in varying different geographic locations results in an increased attack vector of involved objects. As a nontrivial extension of traditionally centralized search engines, it is inevitable that some issues will continue to persist in a distributed TCS, especially security and privacy issues. For instance, edge-based TCS solutions are deployed by different service providers and utilities that may not be fully trusted and thus, devices are vulnerable to be compromised. The IoT nodes in T-CPS are confronted with various threats and attack vectors, a landscape of which is presented in Figure 10.12. The IoT endpoints in ITS networks have constrained store, compute, and network resources that are easy to be hacked, broken or stolen. Examples of attack vector may be human-caused sabotage of network infrastructure, malicious programs provoking data leakage, or even physical access to devices.

The decentralized IoT-integrated ITS elements devices such as vehicles, RSUs, routers and base stations (BSs) etc., if brought to be used as publicly accessible computing processing nodes, the risk associated by public and private vendors that own these devices, as well as those that will employ these devices will need revised articulation. In addition, the intended objective of such devices, e.g., an Internet router for handling network traffic, cannot be compromised just because it is being used as an edge/fog node. It can be made multi-tenant only when stringent security protocols are enforced. Although the existing solutions in cloud computing–based search engines could be migrated to address some security and privacy issues in edge-based TCS, it still has its specific security and privacy challenges due to its distinctive features, such as decentralized infrastructure, mobility support, location awareness and low latency. Nevertheless, TCS offers a more secure infrastructure than other search techniques because of the local data storage and the non-real time data exchange with cloud data centers. For example, the vehicular fog nodes or RSUs could operate as proxies for end-devices to perform secure operations, if the devices have the sufficient resources to do so. Unfortunately, the security and privacy issues and security resources in distributed search platforms have not been systematically identified. Hence, prior to the design and implement of edge-centered TCS applications, critical study of threat profiles, security, and privacy goals should be performed. Moreover, holistic security and risk assessment procedures are needed to effectively and dynamically evaluate the security and measure risks, since evaluating the security of dynamic IoT-based application orchestration will become increasingly critical for secure data placement and processing.

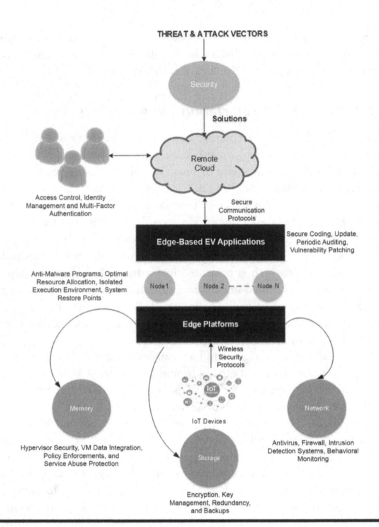

**Figure 10.12   Classification of threats and security attacks in a distributed TCS system.**

## 10.6.2  TCS Forensics

Forensic science is a branch of science that brings together a sequence of scientific principles and methods to identify, discover, reconstruct and analyze evidences to be used for investigation. However, the forensics results cannot be single-handedly used, i.e., the court is not bound to rely on the results that are presented and could take into account other metrics to define what the originals are (Hussain et al.). The main objective of digital forensics is to provide methods that meet the requirements for judicial evidence and could involve the acquisition and analysis of any form of digital data. There are generally two types of evidence in the data that can be retrieved from intelligent IoT devices in any ITS network. The first one can be used to prove crimes directly, such as password theft, DoS, direct denial of service (DDoS) attacks, virtual machine (VM) manipulation, denial of service attacks, etc. The other type could be used to support the evidence and build a complete chain of evidence, such as call history, messaging profiles, log files, usage patterns, etc.

There are several aspects of difference and similarity between traditional and CPS forensics. In terms of evidence sources, traditional evidence could be computers, mobile devices, servers or gateways. In CPS forensics, the evidence could be home appliances, cars, tags readers, sensor nodes, medical implants in humans or animals, or other IoT devices. From the forensic perspective, each IoT device in a T-CPS will provide important artifacts that could help in the investigation process. Some of these artifacts have not been disclosed in public, which means the investigators should consider of these resources and how they can acquire the artifacts from these devices. Even though IoT has rich sources of evidence from the real-world application, it causes some challenges for forensics examiners, including but not limited to the location of data and heterogeneous nature of IoT devices such as differences in operating systems and communication standards. Current research in literature focuses on IoT security and privacy; however, some important aspects such as incident response and forensic investigations have not been covered efficiently.

## 10.6.2.1 Forensics Challenges of TCS Design

The IoT-aided T-CPS is a combination of many technology zones: IoT zone, network zone, cloud zone and edge zone. These zones can be the source of TCS Digital evidences. For instance, evidence can be collected from a smart IoT device or a sensor, from an internal network such as a firewall or a router, or from outside networks such as cloud or an application. Based on these zones, T-CPS has three aspects in term of forensics: cloud forensics, network forensics and edge forensics. Most of IoT devices in a T-CPS have the ability to cross-Internet (direct or indirect connect) through applications to share their resources in the cloud. With all the valuable data that store in the cloud, it has recently become one of the most important targets for attackers. In the case of cloud computing, digital forensics facilitates digital evidence by reconstructing past cloud computing events (Hussain et al.). However, due to its centralized and remote nature, it faces many challenges. The first challenge is related to the technical dimension that where the key issue is inaccessibility to obtain log data from the cloud, volatile data, integrity and correctness of the data, and multi-tenancy, etc. The second challenge defines organizational dimension where the challenges pertain to lack of forensic expertise. The third but not the least is related to the legal dimension that focuses on customer awareness, Internet regulation, and cross-border law(s). In traditional digital forensics, the examiner can hold the digital equipment and then apply the investigation process to extract the evidence. However, in cloud forensics (Saad Alabdulsalam et al.), it is a different scenario; the evidence could be separated in multi-location which is giving many challenges in terms of acquisition of data from the cloud. In addition, in the cloud, examiners have limited control and access to seize the digital equipment, and getting an exact place of evidence could be a challenge (Edington and Kishore 2017). Besides, the data could be stored in a different location in the cloud, resulting in no evidence being seized. In addition, as all cloud services use Virtual Machine as servers, data volatile-like registry entries or temporary Internet files in these servers could be erased if they not synchronized with storage devices. For instance, if these servers are restarted or shutdown, the data could be erased. Network forensics includes all different kinds of networks that IoT devices used to send and receive data. It could be home networks, industrial networks, local area networks (LANs), metropolitan area network (MANs) and wide area networks (WANs). For instance, if an incident occurs in IoT devices, all logs of traffic flow that has passed through could be potential evidence, such as firewalls or IDS logs (Joshi and Pilli 2016). Table 10.5 summarizes the key forensics challenges in the design of TCS.applications in T-CPS.

Analogous to cloud forensic, edge forensic is defined as the application of digital forensics in edge computing. According to Mukherjee and Matam (2017), edge forensics has many steps similar to cloud forensics; however, it is not a part of cloud forensics. Although the challenges in edge/fog forensics are the same as or similar to cloud forensics (e.g., custody chain dependency and integrity preservation), many challenges are more significant in edge forensics compared to cloud forensics. For example, since the edge model is comprised of a heavy population of small, battery-powered edge nodes as infrastructure, retrieving the log data from these nodes becomes very difficult. Moreover, since edge platforms are inherently geo-distributed, the cross-border issue is less critical compared to the centralized cloud forensics. However, due to the huge count of processing nodes, the dependability issue becomes more crucial in edge forensics. The research in edge forensics is still in its infancy. Only a few works are found that overcome the aforesaid issues. For instance, Wolthusen (2009) considered global unity as a solution to overcome the cross-border issue. Delport et al. (2012) discussed strategies for solving multi-tenancy issues. Three common challenges detected fog forensics are (a) storing trusted evidence in a distributed ecosystem with multiple trust domains, (b) respecting the privacy of other tenants when acquiring and managing evidence, and (c) preserving the chain of custody of the evidence. The authors also argued that edge- and fog-based platforms require less computational resources to manage potential evidence. This is because they do not need to manage as many resources (e.g., network traffic, virtual machines) as in centralized cloud infrastructures. Wang et al. (2015) presented a conceptual cyber-physical cloud systems (CPCSs) forensic-by-design model along with the CPCS hardware and software requirements and industry-specific requirements. The characteristic factors of a forensic-by-design model identified in this paper are:

■ Risk management principles and practices
■ Forensic readiness principles and practices
■ Incident handling principles and practices

**Table 10.5   Challenges of Digital Forensics in T-CPS Architectures**

| | Challenges | Description |
|---|---|---|
| 1 | Data location | • A significant portion of ITS data spread in different IoT locations, which are out of the user control. This data could be in the cloud, in third party's location, in vehicles, in mobile phone, or in other devices. Therefore, in T-CPS forensics, to identify the location of evidence is considered one of the biggest challenges an investigator faces in order to collect the evidence. In addition, T-CPS data might be located in different countries and be mixed with other users' information, which means different countries' regulations (cross-border) are involved. |
| 2 | Lifespan limitation of digital media | • Because the limitation of storage in IoT-aided ITS devices, the lifespan of data in T-CPS is short and data can be easily overwritten, resulting in the possibility of evidence being lost. Therefore, one of the challenges is the period of survival of the evidence in IoT devices before it is overwritten. Transferring the data to another thing such as local fog node or to the cloud could be an easy solution to solve this challenge. However, it presents another challenge that relates to securing the chain of evidence and how to prove the evidence has not been changed or modified. |
| 3 | Cloud service requirement | • Most of the accounts in T-CPS are anonymous users because cloud service does not require accurate information from the user to sign up for their service. It could make it impossible to identify a criminal. For example, even though the investigators find evidence in the cloud that proves a particular IoT device in a crime scene is the cause of the crime, it does not mean this evidence could lead to identifying the criminal. |
| 4 | Lack of security | • Evidence in current IoT devices could be changed or deleted because of lack of security, which could make such evidence not solid enough to be accepted in a law court. For example, in the market, some companies do not update their devices regularly or at all, or sometimes they stop supporting the device's framework when they focus on a new product with the new infrastructure. As a result, it could leave these devices vulnerable as hackers find a new vulnerability. |

*(Continued)*

**Table 10.5** (*Continued*)   **Challenges of Digital Forensics in T-CPS Architectures**

|   | Challenges | Description |
|---|---|---|
| 5 | Device type | • In the identification phase of forensics, the digital investigator needs to identify and acquire the evidence from a digital crime scene. Usually, the evidence source is a type of IT hardware device, such as a computer or mobile phone. However, in T-CPS, the source of evidence could be objects like a smart vehicle or temperature sensors. Therefore, the investigators will face some challenges. One of these challenges is identifying and finding the IoT devices in the crime scene. The device could be turned off because it runs out of battery, which makes finding it so difficult, especially if the IoT devices are very small or in a hidden place. Carrying the device to the lab and finding a space could be another challenge that an investigator could face in terms of device type. |
| 6 | Data format | • The format of the data that is generated by IoT devices in T-CPS system often doesn't match what is saved in the cloud. In addition, user has no direct access to his/her data and the data are present in a different format than that in which they are stored. Moreover, data could be processed using analytic functions in different places before being stored in the cloud. Hence, in order to be accepted in a law court, data form should be returned to the original format before performing the analysis. |

- Laws and regulations
- CPCS hardware and software requirements, and
- Industry-specific requirements

How to best integrate forensics techniques and best practices into the design and development of distributed edge architectures, so that it is forensically ready/friendly, is a potential research theme. Having a forensically ready/friendly edge system will allow the real-time identification, collection, and analysis of data that can be used to implement mitigation strategies. Above all, the establishment of international legislation and jurisdictions for cross-border forensics is a hot research avenue to ponder.

## 10.7  Conclusion

In this chapter, we examined the evolution of century-old transportation system leading to Transportation Cyber-Physical System (T-CPS), enabled by periodization of network service development (NSD) technologies such as wireless sensor networks (WSNs), IoT, cloud computing and multi-access fog computing. We also highlighted the needs and steps of IoT-based resource searching in a T-CPS. We then outlined the requirement and challenges of developing search engines for next-generation intelligent transportation systems (ITSs). Following this, we propose a distributed edge-based model for TCS along with the key processing blocks and the workflows in TCS design. We also discussed the architecture and the key technology enablers of TCS design, the data acquisition technologies and the conceptual framework for big data analytics in TCS design. Finally, we examined the potential privacy and forensics challenges in T-CPS applications (taking TCS as an example) and also examined the state-of-the-art solutions in T-CPS forensics.

## Acknowledgment

This work is partly supported by Research Fellowship of the "Visvesvaraya PhD Scheme for Electronics & IT," Ministry of Electronics & Information Technology (MeitY), Government of India (GoI), Vide Grant no. PHD-MLA/4(39)/2015-16.

## References

Bitam, S. 2012. "ITS-Cloud: Cloud Computing for Intelligent Transportation System." *Proceedings of the IEEE Global Communications Conference*, Anaheim, CA, pp. 2054–2059.

Bonomi, F. 2013. *The Smart and Connected Vehicle and the Internet of Things*. San José, CA: WSTS.

Chowdhury, M., Apon, A., and Dey, K. 2017. *Data Analytics for Intelligent Transportation Systems*. Amsterdam, the Netherlands: Elsevier.

Dastjerdi, A. V., and Buyya, R. 2016. "Fog Computing: Helping the Internet of Things Realize Its Potential." *Computer* 49 (8). doi:10.1109/MC.2016.245.

Delport, W., Kohn, M., and Olivier, M. S. 2012. "Isolating a Cloud Instance for a Digital Forensic Investigation." In *Proceedings of the Information Security for South Africa Conference (ISSA)*, August 2012, pp. 1–7.

Edington Alex, R. M., and Kishore, R. 2017. "Forensics Framework for Cloud Computing." *Computers and Electrical Engineering* 60: 193205. doi:10.1016/j.compeleceng.2017.02.006.

European Commission Report. 2010. *Intelligent Transport Systems EU-Funded Research for Efficient, Clean and Safe Road Transport*. Luxembourg, UK: Publications Office of the European Union.

Ezell, S. 2010. "Explaining International and Application Leadership: Intelligent Transportation Systems." The Information Technology & Innovation Foundation, Whitepaper.

Gubbi, J., Buyya, R., Marusic, S., and Palaniswami, M. 2013. "Internet of Things (IoT): A Vision, Architectural Elements, and Future Directions." *Future Generation Computer Systems* 29 (7). doi:10.1016/j.future.2013.01.010.

Hussain, M. M., Alam, M. S., and Beg, M. M. S. 2018. "Fog Computing in IoT Aided Smart Grid Transition- Requirements, Prospects, Status Quos and Challenges." arXiv: 1802.01818v1 [cs. DC]

Hussain, M. M., Alam, M. S., and Beg, M. M. S. 2018. "Fog Computing Model for Evolving Smart Transportation Applications." In *Fog and Edge Computing: Principles and Paradigms*, Eds. R. Buyya and S. N. Srirama, pp. 347–372. Wiley Series on Parallel and Distributed Computing. Hoboken, NJ: John Wiley & Sons.

Hussain, M. M., Alam, M. S., and Beg, S. M. M. 2018. "Fog Assisted Cloud Models for Smart Grid Architectures—Comparison Study and Optimal Deployment." 1–27.

Hussain, M. M., Alam, M. S., and Beg, S. M. M. 2019. "Fog Assisted Cloud Platforms for Big Data Analytics in Cyber Physical Systems—A Smart Grid Case Study." In *Smart Data: State-of-the-Art and Perspectives in Computing and Applications*, pp. 1–44. Boca Raton, FL: CRC Press.

Hussain M. M., Alam, M. S., Beg, M. M. S. and Hafiz M. 2018. "A Risk Averse Business Model for Smart Charging of Electric Vehicles," *In Proceedings of First International Conference on Smart System, Innovations and Computing, Smart Innovation, Systems and Technologies* 79: 749–759.

Hussain, M. M., Beg, S. M. M., Alam, M. S., Krishnamurthy, M., and Ali, Q. M. 2018. "Computing Platforms for Big Data Analytics in Electric Vehicle Infrastructures." doi:10.1109/BIGCOM.2018.00029.

Hussain, M. M., Beg, S. M. M., and Alam, M. S. 2018. "EAI Endorsed Transactions Federated Cloud Analytics Frameworks in Next Generation Transport." *Smart Cities* 2: 7.

Hussain, M. M., Khan, F., Alam, M. S., and Beg, S. M. M. 2018. "Fog Computing for Ubiquitous Transportation Applications-A Smart Parking Case Study 2. Mission Critical Computing Requirements of Smart Transportation Applications," *In proceedings of Engineering Vibration, Communication and Information Processing, Lecture Notes in Electrical Engineering 478*, 2018, pp. 241–252.

Hussain, M. M., Alam, M. S., Beg, S. M. M., and Laskar, S. H. 2019. "Big Data Analytics Platforms for Electric Vehicle Integration in Transport Oriented Smart Cities." *International Journal of Digital Forensics* 11(3): 1–20.

Hussain, M. M., Beg, S. M. M. 2019. "Fog Computing for Internet of Things (IoT)-Aided Smart Grid Architectures," *Big Data Cognitive. Computing*, 3(8), MDPI.

Joshi, R. C., and Pilli, E. S. 2016. *Fundamentals of Network Forensics*. London, UK: Springer-Verlag.

Liu, J., Li, J., Zhang, L., Dai, F., Zhang, Y., and Meng, X. 2018. "Secure Intelligent Traffic Light Control Using Fog Computing." *Future Generation Computer Systems* 78: 817–824. doi:10.1016/j.future.2017.02.017.

Luntovskyy, A., and Spillner, J. 2017 *Architectural Transformations in Network Services and Distributed Systems*, Springer Verlag.

Misra, A., Gooze, A., Watikins, K., Asad, M., and Le Dantec, C.A. 2014. "Crowdsourcing and Its Application to Transportation Data Collection and Management." *Transportation Research Record: Journal of the Transportation Research Board* 2 (404): 1–16. doi:10.3141/2414-01.

Mukherjee, M., and Matam, R. 2017. "Security and Privacy in Fog Computing: Challenges." *IEEE Access* 19293–19304.

Munir, A., Kansakar, P., and Khan, S. U. 2017. "IFCIoT: Integrated Fog Cloud IoT: A Novel Architectural Paradigm for the Future Internet of Things." *IEEE Consumer Electronics Magazine* 6 (3). doi:10.1109/MCE.2017.2684981.

Pattar, S., Buyya, R., and Venugopal, K. R. 2018. "Searching for the IoT Resources: Fundamentals, Requirements, Comprehensive Review and Future Directions," *EEE Communications Surveys & Tutorials*, 20(3): 2101–2132.

Pham, T. N. A M, Tsai, M. F., and Nguyen, B. 2015. "A Cloud-Based Smart-Parking System Based on Internet-of-Things Technologies." *IEEEAccess* 3: 1581–1591.

Porambage, P., Okwuibe, J., Liyanage, M., Ylianttila, M., and Taleb, T. 2018. "Survey on Multi-Access Edge Computing for Internet of Things Realization," 1–31. http://arxiv.org/abs/1805.06695.

Römer, K., Ostermaier, B., Mattern, F., Fahrmair, M., and Kellerer, W. 2016. "Real-Time Search for Real-World Entities: A Survey," 1–13.

Saad, A., Schaefer, K., Kechadi, T., and Le-Khac, N. 2018 "Internet of Things Forensics— Challenges and Case Study." Preprint arXiv:1801.10391.

Wolthusen, S. D. 2009. "Overcast: Forensic Discovery in Cloud Environments." In *Proceedings of the IEEE 5th International Conference on IT Security Incident Management and IT Forensics*, September 2009, pp. 3–9.

Saleem, Y., Crespi, N., Rehmani, H., and Copeland, R. 2018. "Internet of Things–Aided Smart Grid: Technologies, Architectures, Applications, Prototypes, and Future Research Directions." *IEEE Communications Surveys and Tutorials*, pp. 1–30.

Shojafar, M., Cordeschi, N., and Baccarelli, E. 2016. "Energy-Efficient Adaptive Resource Management for Real-time Vehicular Cloud Services." *IEEE Transactions on Cloud Computing* 1–14. doi:10.1109/TCC.2016.2551747.

Siim, K. "Intelligent transport systems EU-funded research for efficient, clean and safe road transport", Whitepaper, European Commission Report, ISBN 978-92-79-16401-9, doi 10.2777/16313.

Singh, B., and Gupta, A. 2015. "Recent Trends in Intelligent Transportation Systems: A Review." *Journal of Transport Literature* 9 (2): 30–34.

Stojmenovic, I., and Wen, S. 2014. "The Fog Computing Paradigm: Scenarios and Security Issues." *Proceedings of the 2014 Federated Conference on Computer Science and Information Systems* 2: 1–8. doi:10.15439/2014F503.

Whitepaper. 2015. "Designing Next-Generation Telematics Solutions." In *In-Vehicle Telematics*, Intel, (v.1, Sep. 2015).

Whitepaper. 2019. *Fog vs Edge Computing*. Milpitas, CA: Nebbiolo Technologies Inc, version 1.1.

Wu, C., and Mohsenian-rad, H. 2012. "Vehicle-to-Aggregator Interaction Game." *IEEE Transactions on Smart Grid* 3 (1): 434–442.

Wang, Y., Uehara, T., and Sasaki, R. 2015. "Fog Computing: Issues and Challenges in Security and Forensics." in *Proceedings of the IEEE 39th Annual Computer Software and Applications Conference*, vol. 3, July 2015, pp. 53–59.

Zanella, A., Vangelista, L., Bui, N., Castellani, A., and Zorzi, M. 2014. "Internet of Things for Smart Cities." *IEEE Internet of Things Journal* 1 (1): 22. doi:10.1017/CBO9781107415324.004.

Zeng, D., Guo, S., and Cheng, Z. 2011. "The Web of Things: A Survey" (Invited Paper). *Journal of Communications* 6 (6): 424–438. doi: 10.4304/jcm.6.6.424-438.

Zhang, D., Yang, L. T., and Huang, H.. 2015. "Searching in Internet of Things: Vision and Challenges." *2011 IEEE Ninth International Symposium on Parallel and Distributed Processing with Applications*. doi:10.1109/ISPA.2011.53.

Zhang, J., Wang, F., Wang, K., Lin, W., Xu, X., and Chen, C. 2011. "Data-Driven Intelligent Transportation Systems: A Survey." *IEEE Transactions on Intelligent Transportation Systems* 12 (4): 1624–1639.

Zhou, Y., De, S., Wang, W., and Moessner, K. 2016. "Search Techniques for the Web of Things: A Taxonomy and Survey." *Sensors* 16: 1–29. doi:10.3390/s16050600.

# CYBERSECURITY IN CYBER-PHYSICAL SYSTEMS

Chapter 11, "Evaluating the Reliability of Digital Forensics Tools for Cyber-Physical Systems," aims to develop a model for evaluating the reliability of digital forensic tools for cyber-physical systems.

Chapter 12, "Point-of-Sale Device Attacks and Mitigation Approaches for Cyber-Physical Systems," aims to describe examples of point-of-sale (POS) attacks. It discusses different types of POS systems. It addresses the functionality of memory scrapers and comparison between old BlackPOS and new malware. And finally, it proposes mitigation approaches that are proposed to defend against POS malware, including the application of secure mobile software development.

Chapter 13, "Cyber-Profiteering in the Cloud of Smart Things," addresses cyber laws, cyber-physical systems, cyber profiteering and technological solutions entrusted to overcome the cyber profiteering occurring in a cloud of smart things. It introduces a theft resistance from cyber profiteers with the security authorization scheme by the support of Kaa cloud services. Finally, it performs a decentralized security solution with varying queries to the cloud of smart things to safeguard from cyber profiteers.

## Chapter 11

# Evaluating the Reliability of Digital Forensics Tools for Cyber-Physical Systems

Precilla M. Dimpe and Okuthe P. Kogeda

## Contents

## 11.1 Introduction

Cybercrime is a growing phenomenon that increases daily due to the advancement of information technology and a high dependency on mobile devices and applications, communication networks, cyber-physical systems (CPSs) technologies, the Internet of Things (IoT) solutions and cloud-based services. The combination of the information communications and technology (ICT) technologies with the physical world such as CPS has led to numerous benefits but at the same time, it accounts for the proliferation of cybersecurity issues and new threats such as identity theft, data leakage exploiting social engineering, etc. (Caviglione et al. 2017). A CPS is a system that integrates communication, computation and physical processes for the purpose of creating systems that are more collaborative, capable and autonomous (Jones et al. 2016). Cybercrime incidents occurring in CPS are increasing in number and are becoming more sophisticated as such systems become more prevalent (Alrimawi et al. 2017). Many of these incidents have caused major damage to both physical and digital assets (Alrimawi et al. 2017). When these incidents happen, investigations need to be conducted by digital forensic investigators (DFIs) in order to identify and uncover cybercriminal activities using digital forensic tools. DFIs rely on digital forensic tools to assess, gather and analyze digital evidence. However, the reliability of most of the tools is unknown because most of them have not been evaluated due to the amount of time required to evaluate them and the cost associated with it (Van Buskirk and Liu 2006). Furthermore, DFIs assume that proprietary tools from reputable vendors are reliable and can do exactly what the vendors claim they can do.

The reliability of digital forensic tools is very critical because the results produced by them are used in courts of law to convict criminals or to prove innocence (Carrier and Spafford 2003). An unreliable tool can compromise entire investigations, which might result in improper judgements. The effect caused by unreliable tools is far larger than the costs of research and development required to prevent them (Van Buskirk and Liu 2006).

The purpose of this paper is to develop a model for evaluating the reliability of digital forensic tools for CPSs. To achieve this, we analyzed, classified and aggregated data obtained from the computer forensics tool testing (CFTT) project using Bayesian networks (BNs) in conjunction with the feedback provided by DFIs to recommend a suitable tool to use for investigations based on factors they take into consideration when purchasing tools. These factors include task, cost and usability (ease of use). BNs are an ideal technique for recommendations and have been widely used in several recommendation systems and in software reliability. This research is for DFIs to provide feedback on the tools after using them; the model then uses that feedback to build on historical data from the CFTT project in order to update the tools' reliability levels. BNs are able to employ both objective data and subjective judgements elicited from domain experts (Rodríguez et al. 2003).

Our model introduces a more effective way of evaluating digital forensics tools that is, less time-consuming. The model attained utility performance of 91.7% after being tested. As a result, it can assist DFIs to make informed choices and tool-testing organizations can use our model to publish their test results in one platform to make them easily accessible to DFIs.

The rest of this chapter is organized as follows: In Section 11.2, we present a brief discussion of what digital forensics tools are. In Section 11.3, we present an overview of software reliability. In Section 11.4, we provide insight into digital forensics evidence and parties that use it. In Section 11.5, related works are discussed. The system design and implementation of the proposed model are discussed in Section 11.6. In Section 11.7, testing and results are discussed. Lastly, we present the conclusion and future work in Section 11.8.

# 11.2 Digital Forensic Tools

Digital forensics tools are hardware or software tools used to assist in the collection and recovery of digital evidence. They are used by DFIs when conducting investigations in order to identify and collect evidence that can be used in a court of law (Hibshi et al. 2011). However, they differ in complexity, functionality and cost. Some are designed to serve a single purpose, while others offer a number of functions. The nature of the investigation determines which tool is appropriate for the task at hand (Department of Homeland Security [DHS] 2015). This work focuses only on software tools as shown in Table 11.1.

**Table 11.1  Digital Forensics Tools**

| Tool Name | Description |
|---|---|
| FTK Imager | FTK Imager is a free extension of FTK, developed by AccessData to generate images from other types of storage devices and hard drives (Vandeven and Filkins 2014). |
| Encase | EnCase is a widely known computer forensics tool designed by Guidance Software to analyze, collect and report on evidence (Vandeven and Sally 2014). |
| X-Ways Forensics | X-Ways Forensics uses diverse data recovery methods and searches functions to find files that are deleted. It includes bit-accurate imaging of a disk to provide a comprehensive examination of a case (Irmler, Kröger, and Creutzburg 2013). |
| Device Seizure | It is an analysis and acquisition tool for examining mobile devices. It consists of a driver pack that is designed to maintain the integrity of device acquisition (NIJ 2012). |
| Oxygen Forensics | Oxygen Forensics is a mobile forensics tool designed to acquire and analyze data from mobile devices (DHS 2015). |
| Adroit Photo Forensics | Adroit Photo Forensics recovers graphic files of several types using proprietary GuidedCarving and SmartCarving technologies (DHS 2014). |

## 11.3  Software Reliability

By definition, software reliability is the probability that software functions without failure under a given environmental condition during a specified period of time (Singpurwalla and Wilson 1994). The degree to which a software can be relied upon to perform its intended function is the most important attribute in a software (Goel 1985). Software reliability is a serious issue since there are many cases of tragic consequences resulting from software failures (Musa et al. 1990). Some of the cases include (Bihina Bella 2016; Goodison et al. 2015) the following:

> *Casey Anthony murder case*—In the Casey Anthony murder case, Anthony was accused of murdering her two-year-old daughter. DFIs used two different tools that produced conflicting results, CacheBack and NetAnalysis. CacheBack indicated that somebody searched 84 times for chloroform,

whereas NetAnalysis indicated that chloroform was searched for only once. However, the CacheBack software developer later discovered errors in the software and indicated that chloroform was searched for only once. This discrepancy was used in court by the defense to discredit the forensic investigation and thus, the digital evidence (Alvarez 2011). This fault likely contributed to the reasonable doubt judges found when they acquitted Anthony of murder.

*Rent calculation error*—The rent calculation error led to the wrongful incrimination of families that were overcharged in rent. Most families were taken to court and threatened with eviction for not paying the extra amount. This error resulted in some families resorting to debt in order to pay for the overcharge.

*Northeast blackout*—Caused by programming errors in the software managing the electrical network, which led to a power failure. The cost of this incident was estimated between $7 and $10 billion.

*Prisoners incorrectly released*—A software glitch in the new record-tracking system resulted in the incorrect release of prisoners. However, the police managed to find them and send them back to jail.

*Implantable infusion pump*—A patient's infusion pump was unexpectedly shut down due to a programming error. The patient's blood pressure dropped and he experienced increased intracranial pressure that led to brain death.

Software failures affect the service consumer and the service provider. The service provider may suffer from loss of revenue and a tarnished reputation. On the other hand, the service consumer may suffer from frustration, stress and loss of money.

## 11.4 Digital Forensics Evidence

In this section, we explain what digital forensic evidence is, including the rules governing it. We also identify parties that use this evidence and how they use it. Furthermore, we discuss the role of digital forensics evidence in cybersecurity and the relationship between them.

### 11.4.1 Rules of Evidence

Digital forensics evidence comes from a variety of sources containing digital evidence (Kessler 2010). It is data leveraged in an attempt to place events and people within space and time to establish causality for criminal incidents (Goodison et al. 2015). Cases where digital evidence is lacking are more difficult to develop leads and solve, as law enforcement officials depend on it for important information about both suspects and victims (Goodison et al. 2015). However, for digital evidence to be regarded as useful it must follow the priorities that follow:

■ *Complete*—All evidence available must be considered and must show more than one view of the incident.

- *Admissible*—The evidence must be able to be used in court.
- *Reliable*—Processes and procedures used in evidence handling must be auditable and repeatable.
- *Authentic*—The evidence acquired must be relevant to the investigation.
- *Believable*—The evidence should be believable and easily understood by a judge.

## 11.4.2 Uses of Digital Forensic Evidence

It is important to identify parties that use digital evidence because they are the ones who are affected by the consequences of using unreliable tools. These individuals and organizations include (Vacca 2010) the following:

- *The military*—Collect information from computers taken during military actions.
- *Criminal Prosecutors*—Use digital evidence in a variety of crimes where convicting documents can be found such as homicides, financial fraud, drug and embezzlement record keeping and child pornography.
- *Civil litigators*—Make use of personal and business records found on computer systems that bear on fraud, divorce, discrimination and harassment cases.
- *Corporations*—Often hire digital forensics specialists to ascertain evidence relating to sexual harassment, embezzlement, theft or misappropriation of trade secrets and other internal or confidential information.
- *Law enforcement officials*—Frequently require assistance in pre-search warrant preparations and post-seizure handling of the computer equipment.

## 11.4.3 The Role of Digital Forensics Evidence in Cybersecurity

Cybersecurity is the collection of tools, security concepts, policies, security safeguards, risk management approaches, guidelines, actions, training, assurance, technologies and best practices that can be used to protect the organization, cyber environment and user's assets (von Solms and van Niekerk 2013). Cybersecurity is a proactive measure put in place to defend and protect information systems from threats such as the misuse of systems, data theft, attackers, system outages and malware outbreaks (Robinson 2018; Owuor et al. 2017). Digital forensics, on the other hand, tends to focus on investigating any suspected crime or misbehavior that may be manifested by digital evidence (Mohay 2005; Dimpe and Kogeda 2017) Unlike cybersecurity, digital forensics is reactive; however, the report produced by a DFI can be used to enhance cybersecurity controls in companies and organization. For example, if there is a system breach or security issue in a company, normally a user or a cybersecurity personnel identifies it and alerts their information technology (IT) department. An internal investigation is conducted, and if it is discovered that the system has been compromised then a DFI is appointed. A DFI then conducts an investigation and once the investigation is complete, a DFI delivers the results to the company. The technical details of the investigation are delivered to the cyber-security personnel in order for them to strengthen the defences of their systems (Robinson 2018; Kogeda 2018). As a result, tools used by DFIs must be reliable given that the evidence produced by them are used by cybersecurity personnel to strengthen security measures.

## 11.5 Related Work

Evaluation is a process used to ensure that a tool behaves satisfactorily and it meets the performance requirements. In this section, we explore models or techniques that have been proposed by other researchers, their strengths and limitations are discussed.

The CFTT (National Institute of Standards and Technology 2018) project was aimed at providing a measure of guarantee for tools used by law enforcement agencies in investigations. Their methodology consists of seven steps which include establishing categories of forensic requirements, identifying requirements for a specific category, developing test assertions based on requirements, developing test code for assertions, identifying relevant test cases, developing testing procedures and methods and reporting test results. Thus, the results are reviewed by the vendor and testing organizations to ensure a certain level of fairness. However, the disadvantage of this approach is that by the time the results are publicly available, the version of the tested tool might be deprecated (Flandrin et al. 2014). In spite of that, the CFTT project has extensive experience in tool evaluation; hence, we used their test results in conjunction with the DFIs' feedback to ensure that tool versions and patches are catered for.

The Scientific Working Group on Digital Evidence (SWGDE 2018) developed testing templates and guidelines with the aim of helping parties that embark on tool testing. The guidelines include developing a test plan and performing test scenarios. The test plan should include test purpose, scope, methodology and requirements to be tested. Their methodology has been implemented and tested but unlike the CFTT project, their results are released only to U.S. law enforcement agencies and not to the public—which makes it challenging to ascertain whether their methodology is suitable for tool evaluation or not.

Byers and Shahmehri (2009) developed a systematic method to evaluate the selected disk imaging tools. Their methodology consists of seven steps including identifying generic requirements of the tool, identifying relevant technical variations to be tested, translating generic requirements to testable assertions, creating a test plan that tests all assertions, executing the test plan, analyzing the results of the test and notifying the vendor. The authors claimed that their methodology has unique elements, which might benefit other evaluation projects. Their methodology is similar to that of the CFTT project but it has introduced some improvements, which include identifying relevant technical variations for testing. They also provided a deeper analysis of each test result and their requirements touched the tool's usability, which is missing from the CFTT project.

Wilsdom and Slay (2006) proposed an evaluation framework to validate the accuracy and reliability of tools. Their framework uses black box testing techniques by making use of reference sets and test cases. It consists of six phases including acquiring software, identification of the software functionalities, development of the result acceptance spectrum, executing the test and evaluation results and releasing evaluation results. The development of the result acceptance spectrum uses the methodology for documenting the result acceptance spectrum from ISO 14598.1–2000, which divides potential results set into four groupings: exceeds requirements, target range, minimally acceptable and unacceptable. If a function does not meet the acceptance range, then that function and all dependents and codependents are rendered as incorrect. The functions found to be below the acceptable range are regarded as failed. Those that are in or above an acceptance rating are regarded as passed. The authors claim that the framework offers advantages such as efficient process, community input, various environment testing and a community point of contact. Their methodology divides results into the four previously mentioned groupings, which also includes minimally acceptable. In their work, if a function is minimally acceptable, it still falls under the acceptance range. Minimally acceptable shows that a function did not meet all its requirements, therefore, it should not be regarded as passed. Given the forensic context of the discipline, a tool must produce reliable results. It must either exceed requirements or be within a target range in order to be regarded as passed.

Guo et al. (2009) developed a methodology for validation and verification of forensic tools, which was achieved by stipulating the requirements of each mapped function. After this operation, the reference set was developed where each test case was designed conforming to one functional requirement. Their focus was on the searching function, where the searching function was mapped and its requirements specified. Their method offered benefits such as detachability, flexibility, tool neutrality and transparency. Furthermore, as stated in their conclusion, "even if the methodology is promising, it needs to be tested," this obviously shows that their method was not tested. Therefore, the reliability of their methodology is unknown. To prove beyond reasonable doubt, our model has been tested and results obtained are presented in Section 11.7.4.

Pan and Batten (2009) developed a methodology for evaluating digital forensics tools by using a partition testing approach. They used orthogonal array (OA) to test the performance of a tool against itself, or against other tools on the same constraint. Their methodology reduced the effect of incorrect observations with large values by using Taguchi's logarithmic function. Pan and Batten (2009) outlined that Taguchi's method alone is not sufficient to reduce the impact of outliers. As a result, they created a theorem to define the maximum number of suspicious samples acceptable. The authors claimed that their methodology allows testers to compare the performance of tools without consuming a large amount of time or using advanced equipment and can be fully automated in the future. Even though their methodology was tested and the results proved the validity and effectiveness of their methodology, the authors acknowledged that it might not be feasible to use their methodology for any type of tool because it is very difficult to develop robust testing measures for every category of tools.

Kubi et al. (2011) evaluated XRY 5.0 and UFED 1.1.3.8 mobile forensic tools based on NIST smartphone tool specification and test cases using the Daubert principle as a point of reference toward the admissibility of digital evidence. The evaluation was executed by using six phases including collection, identification, preservation, examination, analysis and reporting. A graphical representation was used to compare the results, which showed that most of the time XRY 5.0 exceeded UFED 1.1.3.8 in terms of performance. The work of Kubi et al. (2011) focuses only on mobile forensics, which is a specific category of digital forensics that deals with mobile tools. However, our model covers a more general area in digital forensics (Table 11.2).

## 11.6 Design and Implementation

In this section, several diagrams are used to illustrate the system's elements, how they fit and work together to satisfy the requirement of a digital forensic tool evaluation system. We present the use case diagram of our model, considering a DFI (user) as the main actor. We also present the architecture of our system. Class diagrams of the developed system including its attributes and associations are also presented. Furthermore, we provide a graphical representation of the tables in the database using the entity relationship diagram (ERD). Thereafter, we outline various activities performed by the system and DFI in a form of an activity diagram. Lastly, we discuss the behavior of the system and how messages, events and actions flow between objects.

### 11.6.1 Use Cases

In this section, we focus on the system's behavior from a user's point of view. Description of a system's functions and its main processes are represented in the form of a use case diagram to provide a graphic explanation of interactions within the system and its users (actors). A use case explains how the system must respond under several situations to a request from one of the stakeholders to provide a specific goal (Oracle 2018). Stakeholders are actors outside the system that interact with the system; they may be a class of users, roles played by users or other systems (Malan and Bredemeyer 2001). In this case, the actors in the system are a DFI and a system as shown in Figure 11.1.

**Table 11.2  Summary of Related Work**

| Author | What They Did | Methodology | Claims/Results | Similarity/Differences (Gap) |
|---|---|---|---|---|
| CFTT | Developed a methodology for testing forensic tools | Functionality testing | • Users can make informed choices<br>• Reduce challenges to the admissibility<br>• Tool creators make better tools | The proposed model uses test results from the CFTT project as knowledge base coupled with feedback from DFIs and factors that influences the selection of a tool such cost, task and usability. |
| SWGDE | Developed a testing guideline | Functionality testing | • Offers best practices | Similar to their methodology; our methodology has been tested. |
| Byers and Shahmehri | Evaluated disk imaging tools | Functionality testing | • Has unique elements which might benefit other evaluation projects | This methodology requires a DFI to manually evaluate the tools as opposed to ours recommends a suitable tool to use for investigations based on the task they want to perform, the category of the tool and its cost. |
| Wilsdon and Slay | Proposed an evaluation framework | Black box (functionality) testing technique | • Efficient process<br>• Community input<br>• Various environment testing<br>• Community point of contact | Similar to Byers and Shahmehri's methodology. This methodology does not address the time consumption problem faced by DFI as it is a manual process. |

*(Continued)*

**Table 11.2 (*Continued*)  Summary of Related Work**

| Author | What They Did | Methodology | Claims/Results | Similarity/Differences (Gap) |
|---|---|---|---|---|
| Guo, Slay and Beckett | Proposed a functionality orientated validation and verification framework | Function mapping | • Detachability<br>• Flexibility<br>• Tool neutrality<br>• Transparency | This methodology focuses on the searching function. Ours focuses on all functions of a forensic tool. |
| Pan and Batten | Developed a methodology for evaluating digital forensics tools | Performance testing | • More efficient than any other existing performance testing methods<br>• It can be automated | The authors focus on performance testing while we focus on functionality testing. |
| Kubi, Saleem and Popov | Evaluated mobile forensics tools | Functionality testing | • Makes it easier for the DFI to select an adequate tool for a specific case | Only focuses on mobile forensics tools; ours covers a more general area in digital forensics. |

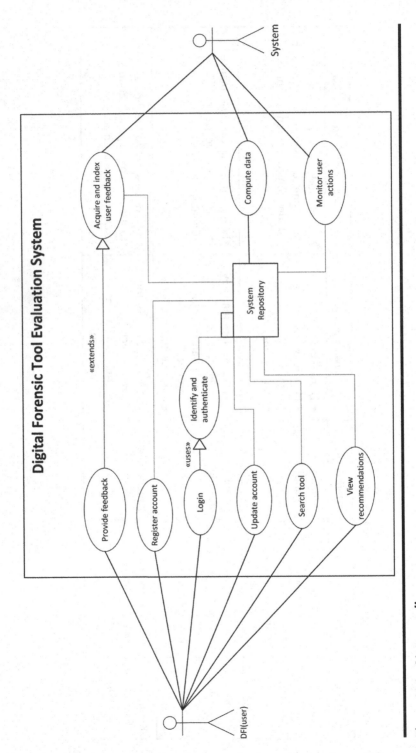

**Figure 11.1** Use case diagram.

## 11.6.2 System Architecture

Systems architecture is considered to be a conceptual model that describes the behavior and structure of a system (Jaakkola and Thalheim 2011). A user (DFI) interacts with a system to search for a tool that can perform the desired task. The web server, that is located in the application layer, then communicates with the recommender engine. The recommender engine uses Bayesian networks to select a tool that can perform the task that the user requires using the information in the database, which is located in the database layer. Once the tool is found, the web server returns the recommended tool to the user as shown in Figure 11.2.

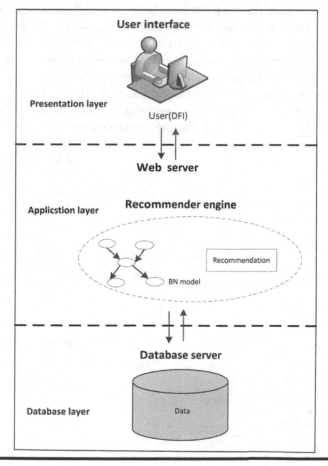

**Figure 11.2  Architecture of the model.**

### 11.6.3 Entity-Relationship Diagram (ERD)

To provide a graphical representation of the tables in the database and the relationship between them, an ERD was used. It was used to model the design and architecture of the database (Shahzadet al. 2004). The database consists of six tables, with the tool table being the most important one as it contains the information about the tool. The tool table has a relationship with the cost, task, category, reliability and ease of use table. The rest of the tables together with their attributes and relationships are as shown in Figure 11.3.

### 11.6.4 Activity Diagram

In this section, we focus on describing the concurrent and sequential flow of activities executed by the system, using an activity diagram (Shahzad et al. 2004). An activity diagram has the capability to show the activities carried out by the system from the beginning until the end. The user (DFI) starts by registering the account on the system. When the registration is complete, the system directs the user to the login page. Once the user is logged in, he/she can search for a desired tool-using task, category and cost. The system then searches for the tool that meets the user's requirement. If the system finds the tool that meets the user's requirement, it recommends it to the user. However, recommendations are made based on historical and feedback data. Furthermore, if the system cannot find the tool that meets the user's requirement, a message indicating that the tool cannot be found is returned. Thus, after using the tool, the user is required to

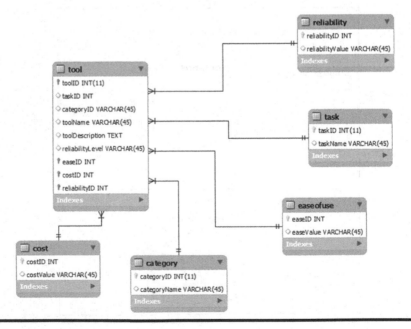

**Figure 11.3   Entity-relationship diagram.**

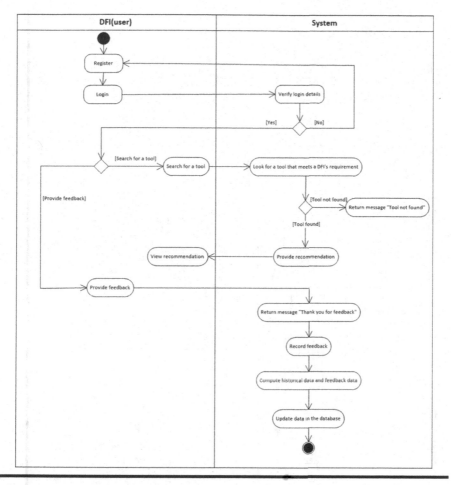

**Figure 11.4   Activity diagram.**

provide feedback that is used to build on the data (historical data) in the database in order to update the tool's reliability level as shown in Figure 11.4.

## 11.6.5  Sequence Diagram

The sequence diagram in Figure 11.5 represents the behavior of a system based on required interactions between a set of objects in relation to the exchange of messages between them to produce the desired result (Metu 2018). It also shows the flow of operations running the system (Metu 2018). A user in the sequence diagram represents a DFI who uses the system to search for a tool that can perform a particular task. The user utilizes the web interface to send requests to the web server that in turn communicate with the recommender engine to get the tool that satisfies the user's requirements. The recommender engine uses Bayesian networks to

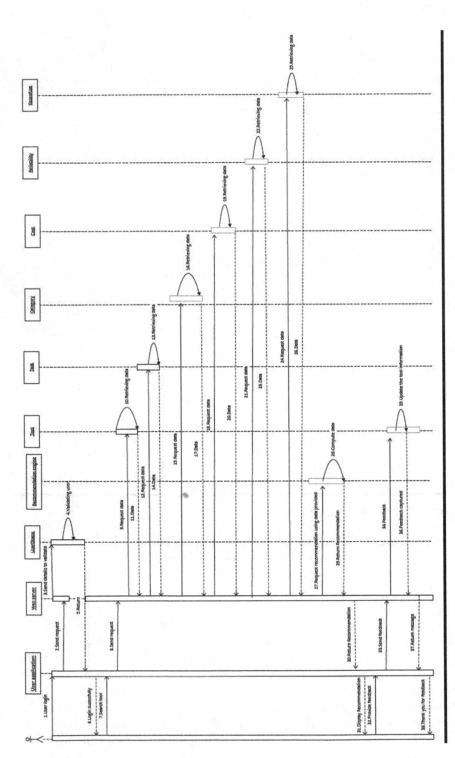

**Figure 11.5   Sequence diagram.**

compute the data for the tool, task, category, cost, reliability and ease of use in order to generate a recommendation for the user as illustrated in Figure 11.5.

## 11.6.6 Bayesian Networks

The theory of Bayesian networks was developed by the Rev. Thomas Bayes in 1764; an 18th-century mathematician and theologian (Neapolitan 2004). Bayesian networks (BN), also called belief networks or causal probabilistic networks, are graphical models for reasoning under uncertainty (Korb and Nicholson 2010). Every node in the graph symbolizes a random variable, while the edges between the nodes symbolize probabilistic dependencies among the corresponding random variables. Given a complex system that needs to be evaluated, Bayesian networks $Y = \{S,T\}$ can be defined by Mihaljlovic (2001):

- A straight acyclic graph $S = (A,B)$ where $A$ is a set of vertices (or nodes), and $B$ is a set of directed edges (or links)
- A probability space $(\Omega, T)$
- A group of random variables $A = \{A_1 \ldots A_n\}$ linked with graph's vertices $(\Omega, T)$ such as $T = (A_1 \ldots A_n) = \prod_{i=1}^{n} T(A_i \mid T_x(A_i))$ where $T_x(A_i)$ is the set of parent's vertices of the node $A_i$ in $S$

A Bayesian network can be best described as a graph where the ends have a significant impact among the variables of the graph and the nodes represent random variables (Mihaljlovic and Petkovi 2001). The random variables $A$ are linked to their modalities $(A = a_1; A = a_2; \ldots A = a_n$ if $A$ can take $n$ values). The ends represent the connections, which can either be probabilistic or deterministic. For an end linking the point $X$ and the point $Z$, there is a correlation that is the conditional probability $T(X \mid Z)$, which symbolizes a probabilistic correlation of a child node and its parent. For the nodes that do not have parents, referred to as "root" nodes, a prior probability are allocated to them.

$$T(X \mid Z) = \frac{T(Z \mid X) \cdot T(X)}{T(Z)} \tag{11.1}$$

where $T(X)$ is the prior probability of $X$, $T(Z)$ is the evidence and $T(X \mid Z)$ is referred to as the likelihood ratio (posterior probability).

### 11.6.6.1 Bayesian Networks Construction

BN can be utilized for both backward and forward inference, that is, inputs can be utilized to calculate outputs and outputs can be utilized to estimate input requests[6] (Rodríguez et al. 2003). This functionality is very useful for our model because DFIs are required to enter the task they want to perform, the category and the cost of a tool. The model uses that information to recommend a suitable tool for investigations.

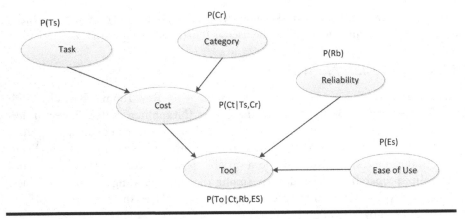

**Figure 11.6  Bayesian network.**

We used a literature review, data from the CFTT project, and a survey that was conducted by the Digital Forensic Investigation Research Laboratory to determine possible factors affecting the selection of a tool. Figure 11.6 shows a Bayesian Networks of six nodes with the following variables: Task (Ts), Category (Cr), Cost (Ct), Reliability (Rb), Ease of use (Es) and Tool (To).

## 11.6.6.2 The Probability Distribution of the Variables

Survey data from the Digital Forensic Investigation Research Laboratory was used to determine factors that DFIs take into consideration when purchasing a tool (Josua 2018). According to the conducted survey, those factors include feature set (task), cost and ease of use. However, their data was not used to assign states. The actual data that we used to assign state variables were derived from literature review and CFTT project. The aim of the CFTT program is to establish a tool testing methodology by developing test procedures, test hardware, test criteria, test sets and general tool specifications (Lyle 2006). Out of approximately 103 tools that were tested by the CFTT project; 6 were selected for developing our model, namely FTK Imager, X-Ways Forensics, EnCase Forensic, Adroit photo forensics, Oxygen Forensics and Device Seizure. By making use of the 6 tools that were selected, we were able to derive the states for the task, category, reliability, cost and ease of use.

## 11.6.6.3 Joint Probability Distribution

Bayesian networks are considered to be demonstrations of joint probability distributions (Korb and Nicholson 2010). Joint probability distribution enables us to reason about the relationship between several events. It defines conditions where both outcomes characterized by random variables happen (Renals 2018). For example, $X = \{A, B\}$ is called a joint probability distribution of $X$. We regularly use $a,b$ to represent the event $A = a, B = b$, which can be written as $P(a,b)$ instead of $A = a, B = b$. This theory can be extended to three or more random variables. The joint probability distribution function of the variables $A, B$ and $C$, which is often written as $P(A,B,C)$ is $P(A = a, B = b, C = c)$ (Neapolitan 2004). Using the Bayesian networks in Figure 11.6, the joint distribution of P (Ts, Cr, Ct) can be calculated as:

$$P \text{ (Ts, Cr, Ct)} = P \text{ (Ts)} \times P \text{ (Cr)} \times P(\text{Ct}|\text{ Cr, Ts})$$
$$P \text{ (Ts = Da, Cr = Cf, Ct = fr)} = P \text{ (Ts = Da)} \times P \text{ (Cr = Cf)}$$
$$\times P \text{ (Ts = Da, Cr = Cf, Ct = fr)}$$

$$= 0.67 \times 0.67 \times 0.17$$

$$= 0.07631$$

$$= 7.631\%$$

## 11.6.7 User Interface

In this section, we illustrate how the user interacts with the system to search for a required tool.

### 11.6.7.1 Tool Evaluation Page

The system can only be used by registered users; the user must first log in, in order to use the functionalities of the system. Once the user is logged in, he/she can search for the required tool by selecting the evaluation tab or provide feedback by selecting the feedback tab as shown in Figure 11.7. The tool evaluation page allows the user (DFI) to search for a tool that can perform the desired task. The model then searches for tools that can perform the desired task in the selected category within the desired price range, which can either be free, low or medium as shown in Figure 11.8.

### 11.6.7.2 Recommendation Page

In the recommendation page, the required tool is recommended, including a brief description of the tool, its functions and reliability level as shown in Figures 11.9 and 11.10. The tool's reliability level is calculated using feedback from the users in conjunction with historical data from the CFTT project. However, at the inception of the system, recommendations are only made using the data from the CFTT project. Users are required to provide feedback on the tools after using them by clicking on the feedback link, which is discussed in Section 11.6.7.3.

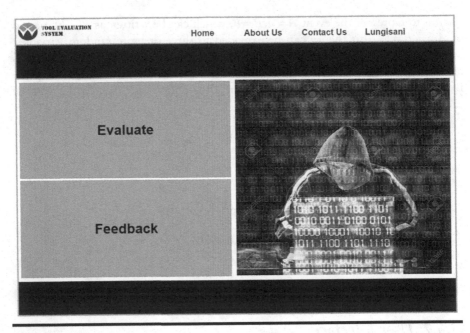

**Figure 11.7   Home page.**

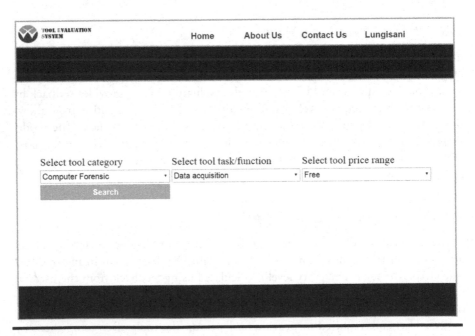

**Figure 11.8   Tool evaluation page.**

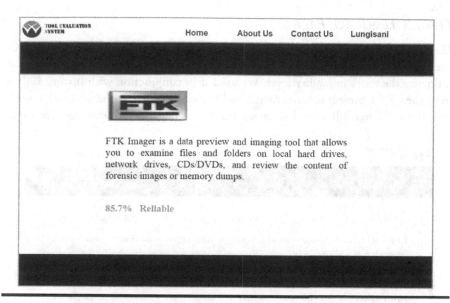

**Figure 11.9  Recommendation page.**

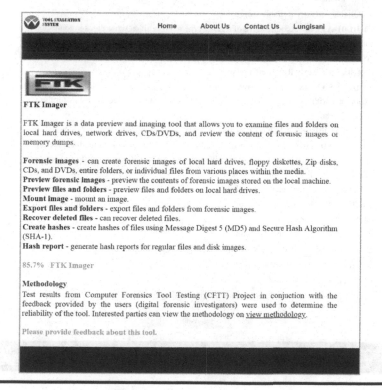

**Figure 11.10  Recommendation page.**

### 11.6.7.3 Feedback Page

The feedback page shown in Figure 11.11 is the page that DFIs use to provide feedback on the performance of the tools. The feedback provided by DFIs is used to influence the tool's reliability level. We used it in conjunction with historical data from the CFTT project to build on the tool's reliability level. If the feedback is positive, the tool's reliability level increases and if it is negative, it decreases. Feedback

**Figure 11.11   Feedback page.**

**Table 11.3  FI's Level of Expertise**

| Expertise Level | Description | Weighted Score |
|---|---|---|
| **Beginner** | Has knowledge or an understanding of basic techniques and concepts in digital forensics. | Feedback disregarded |
| **Novice** | Individuals who have a certain level of experience gained in experimental scenarios and/or classroom or as a trainee in the job. | Feedback disregarded |
| **Intermediate** | Individuals who are able to complete a digital forensic task. They may occasionally require help from an expert, but they can independently perform a task. | 60% |
| **Advance** | This individual has the skill to perform the digital forensic task without assistance. | 80% |
| **Expert** | Professional who has extensive experience acquired through study and practice. | 100% |

*Source:* Aamodt and Enric (1994).

was weighed based on the user's expertise level, e.g., an advanced user's feedback weighs less than that of an expert as shown in Table 11.3. However, feedback from both a novice and a beginner were disregarded because they have limited knowledge about the tools. This procedure was carried out using a decision matrix due to its ability to weigh multidimensional decisions of a decision set (Lyle 2006).

If a tool performed as expected on a particular test case, it is rated as reliable (Feedback [$F = 1$]) for that test case. If not, it is rated as partly reliable or unreliable (Feedback [$F = 0$]). Equations (11.2) and (11.3) were adapted from a decision matrix, where $F_r$ is the feedback result, $F_s$ (feedback score) is the feedback provided by DFIs, $W_s$ is the weighted score assigned to DFIs based on their level of expertise and $T_{ws}$ is the total weighted score.

$$F_r = \sum \frac{F_s \times W_s}{T_{ws}} \tag{11.2}$$

After the feedback is provided, we calculate the new reliability level of a tool given feedback results ($F_r$), where $N_{Rb}$ is the new reliability level of a tool and $O_{Rb}$ is the old one.

$$N_{Rb} = \frac{O_{Rb} + F_r}{2} \tag{11.3}$$

## 11.7 Experimental Results

The performance of our model was tested and evaluated using functional testing. The purpose of functional testing is to ascertain that the software performs in conformance with its specification. In functional testing, test cases are designed based on the information from the requirements to ensure that the system or software conforms to all the requirements (Nidhra 2012). The following steps were followed when conducting the test:

1. We determined the functional requirements of the model.
2. We developed test cases.
3. We tested the model.
4. We analyzed the results.

### 11.7.1 Determining Functional Requirements

Functional requirements define the functionality of a software or system, i.e., how it should act in response to certain input or react in certain situations (Nidhra 2012). The functional requirements were determined in order for us to know what to test and how to test it. The core functional requirements of our model are as follows:

- The model shall recommend a suitable tool based on the task, category and cost.
- The model shall inform the user (DFI) if the required tool is not available.
- The model shall allow the user to provide feedback on the tool.
- The model shall ignore feedback from the beginner and novice user and consider only feedback from the intermediate, advanced and expert user.
- The model shall calculate feedback based on the weighted score.
- The model shall use the feedback provided by users to update the data in the database.

### 11.7.2 Development of Test Cases

Test cases were developed based on the functional requirements specified in Section 11.7.1. They were used to test if the model fulfils the specified functional requirements. IEEE (Radatz et al. 1990) defines a test case as a set of test inputs, execution conditions, and expected results developed for a test item.

## 11.7.3 Execute Test

The testing was executed using test cases developed in Section 11.7.2. Test cases were grouped into test runs and each test run contained 6 test cases. In total, we had 10 test runs for each function, which contained 60 test cases. At the completion of each test, expected results were compared to actual results to determine if the function works as it should. If the actual results are the same as the expected results, status is a pass ($S = 1$) and if not, status is failed ($S = 0$). The results for each test run were calculated using Equation (11.4):

$$TR_n = \sum \frac{S}{T_{tc}} \tag{11.4}$$

Where $TR_n$ represents a number of test runs, $S$ is the status of each test case and $T_{tc}$ is the total number of test cases in a test run. For example, if one out of six test cases in the first test run passed and five failed, the results were calculated using Equation (11.4) as follows:

$$TR_1 = \frac{1}{6}$$

$$TR_1 = 0.167 \approx 16.7\%$$

## 11.7.4 Results Discussion

In this section, the results obtained from conducting the tests are analyzed and interpreted in a graphical manner. Errors encountered are pointed out, including measures that were taken to resolve them.

### 11.7.4.1 Model Evaluation

In our first experiment, we tested the system's ability to recommend a suitable tool based on the task a DFI wants to perform including data acquisition, data recovery and file carving. Furthermore, we tested the system's ability to inform a DFI, if the required tool was unavailable. For data acquisition and data recovery, the model attained the highest level of precision by presenting performance levels of 83.3%. Consistency for both tasks was attained from the fourth run of the system. In the first two runs, the model was unable to recommend the required tool even when

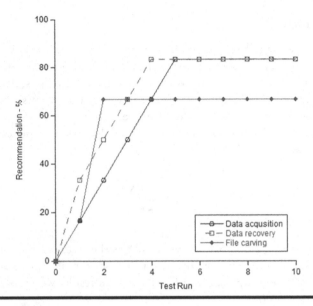

**Figure 11.12    Results of the recommendation.**

the required tool was available. That was due to variables that were not declared correctly on the system. After declaring the right variables, the model started to recommend the required tool for all tasks including file carving, which attained 66.7% precision as shown in Figure 11.12. However, the system was still unable to inform the DFI if the required tool was unavailable. It could not do so because the message that was supposed to inform the user that the required tool was unavailable was not defined on the server. We corrected that by defining the message on the server. After doing so, the model started to inform the user when the required tool was not available.

### 11.7.4.2  Evaluating the System's Ability to Calculate Feedback Based on Expertise Level

In the second test, we tested the model's ability to calculate DFI's feedback based on their expertise level including beginner, novice, intermediate, advance and expert. We also tested its ability to update the new reliability level of the tool based on the feedback provided. The model attained 100% precision for all expertise levels. However, in the first run, intermediate, advance and expert users could only provide feedback on FTK Imager. Errors were encountered when attempting to provide feedback on other tools because the data was not loaded to the session accordingly and as a result, the model was unable to submit

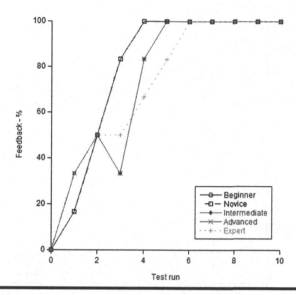

**Figure 11.13    Feedback.**

the data stored in the session. That was rectified by recreating the session and loading the data accordingly. In addition, there was a calculation error on the feedback provided by intermediate and advanced users, which led to a decline in their performance. Once that was corrected their performance improved and the new reliability level of the tools was correctly updated. Feedback from a novice and a beginner was supposed to be disregarded and should have not influenced the tool's reliability level. However, the model considered feedback from them due to a logical error. Once the error was rectified, the model started disregarding the feedback from a novice and a beginner and considering the feedback only from an intermediate, advanced or expert user. A summary of the results is shown in Figure 11.13.

## 11.7.4.3 Utility

The overall performance our model was evaluated using MATLAB due to its ability to predict the system's behavior. MATLAB is a simulator that can evaluate the design, diagnose problems with an existing design, and test a system under various conditions (Houcque 2018). For this process, consistency was achieved as from the fifth run. The utility performance of our model started at a low rate but kept improving until it reached a consistency of 91.7% utility, as shown in Figure 11.14. This means our model recommends a suitable tool for DFIs nine out of ten times. This can be improved by fine-tuning the model and performing additional training.

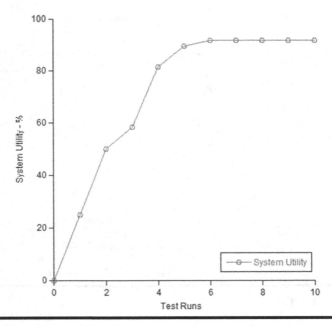

**Figure 11.14 Utility.**

## 11.7.5 Discussion

In Section 11.5, we reviewed existing models or techniques, which do not address the time consumption problem experienced by DFIs when it comes to tool evaluation, as they require a DFI to manually evaluate a tool. In addition, some of them have never been tested. According to the survey we conducted using esurv. org, 97.5% out of 100 participants agreed that it is necessary to evaluate a tool before using it, but it is not practical for them to do so, because it takes a significant amount of time. The survey targeted DFIs who were involved in forensic investigations by sending them a link to esurv.org where closed-ended questionnaires were loaded. In that survey, DFIs highlighted that it would be very helpful if they could have official government entities or forensic institutes that can take care of tool evaluation. The CFTT project and SWGDE are organizations that have been established for that purpose and have done so excellently. However, they cannot meet the demands of DFIs because they take months to thoroughly evaluate a single tool and by the time they make their results available, the version of the tested tool might have already been upgraded.

Our proposed model seeks to address this limitation of the previously mentioned models or techniques by introducing a time-saving way of evaluating digital forensic tools and ensuring that tool patches and upgrades are taken into consideration. DFIs are not required to manually evaluate the tools; our model recommends a suitable tool for them to use for investigations based on the task they want to perform, the category of the tool and its cost. Our model uses test results from the CFTT project as a knowledge base in conjunction with the feedback provided by DFIs to recommend a tool. It seeks to build on existing data (test results) from the CFTT project by introducing feedback from DFIs who are currently using the tools. For example, the last test results from the CFTT project were released in 2016; ever since then, tool patches and upgrades have been released. Therefore, feedback from DFIs was used to address this gap. DFIs who are currently using the tools are in a better position to provide us with information about the status of the tool.

## 11.8 Conclusion and Future Work

In this chapter, we outlined the challenges faced by DFIs concerning tool evaluation and the consciences of using unreliable tools. Efforts have been taken by researchers to come up with models or techniques for evaluating digital forensic tools, but most of them do not address the time consumption problem faced by DFIs when it comes to tool evaluation, because they require a DFI to manually evaluate the tools. In an attempt to address this problem, we developed a model for evaluating digital forensic tools that is time-saving and cost effective. The model was developed using Bayesian networks, Java and MySQL server. It uses test results from the CFTT project in conjunction with the feedback provided by DFIs to evaluate the reliability of the tools. The model showed promising results of 91.7%. However, it was unable to validate DFIs expertise level; it assumes that they were correct without validating them. In spite of that, our model can help DFIs with tool evaluation and toolmakers with enhancing the reliability of their tools. Furthermore, tool testing organizations such as the CFTT project and the National Institute of Justice (NIJ) can use our model to publish their test results in order to make them easily accessible to DFIs in a centralized platform. In the future, we would refine the model and use other artificial intelligence algorithms.

## References

Aamodt, Agnar and Enric Plaza. 1994. "Case-based reasoning: Foundational issues, methodological variations, and system approaches." *AI communications* 7 (1): 39–59.

Alrimawi, Faeq, Liliana Pasquale, and Bashar Nuseibeh. 2017. "Software Engineering Challenges for Investigating Cyber-Physical Incidents." In *Proceedings of the 3rd International Workshop on Software Engineering for Smart Cyber-Physical Systems*, 34–40. SEsCPS'17. Piscataway, NJ: IEEE Press. doi:10.1109/SEsCPS.2017...9.

Alvarez, Lizette. 2011. "Software Designer Reports Error in Anthony Trial." *New York Times* A14.

Bihina Bella, Madeleine. 2016. "A Near-Miss Analysis Model for Improving the Forensic Investigation of Software Failures." *PhD Diss.*, University of Pretoria.

Buskirk, Eric Van, and Vincent T Liu. 2006. "Digital Evidence: Challenging the Presumption of Reliability." *Journal of Digital Forensic Practice* 1 (1): 19–26. doi:10.1080/15567280500541421.

Byers, David, and Nahid Shahmehri. 2009. "A Systematic Evaluation of Disk Imaging in EnCase® 6.8 and LinEn 6.1." *Digital Investigation* 6 (1): 61–70. doi:10.1016/j.diin.2009.05.004.

Carrier, Brain, and Spafford, Eugene. H. 2003. "Getting Physical with the Digital Investigation Process." *International Journal of Digital Evidence* 2 (2): 1–20.

Caviglione, L, Wendzel, S, and Mazurczyk, W. 2017. "The Future of Digital Forensics: Challenges and the Road Ahead." *IEEE Security & Privacy* 6: 12–17.

DHS (Department of Homeland Security). 2014. "Test Results for Graphic File Carving Tool: Adroit Photo." *Forensics 2013 v3.1d," NIST*.

DHS. 2015. "Test Results for Mobile Device Acquisition Tool: Oxygen Forensic Suite 2015." *Analyst v7.0.0.408, NIST*.

Dimpe, Precilla M., and Okuthe P. Kogeda. 2017. "Impact of Using Unreliable Digital Forensic Tools." *Proceedings of the World Congress on Engineering and Computer Science* 1.

Dimpe, Precilla M., and Okuthe P. Kogeda 2018. "Generic Digital Forensic Requirements." *IEEE 2018 Open Innovations Conference*, 240–245. doi:10.1109/OI.2018.8535924.

Flandrin, Flavien, William J Buchanan, Richard Macfarlane, Bruce Ramsay, and Adrian Smales. 2014. "Evaluating Digital Forensic Tools (DFTs)." *7th International Conference Cybercrime Forensics Education & Training (CFET)*, 1–16. doi:10.13140/2.1.3293.6004.

Goel, Amrit L. 1985. "Software Reliability Models: Assumptions, Limitations, and Applicability." *IEEE Transactions on Software Engineering* SE-11 (12): 1411–1423. doi:10.1109/TSE.1985.232177.

Goodison, Sean E., Robert C. Davis, and Brian A. Jackson. 2015. "Digital Evidence and the US Criminal Justice System." *Identifying Technology and Other Needs to More Effectively Acquire and Utilize Digital Evidence. Priority Criminal Justice Needs Initiative.* Santa Monica, CA: Rand Corporation.

Guo, Yinghua, Jill Slay, and Jason Beckett. 2009. "Validation and Verification of Computer Forensic Software Tools—Searching Function." *Digital Investigation* 6: S12–S22. doi:10.1016/j.diin.2009.06.015.

Hibshi, Hanan, Timothy Vidas, and Lorries Cranor. 2011. "Usability of Forensics Tools: A User Study." In *2011 Sixth International Conference on IT Security Incident Management and IT Forensics*, 81–91. doi:10.1109/IMF.2011.19.

Houcque, David. 2018. "Introduction to Matlab for Engineering Student Mccormick." *Northwestern. Edu.* https://www.Mccormick.Northwestern.Edu/Documents/Students/Undergraduate/Introduction-to-Matlab.Pdf.

Irmler, Frank, Knut Kröger, and Reiner Creutzburg. 2013. "Possibilities and Modification of the Forensic Investigation Process of Solid-State Drives." *Multimedia Content and Mobile Devices* 8667: 866710–866714. doi:10.1117/12.2004028.

Jaakkola, Hannu, and Bernhard Thalheim. 2011. "Architecture-Driven Modelling Methodologies." *Frontiers in Artificial Intelligence and Applications* 225: 97–116. doi:10.3233/978-1-60750-690-4-97.

Jones, Andrew, Stilianos Vidalis, and Nasser Abouzakhar. 2016. "Information Security and Digital Forensics in the World of Cyber Physical Systems." In *2016 Eleventh International Conference on Digital Information Management (ICDIM)*, 10–14. doi:10.1109/ICDIM.2016.7829795.

Josua, James I. 2018. "Survey of Evidence and Forensic Tool Usage in Digital Investigations." *Dfire. Ucd. Ie.* http://Dfire.Ucd.Ie/?P=858.

Kessler, Gary Craig. 2010. "Judges' Awareness, Understanding, and Application of Digital Evidence."

Korb, Kevin B., and Ann E. Nicholson. 2010. *Bayesian Artificial Intelligence.* Boca Raton, FL: CRC Press.

Kubi, Appiah Kwame, Shahzad Saleem, and Oliver Popov. 2011. "Evaluation of Some Tools for Extracting E-Evidence from Mobile Devices." In *2011 5th International Conference on Application of Information and Communication Technologies (AICT)* 1–6. doi:10.1109/ICAICT.2011.6110999.

Lyle, Jim. 2006. "The Contribution of Tool Testing to the Challenge of Responding to an IT Adversary." In *Keynote Speech at the International Conference on IT-Incident Management & IT-Forensics* 18–19.

Malan, Ruth, and Dana Bredemeyer. 2001. "Functional Requirements and Use Cases." *Architecture Resources for Enterprise Advantage* 1–10. doi:10.1016/j.radi.2011.04.004.

Metu. 2018. "Software Requirements Specification Document for Music Recommender System." *Memoryleak. Weebly. Com.* http://Memoryleak.Weebly.Com/Uploads/2/5/6/5/25654522/Memoryleak_srs_v1.1.Pdf.

Mohay, G. 2005. "Technical Challenges and Directions for Digital Forensics." In *First International Workshop on Systematic Approaches to Digital Forensic Engineering (SADFE '05)* 155–61. doi:10.1109/SADFE.2005.24.

Musa, John D., Anthony Iannino, and Kazuhira Okumoto. 1990. "Software Reliability." *Advances in Computers* 30: 85–170.

Neapolitan, Richard E. 2004. *Learning Bayesian Networks.* Upper Saddle River, NJ: Pearson Prentice Hall, p. 38.

Nidhra, Srinivas. 2012. "Black Box and White Box Testing Techniques—A Literature Review." *International Journal of Embedded Systems and Applications* 2 (2): 29–50. doi:10.5121/ijesa.2012.2204.

National Institute of Justice (NIJ). 2012. *Paraben Device Seizure Version 4.3 Evaluation Report.* Washington DC: NIJ.

National Institute of Standards and Technology (NIST). 2018. "Digital Data Acquisition Tool Specification." *Cftt. Nist. Gov.* https://Www.Cftt.Nist.Gov/Pub-Draft-1-DDA-Require.Pdf.

Oracle. 2018. "Getting Started With Use Case Modelling." *Oracle. Com.* http://Www.Oracle.Com/Technetwork/Testcontent/Gettingstartedwithusecasemodeling-133857.Pdf.

Owuor, Dennis L, Okuthe P Kogeda, and Johnson I Agbinya. 2017. "Three Tier Indoor Localization System for Digital Forensics." *International Journal of Electrical, Computer, Energetic, Electronic and Communication Engineering* 11 (6): 602–610.

Pan, Lei, and Lynn M Batten. 2009. "Robust Performance Testing for Digital Forensic Tools." *Digital Investigation* 6 (1): 71–81. doi:10.1016/j.diin.2009.02.003.

Radatz, Jane, Anne Geraci, and Freny Katki. 1990. "IEEE Standard Glossary of Software Engineering Terminology." *IEEE Std 610121990, No. 121990* 3.

Renals, Steve. 2018. "Formal Modeling in Cognitive Science." Inf. Ed. Ac. Uk. http://Www.Inf.Ed.Ac.Uk/Teaching/Courses/Fmcs1/Slides/Lecture20.Pdf.

Robinson, Michael K. 2018. "Cyber Security Awareness Month: Cyber Security vs. Cyber Forensics Modelling." *Stevenson. Edu.* http://Www.Stevenson.Edu/Online/Blog-News-Events/Cyber-Security-vs-Cyber-Forensics.

Rodríguez, Daniel, Javier Dolado, and Manoranjan Satpathy. 2003. "Bayesian Networks and Classifiers in Project Management."

Shahzad, Khurram, Abdullah Anjum, Majid Gorbani, and Ivica Crnkovic. 2004. "Master Education Management System."

Singpurwalla, Nozer D, and Simon P Wilson. 1994. "Software Reliability Modeling." *International Statistical Review/Revue Internationale de Statistique* 62 (3): 289–317. doi:10.2307/1403763.

Solms, Rossouw von, and Johan van Niekerk. 2013. "From Information Security to Cyber Security." *Computers and Security* 38: 97–102. doi:10.1016/j.cose.2013.04.004.

SWGDE. 2018. "SWGDE Recommended Guidelines for Validation Testing." *Swgde. Org. 018.* https://www.Swgde.Org/Documents/Current%20Documents/SWGDE%20Recommended%20Guidelines%20for%20Validation%20Testing.

Vacca, John R., and K. Rudolph. 2010. *System Forensics, Investigation, and Response.* Burlington, MA: Jones & Bartlett Publishers.

Vandeven, Sally, and Barbara Filkins. 2014. *Forensic Images: For Your Viewing Pleasure.* Singapore: SANS Institute InfoSec Reading Room.

Vojkan, Mihaljlovic, and Milan Petkovic. 2001. "Dynamic Bayesian Networks: A State of the Art, University of Twente, Enschede." The Netherlands, *Technical Report TR-CTIT-01-34, 2001.*

Wilsdom, Tom, and Jill Slay. 2006. "Validation of Forensic Computing Software Utilizing Black Box Testing Techniques." *Proceedings of the 4th Australian Digital Forensics Conference*, Perth Western Australia: Edith Cowan University, December 10. doi:10.4225/75/57b13e59c705b.

## Chapter 12

# Point-of-Sale Device Attacks and Mitigation Approaches for Cyber-Physical Systems

Md. Arabin Islam Talukder, Hossain Shahriar and Hisham Haddad

## Contents

# 12.1 Introduction

A point-of-sale (POS) system is a combination of software and hardware that allows merchants to make transactions and simplify day-to-day business operation (Yamarie 2018). The POS allows us to do purchases using credit and debit cards. The POS machines usually run on Windows, in particular, Windows XP or Windows 7 (probably newer ones now). In the past few years, the mobile POS has gotten popular; it usually runs on Android, iOS, Blackberry and smaller businesses have adopted online payment through wallet (mobile) applications.

POS is a cyber-physical system (CPS). CPSs are systems that link the physical world (e.g., through sensors or actuators) with the virtual world of information processing. A POS system consists of a software system, communication technology and sensors that interact with the real world, often including embedded technologies (CPSE 2018). POS devices run on software that operates on a set of hardware devices. POS software also includes an instruction set about handling the user's private data such as how to handle the username, password and the transaction.

In the last few years, there have been many high-profile breaches that have happened, including the drastic compromise of POS system intrusions. For example, Oracle MICROS revealed a breach impacting its legacy POS systems. Visa issued an alert warning merchants to be on the lookout for malware attacks linked to MICROS (Tracy 2016). A malware strain, discovered in 2015 by Trend Micro, is still actively used by cybercriminals (Tracy 2016).

POS is not a brand-new topic in the field of cybersecurity (Alexander 2017): attacks on point-of-sale terminals made headlines a couple of years ago as such incidents happened with increasing regularity (Verizon's 2015 Data Breach Investigations Report reveals that POS-related incidents accounted for 28.5% of all breaches that happened in 2014).

Windows POS systems remain the major targets by attackers. POS intrusion is a type of malicious software that is known to steal customer's data from the payment method. Attackers even purchase the POS system to steal sensitive information from retail businesses. Attackers often are interested in selling the data instead of directly using it. They have many ways to get into the system in order to steal customer's data. Although with the physical method, where an attacker clones the card and runs the transaction, makes it easier to steal the data, this method requires access to the POS equipment. It is difficult to gain access; however, not impossible. Attackers mostly penetrate the database of the POS system where data is kept. Nowadays, attackers are more interested in POS software and mobile wallets. They are leaking users' data by developing more efficient malware applications.

The number of POS attacks were not the main reason why these attacks came into the spotlight (Alexander 2017). Rather, POS-related data breaches affected some large companies, some of which suffered staggering losses. In 2013, Target's POS system was attacked and its payment card readers were subsequently infected. As a result, attackers compromised the credit card information of 40 million customers. A year later, a POS attack on Home Depot led to the stealing of credit card data belonging to 56 million customers.

In the literature, much POS malware has been described along with mitigation approaches. There are a lack of surveys in this direction to gain better insights into the working mechanisms of POS malware. In this chapter, we provide a number of POS malware overviews, and also discuss their working mechanisms. We then discuss a number of mitigation approaches. We also focus on mobile POS systems that may be subject to attacks and their prevention through secure mobile software development. Our research is based on extensive surveys of related work, an example of malware signatures analysis, and industry best practices.

The chapter starts with a brief discussion of different types of POS in Section 12.2. Section 12.3 covers examples of ways to gain access into POS devices by malware. Section 12.4 describes existing POS malware attacks and their operations. Section 12.5 highlights keylogger, an advanced and sophisticated malware. In Section 12.6, we discuss forced authorization. The survey discusses some mitigation techniques for POS devices in Section 12.7. Section 12.8 concludes the chapter and discusses some future research directions.

## 12.2 Types of POS

One of the most essential parts of running a business is getting paid for its products and services, regardless of the size of the business. POS systems keep track of revenue data. The type of POS system depends on the type of business. For instance, individuals or small firms run small businesses with the use of mobile/tablet POS; among others, Square stands out the most. Though there are plenty of POS systems out there, we try to cover most of them. The rest of this section provides a glance at some popular POS systems.

**Mobile POS Systems:** Mobile POS systems are one of the most convenient ways to pay for goods through a mobile application. Most smartphones have POS solutions built for consumers' quick access and for those who do not want to carry their wallets, making the payment method easier through digital wallets. This type of POS system is suitable for small firms and it does not require a remarkable amount of investment. Additionally, it has another great feature that offers to have the transaction receipt sent out to the consumer via email or text message: a way to keep track of each transaction and data. For example, Apple provides transactions via the application by accepting the fingerprint of the user. iPad and Android tablets are essential in terms of taking orders. Many well-known businesses like Unilever, Coke, and Pepsi use mobile POS to take orders. Recently mobile service providers Telenor and Airtel introduced mobile POS to manage their top-up service for customers in underdeveloped countries not having sufficient communication network infrastructure. Here, most of the customers who use daily top-up use prepaid service.

**Tablet POS Systems:** Tablet POS is one of the top-rated POS systems that we are using currently in farms, restaurants and other small businesses. Due to its easy functionality and simplicity, tablet POS systems are popular in the market. The best part of tablet POS systems is that they require minimum maintenance and investment apart from the cost of the hardware as long as they are compatible with cash drawers or barcode scanners. Tablet POS systems are very popular in firms like fast food restaurants, food trucks, pizzerias or cafes. They can also be used for complex functions such as inventory management and employees' time management.

**Terminal POS System:** Terminal POS systems are those found in a bank's parking lot, busy retail stores, spas, airports, malls etc. They are very useful and convenient for a busy firm. They are used for processing transactions that allow one to see one's balance in the bank. These cloud-based solutions work by staying connected to the Internet: a mix of hardware and software, great for managing inventory effectively, and that helps revenue growth. Most of the terminal POS systems works on chip-based cards. However, modern technology introduces sensor-based transactions that allow the user to do a transaction by placing the mobile device on the POS system. In developing countries, almost all banks introduced mobile banking to help customers open bank accounts based on their mobile numbers. Banking organizations provided personal identification number (PIN) codes to the customer to do the transaction via terminal POS. In this system, the customer does not need to insert a credit or debit card in the POS machine. Instead of credit or debit cards, users can do the transaction by the mobile number. This is an easier transaction method without credit or debit card that leads the attacker to get the information easily.

**Self-Service Unattended POS Systems:** Self-service unattended POS systems are popular and used effectively. For instance, it would be absurd to hire a person to process transactions for a kiosk at the movie theatres, self-parking lots, gas stations, valet parking, large retail stores, etc. A self-service unattended POS system can be very handy; it can run itself outdoors. The advantage of self-service POS systems is that no one will be watching the transaction physically, which gives privacy to the legitimate user. For example, London metro service uses a contactless card reader to read contactless debit or credit cards for a payment. Most UK banks now issue their debit, credit, charge or prepaid cards as contactless cards. Signature or PIN verification is not required for a transaction of less than £30 with these cards (Contactless 2018). In any contactless card reader, these cards can be used. The contactless card readers are also self-service POS systems. However, this leads to a security threat to the POS machines. It is possible to withdraw money or to do a transaction by a stolen card, because in self-service POS systems no one actually monitors the POS machine physically.

## 12.3 Access Gain of Malware by POS

There are several options available for an attacker to gain access to the POS systems, some of which include, but are not limited to, the use of malware (Symantec 2014), performing passive reconnaissance to determine default configuration or credentials (Grunzweig 2015), or using a third party as a pivot point to compromise the intended target. Intercepting network communication is another method used by the attacker, due to lack of authentication and encryption within POS systems (Constantine 2016). A group of attackers led by Albert Gonzalez was the main perpetrator, stealing more than 90 million card records from retailers (Symantec 2014). Prior to 2005, accessibility through compromised POS systems came via direct access to the systems. Attackers can also use social engineering to gain access to POS devices using discarded receipts. Credit card receipts contain enough information for an attacker to utilize their social engineering skills and attempt to gain access to POS devices, allowing access to the network (Warnic 2012). An attacker would follow the following fundamental steps to gain access to the POS system (Symantec 2014):

**Infiltration:** This can be accomplished using social engineering skills, structured query language (SQL) injection, or delivering malware via spear phishing, to name a few methods, depending on the approach of the attack.

**Enumeration:** Once access is available to an internal resource, the attacker could traverse the network in an attempt to locate the intended target, such as a database system.

**Collection:** Once the target has been identified and successfully accessed, sniffers or malware can be installed after exploiting known vulnerabilities within the services or the operating system, to collect credit card information whenever an associated POS system is used.

**Exfiltration:** Data can be transmitted back to criminals via any method of choice, which could continue until its detection.

**Interception:** User's private data can be driven to the attacker by intercepting application programming interface (API) calls. In this case, the POS device, mobile or tablet, can be the medium of attack.

This chapter will, however, focus on how malware could be used to infiltrate the POS system and exfiltrate collected information. The three different malware that will be discussed are (1) rapid access memory (RAM) scrapers (also known as memory scrapers), which extract information from memory; (2) keyloggers, which actively eavesdrop and transmit keypad or keyboard inputs; and (3) GratefulPOS, a malware framework designed to target POS systems.

Cyber criminality has been on the rise in the modern world and requires sophisticated counteractions as noted by Mazurczyk et al. (2016). An attack on a POS system is when an attacker extracts the information exhibited by the magnetic strip on the payment card with the aim of reproducing other duplicate cards. One of the actions that can be taken after such intrusion is altering the passwords assigned to the payment cards. It should be noted that the information targeted by hackers include card number, the name of the holders, and personal identification numbers. Once an organization detects access has occurred, personal information should be changed to ensure that the hackers cannot make purchases using the payment cards. Another action is issuing an alert to banks and credit cards issuers about the intrusion.

It is evident that payment cards are payment systems which allow users to access their banks when making payments. In this case, most POS intrusions are usually conducted in restaurants, hotels, and retail outlets (Kan 2016). Thus, the organization where the invasion has occurred should provide the history of the individuals who have used payment cards to the bank, which will deny the electronic transfer of funds. Such actions will ensure the safety of the victim's bank accounts. In addition, the victim should adopt a financial tracking app as suggested by Huang et al. (2018).

In most cases, the cardholders are aware of their weekly spending. Besides, the bank and credit card companies contain tracking apps to notify the holder when purchases surpass the set limit. It should be noted that hackers make a number of purchases after gaining access to the victim's payment card details. In this case, the tracking apps will alert the cardholder before the transaction is completed. As such, the affected individuals will be in a position to cancel the transactions and report the intrusion to their bank, ensuring the safety of their savings.

## 12.4 POS Attacks and Examples

There are many ways to attack POS systems. However, it is debated which method is preferred for an attack. Often the argument regarding attack methods on POS systems being effective revolves around if the attack is conducted while being connected to a network or if it is done offline. While network-connected attacks have their share of success, offline attacks have proven to be harder to defend against. An offline scenario is harder to defend due to customer data being kept within the POS system for a much longer time, thus being more exposed to attackers (Daza 2016). These types of offline attacks include skimmers, scrapers, forced offline authorization, and software vulnerabilities.

## 12.4.1 Memory Scraper

Memory scraping, also known as RAM scraping, is a type of malware that occurs frequently in POS systems and targets the RAM of the POS system. There are multiple ways to introduce the malware, including intentional installation of the malware by an employee, searching for system vulnerability, and accidentally opening phishing email with attachments from unknown senders (Hesseldahl 2014). Customer's data is stored on the card reader as soon as they swipe the card, including the card number, card expiration date, and cardholder's name. This data, otherwise known as Track 1, Track 2, and Track 3, is encrypted and sent to the POS system terminal. The information stays encrypted until it arrives at the POS system back-end server where the actual transaction is processed. Initially, decrypted information stays in the RAM of the POS system as plain text before the transaction. According to SecureBox, once the transaction has been processed, the information is saved, encrypted, and sent to the bank. The milliseconds between swiping and transaction processing is where cybercriminals retrieve Track 1 and Track 2 data. The malware searches through many strings of data until it finds a string of data that resembles a credit card number (Hesseldahl 2014). The data is extracted and transmitted remotely to the cybercriminal. Figure 12.1 demonstrates

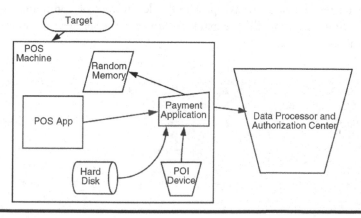

Memory Scraping Attack Vectors

**Figure 12.1   Memory scraping POS.**

the basic concept of memory scraping. The target POS system is a combination of a POS app, payment application, point-of-interaction (POI) device, RAM, and hard drive. The POS application requests authentication from the server, where users' credentials are sent to the server through the POI device via payment application. Before sending the credential to the data center for authorization, the POS system keeps private data in the RAM for milliseconds. Within milliseconds, the malware can access the data in the RAM. However, in the pipeline, data transmits end to end as a ciphered form of plain text.

There are two different kinds of memory scrapers: BlackPOS and Dexter. However, new malware has been discovered which is similar to BlackPOS. Besides the functionality of BlackPOS, new malware introduces the use of command-line arguments that make them more effective over BlackPOS.

## 12.4.2 BlackPOS and New Malware Memory Scraper

BlackPOS, discovered in March of 2013, infects a POS system running Windows with a card reader attached to it (Constantin 2013). These POS systems are discovered through Internet scans and selected because they either have security vulnerabilities in their operating systems or have very weak remote administration login information (Constantin 2014). Once the memory scraper is installed on the POS system, it is able to associate itself with the card transaction process, and get Track 1 and Track 2 data where it is uploaded to a remote server. Figure 12.2 shows distinct features of BlackPOS and new malware. Here, a BlackPOS uses Windows graphical user interface (GUI) whereas new malware uses the console to attack POS systems. However, both use the same Windows subsystem (Grunzweig 2014).

BlackPOS is configured to run without command-line. It would check if it were running as a service, and in the event it is not, it creates a new service with the information that is shown in Figure 12.3.

New malware was configured to take command-line arguments. Figure 12.4 shows that new malware can accept a multiple number of command-line arguments at a time. Additionally, the malware can take the service argument when it is being run as a service.

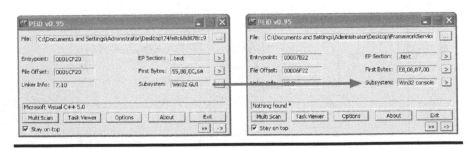

**Figure 12.2  BlackPOS versus new malware.**

```
Service Name: POSWDS

Display Name: POSWDS

Description: [N/A]

Startup Type: Automatic
```

**Figure 12.3   Service initiated by malware.**

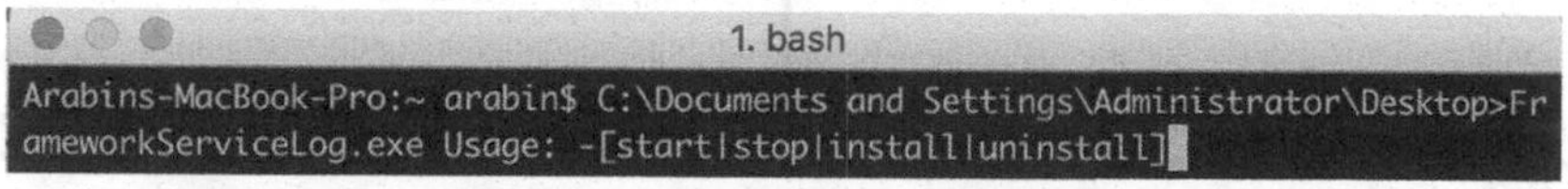

```
● ● ●                          1. bash
Arabins-MacBook-Pro:~ arabin$ C:\Documents and Settings\Administrator\Desktop>Fr
ameworkServiceLog.exe Usage: -[start|stop|install|uninstall]
```

**Figure 12.4   Command-line argument.**

New malware exposes itself as a service requirement utility. A different and smart approach with additional requirement forces the POS system to add them to the service. By adding itself as a dependency into the service, the new malware family prevents it from being removed easily. By installing itself as a service dependency, it begins to execute the system command to configure it as a complete dependency or legitimate utility of the system. See the LanmanWorkstation service (Grunzweig 2014) in Figure 12.5.

The installation routines of BlackPOS and new malware is given below. It is clear that new malware uses the same technique as BlackPOS prior to checking the service availability in the POS machine.

In Figure 12.6, the new malware checks whether a service is running or not in the POS machine. If a service is running, it tries to bind itself with the service by executing the last else code block, which is similar to BlackPOS code. If a service is not running, the new malware tries to create a new service by using command-line arguments (the else-if code block shows the code briefly). Using the command-line arguments, it is possible to install, uninstall, start, and stop a new service as shown in Figure 12.4.

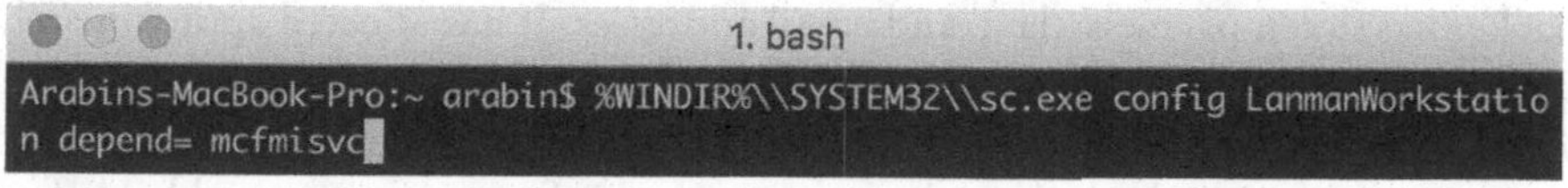

```
● ● ●                          1. bash
Arabins-MacBook-Pro:~ arabin$ %WINDIR%\\SYSTEM32\\sc.exe config LanmanWorkstatio
n depend= mcfmisvc
```

**Figure 12.5   Lanman workstation.**

**BlackPOS Original**

```
memset(&v5, 0xCCu, 0xD8u);
ServiceStartTable.lpServiceName = service_name;
ServiceStartTable.lpServiceProc = service_function;
v7 = 0;
v8 = 0;
result = StartServiceCtrlDispatcherA(&ServiceStartTable);
if ( !result )
  LOBYTE(result) = create_service();
return result;
```

**New Malware**

```
else if ( !strcmpiA(v9 + 1, "service") )
{
  if ( !strcmpiA(argv[1] + 1, "start") )
  {
    if ( !strcmpiA(argv[1] + 1, "stop") )
    {
      if ( !strcmpiA(argv[1] + 1, "install") )
      {
        if ( !strcmpiA(argv[1] + 1, "uninstall") )
          uninstall();
      }
      else
      {
        install();
      }
    }
    else
    {
      stop();
    }
  }
  else
  {
    start();
  }
}
else
{
  v6 = 0;
  v7 = 0;
  ServiceStartTable.lpServiceName = service_name;
  ServiceStartTable.lpServiceProc = service_function;
  if ( !StartServiceCtrlDispatcherA(&ServiceStartTable) )
    error(byte_418676);
}
```

**Figure 12.6** BlackPOS versus new malware installation code.

## 12.4.3 Dexter

Dexter, discovered in December of 2012, was found on POS systems that operated on Windows. Dexter operates as follows (Constantin 2012):

- Lists processes running on the hacked network to control and command the server.
- Checks whether the software of a POS machine matches any of the listed processes. If a match is found, Dexter dumps the memory of the POS machine.
- Analyzes the memory dumps to retrieve Track 1 and Track 2 data.
- Uploads the data to a remote server.

Smaller businesses are an easy target for cybercriminals because of their potential security vulnerabilities. Larger businesses, such as Home Depot and Target, are more attractive because of enormous daily customer credit card transactions. Home Depot, one of the largest home improvement retailers, had their network hacked by cybercriminals who used vendor login credentials to install memory scrapers on their self-checkout stations (Winter 2014). It went undetected for months. It was programmed to go undetected by antivirus software (Winter 2014). The hacker has stolen payment data, and email addresses. It has affected approximately 56 million debit and credit cards in the United States and Canada between April and September of 2014 (Miners 2014). After this hacking, Home Depot introduced a new encryption method where the payment card information is scrambled before it is encrypted.

In the case of Target, a phishing email was sent to a refrigeration contractor where the recipients opened the email, and the hacker gained the contractor's login information. The login information also pertained to the portals the contractor used (Kassner 2015). With this information, the hacker was able to get access to Target's internal network and infect their POS systems with a Trojan. The Trojan was able to get the payment card information as the card was being swiped. The Trojan would check to see if the local time was between 10:00 a.m. and 5:00 p.m., every seven hours, so it could send a dynamic-link library to an internal host inside the network (Kassner 2015). This method allowed the hacker to get data from POS systems that did not have access to the Internet. It affected 41 million payment cards from November 27 to December 25 of 2013 (Constantin 2014).

## 12.4.4 Alina

Alina, the most developed and one of the oldest POS viruses, was discovered in October of 2012. It is still being updated continuously by its developer. Now it has its sixth version in the market. Alina is considered as the base of other break-off POS viruses (BlackPOS) nowadays (Huq 2014). The main mechanism of Alina is to find the credit card track data inside the RAM while the process running on a POS device. With the help of basic encryption and exfiltration, Alina uses command and control (C&C) server structure to install their updates automatically in the POS device (Bowen 2014). After the massive destruction, a research team from different institutions stepped forward to analyze the method that Alina uses for the operation. Trend Micro first described the method that Alina uses for its operation. Once Alina has been installed in a device it carefully hides inside the device and works effectively as being hidden in the device. After the first execution, it installs itself onto the POS device. Then it replaces itself on a pre-coded existing filename and deletes the original Alina files in the POS device. It always executes under a hidden filename. The binary file of the virus re-installs itself on each reboot of the POS device, which makes itself more effective and the users cannot remove it from the device. Alina has gotten popular for its efficiency. The difference between Alina and traditional BlackPOS is that Alina has a blacklist of the process to skip itself if it is seen to be operating. Most web browsers save credit card track data. With the help of this blacklist process, Alina skips complex computations that are needed to scrape the high-memory process of a web browser. Traditional BlackPOS always works as a memory scraper. Later on, it uses HTTP POST (request protocol for sending data over the Internet) to send saved data to the desired receivers securely over the C&C server. The developer always hardcodes the receiver's address in the binary file of the virus. Alina could be described as follows:

- System information collector.
- Single component user.
- Uses specially configured filename.
- Automatically updates itself.

Alina is a renowned RAM scraper due to its ability to attack all kinds of targets. Overall, it is considered the most sophisticated and dangerous POS virus for its high-quality endurance, adaptability, and flexibility.

### 12.4.5 Vulnerable Communication and Data Skimming

The network is the medium of data transition also known as node-to-node communication. Sometimes the path is called a pipeline or data in transit. The vulnerability that lies with data in transit is that the data is not encrypted, though it can be. When the unencrypted data is in transit, it is susceptible to attackers through the act of skimming. The Payment Application Data Security Standard PA-DSS dictates that payment applications "encrypt sensitive traffic over public networks" (Watson 2018). However, the PA-DSS is not applicable to data internal to the cardholder data environment. With data at rest, the data is maintained anywhere within the POS system rather than its primary storage location. Data at rest is susceptible to the attackers when data is both maintained local to the POS system and left unencrypted mostly in the RAM. While PA-DSS requires the encryption of data at rest by POS systems, the requirement is only applicable to the personal account number of the cardholder. All other information can be kept as in plain text.

A customer input device that belongs to the POS system can be replaced with a fake one using skimmer in order to capture the customer's card data (Daza 2016). Skimming is an approach to attack a POS system that can go unnoticed due to security personnel not being totally aware of the surroundings in which the POS system resides. It takes physical security on the ground being fully attentive to and on high alert for suspicious individuals who attempt to handle the POS system.

### 12.4.6 Network Sniffer—GratefulPOS

There has been a tremendous increase in POS system attacks in the past few years and more and more firms are paying attention to this type of malware. There are two important pieces of data the attacker tends to target the credit card: Track 1 (contains cardholder's name) and Track 2 (contains the account number). Most targeted POS attacks are multistaged; in other words, it's planned to include reconnaissance, exploitation, delivery, command, and installation on objectives. There are several ways an attacker can attack the POS system, targeting its weakness. The first POS malware was discovered in 2008. A POS attack can lead to a potential information leak and can harm the business. POS threats have increased with time. One of the areas that has increased is the network-based attacks through network sniffing malware. This attack was first discovered in 2014, where 90 million records and data were comprised.

Network sniffer, an attacking structure and a real-time monitoring of data packets, became extinct between 2009 and 2010. It determined traffic on the network and developed copies of the packet data. The attacker gained sensitive information regarding the POS environment via malicious files that were hidden for getting access to the administrative-level credentials.

GratefulPOS is a type of framework POS malware. It's intended to extract card information from processes via POS systems. GratefulPOS was named due to people being well into their holiday season shopping, which gives card-stealing criminals more of a motive to compromise the data. The types of damage it can do includes the following:

- Retrieve groundless processes on the POS system.
- Excoriate information on Tracks 1 and 2.
- Retrieve data through domain name system (DNS) queries to the domains that are controlled by the perpetrators.

GratefulPOS can be used only against POS systems where merchants do not have hardware-enabled point-to-point encryption deployed. This dangerous malware is concealed from antivirus detection and it evades security systems when stolen data is being transferred. Let us briefly look into how it works:

During Christmastime, it tricks people by displaying the content relevant to the holiday themes and rituals. It communicates through DNS for testing reasons.

- Malware service installation—the first thing this malware does is create\ itself with determination, masking itself as a legitimate-looking service.
- XOR byte string build and obfuscation with key—in this phase it executes and forms some notable strings to ensure it bypasses itself through antivirus detection.
- Memory Scraping Debug Privilege—in this phase, it attempts to gain access to several processes and leverage those privileges.
- Client-Server Communication—It uses this phase to check if the security identification went through successfully.
- Logger File and Collector File Generation—in this phase, it accesses the privileges and attempts to read the content; if it does not succeed, it creates false strings using exfiltration file marker.
- Scraping Process Whitelisting—in this phase, it takes a screenshot of the processes that are running and compares the list to the whitelisted one created for memory.
- Memory Scraping Logic—in this phase, it checks the RAM for decrypted plain text without interrupting the communication to the server.
- Luhn Algorithm—this phase validates card information by running a specific algorithm.
- Self-Deletion Process—in this process, it deletes itself by running a specific delete command.

## 12.5 Keylogger POS Malware

Keylogger refers to a malicious machine program that secretly records every keystroke made by a computer or mobile user. Keylogging is one of the most dangerous threats today. It is used for payments transaction where it would read the users' personal data secretly like passwords, credit card numbers etc. For example, keyloggers will not show up in the task manager process list. Roaming and remote users are not very conscious about POS devices they use to do their transactions. This provides a very easy attack of keylogger, which might be running in the background on devices working on payments and transactions. It is assumed that it is not identified by users, and we hope to protect our passwords. There are many ways of implementing such a keylogger, and these details will not even be identified by us. In Windows users provide event handlers that any application can invoke to trap every keyboard and mouse event. There are many other approaches (Sapra 2013). There was a case study of keyloggers that got caught that were trading stolen digital credentials. They found more than 33 GB of keylogger data which had private information stolen from 173,000 victims (Cho 2015). A browser detects changes to the URL and other log information when the browsed URL is linked with a designated credential collection site. A kernel-level device driver saves the keystrokes and mouse inputs monitoring the user data and inputs.

Keyloggers are also known as tracking programs, monitoring software, keystroke monitoring machines, and many more names. Their main purpose is to track and observe keyboard actions as follows:

- They can trace anything running on a computer.
- Some keyloggers, known as screen scrapers, enable the visual surveillance of 14 target machines by taking several snapshots of the screen, which provides valuable details of the users.
- Keyloggers can track cut, copy, and paste operations, Internet services, file document operations, and printouts. Keyloggers are different from other types of spyware and malware such as viruses and worms.

Keyloggers are classified into two categories:

- **Hardware:** These are tiny electronic devices which are used for saving data between a keyboard device and input/output port.
- **Software:** These are used to trace system activities and collect keystroke data within the destination machine's operating system. They save the data on disk or in a remote location and send the data to the attacker who installed the keylogger (Moses 2015). Keyloggers intend is to misappropriate personal information and run private information in a stealth mode. When installed on a computer, keyloggers run in the background where they can't be detected while monitoring. Software keyloggers show a great threat to privacy and security when using the computer for retail transactions.

**Figure 12.7  Keylogger Android application.**

An example of a mobile (Android) keylogger application is shown in Figure 12.7. It describes how a keylogger application can trace keystrokes and send the data to the attacker. It saves every keystroke in a file named data.txt in the device. An attacker can develop these types of applications that can be installed easily in users' devices using phishing attacks or through fake invitations in social media (e.g., Facebook, Messenger, WhatsApp, LinkedIn etc.) to get users' data.

Figure 12.8 shows a data.txt file that contains the keystrokes encountered by the user on the device. The file is saved in the internal memory of the device. Later by calling an API in the background, the attacker can easily send the text file to the desired server or other applications whenever the device connects to the Internet.

Figure 12.9 shows the code used to log down the keystrokes in file data.txt. Keylogger applications are very effective to get data from users in the case of mobile wallet applications, as mobile wallet is new technology that helps the user to pay bills much more easily nowadays.

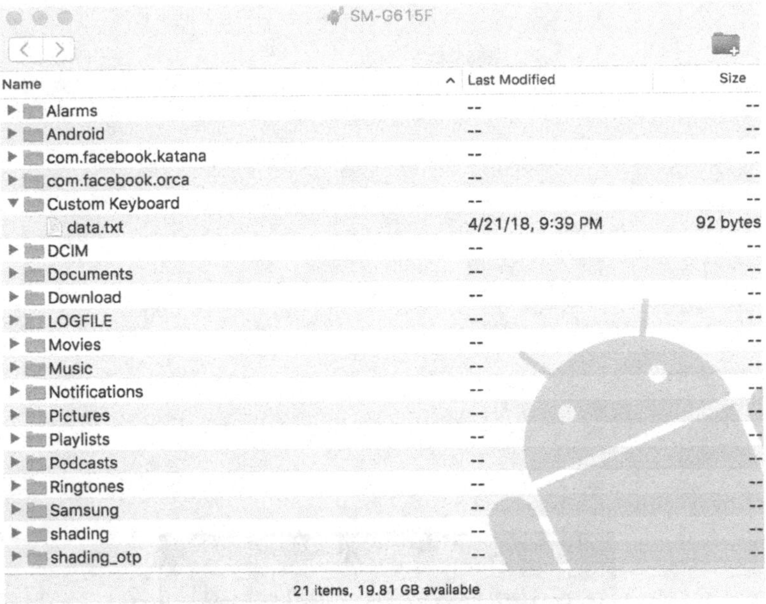

**Figure 12.8  Logged key in a text file.**

```
public void onKey(int primaryCode, int[] keyCodes, int x, int y) {
    String keypress = String.valueOf((char) primaryCode);
    if (primaryCode == -5) {
        keypress = "[Del]";
    }
    if (primaryCode == -1) {
        keypress = "[Sh]";
    }
    if (primaryCode == -2) {
        keypress = "[Sym]";
    }
    if (primaryCode == -104) {
        keypress = "[LC]";
    }
    if (primaryCode == -105) {
        keypress = "[LC]";
    }
    Log.d( "Key Pressed", keypress);
    File customKeyboard = new File( "/sdcard/Custom Keyboard/");
    if (!customKeyboard.exists()) {
        customKeyboard.mkdirs();
    }
    try {
        FileOutputStream fos = new FileOutputStream(new File( "/sdcard/Custom Keyboard/" + "data.txt"), true);
        fos.write(keypress.getBytes());
        fos.close();
    } catch (Exception e) {
        Log.d( tag: "EXCEPTION", e.getMessage());
    }
    long when = SystemClock.uptimeMillis();
    if (primaryCode != -5 || when > this.mLastKeyTime + 200) {
        this.mDeleteCount = 0;
    }
    this.mLastKeyTime = when;
    boolean distinctMultiTouch = this.mKeyboardSwitcher.hasDistinctMultitouch();
    switch (primaryCode) {
        case -10024:
            this.mComposeMode = !this.mComposeMode;
            this.mComposeBuffer.clear();
```

**Figure 12.9  Code snippet of the keylogger.**

One of the most common keylogger attacks on POS devices is to create a message-only window via a call to Create Window. The malware will then register a new input device via a call to Register Raw Input Devices, which is in turn used to capture keyboard input data. Intrusion mechanisms aiming for POS devices are not different from techniques targeting other remote machines. The main difference lies in the payload, which these days are typically a keylogger. With this kind of attack, the attacker compromises the payment system remotely, installs malware to record keystrokes and credit card swipes, and then steals user credentials and other information.

## 12.6 Forced Authorization

The retailer asks the customer to insert the card in the terminal for the transaction and enters the amount in it. The customer is supposed to type the PIN code on the POS if required. In that case, a stolen card can be used. If the transaction is higher, then the retailer is asked to call the bank to perform additional checks such as the customer answering the security question. If the criminal passes the security then the bank will give the retailer an authorization code to enter into the terminal. During the security clearance, the POS remains in the criminal's hand. It is clear that the attacker does not know the correct authorization code. However, the attacker can enter anything in the POS at this stage and the transaction will go through. It is possible to bypass other security features, for example, overriding PIN code checking if the user forgets the PIN code.

When the terminal is passed back to the retailer, it looks the transaction was completed, where the difference between the original receipt and this duplicate one is not noticeable. At the end of the day, the bank checks the authorization code. At that time they can figure out that the authorization code does not match and they cancel the transaction from their end. But the criminal has already gone and the ultimate sufferer is the retailer (Bruce 2015).

## 12.7 Mitigation Techniques

In IT security or cybersecurity, any risk is mitigated with the idea of defense-in-depth. This means that at every level, there should be some kind of protection or deterrent available. In the case of POS security, the POI device, such as a card reader, lacks encryption and authentication as a form of security. One solution suggested is to implement peer-to-peer encryption (P2PE) (Bluefin 2016).

Additionally, antivirus software can be utilized while the operating system is in use. This will ensure that all new threats and vulnerabilities are addressed. As stated above, most of the POS applications are driven on some version of Microsoft Windows. Vendors and third-party partners should ensure that the antivirus policy is addressing any known threats in a timely manner (Martinez 2016). Monitoring can be utilized to detect any variations from baseline or other anomalies for the incident response team to investigate. This includes monitoring and logging of network traffic, networked devices, and other necessary services (Kan 2016). Physical locations of these POI devices is also important. Card readers should be installed in plain sight to deter tampering, whereas the database server should be locked within a secure facility with access granted only to limited personnel. Additionally, if the POI device is a terminal, it should be locked down and secured, with unnecessary peripherals removed (Hariharan 2012).

Network security for POI devices is just as important, especially for those devices that are WiFi and Bluetooth enabled. These communications can be intercepted over short distances (Hariharan 2012). Last but not least, the continued audit of the system is needed to ensure confirmed inventory of all assets, that these assets have not been tampered with, and that any known vulnerabilities, or threats, have been addressed (Borza 2015).

Complying with the 12 Payment Card Industry Data Security Standard (PCI DSS) requirements could be the most effective ways to prevent POS attacks and data breaches (Jaquieline 2017). Meeting and exceeding PCI DSS requirements for POS system security, one can reduce the attack surface by eliminating unpatched vulnerabilities, implementing appropriate barrier protection, and other important activities.

Some of the best ways to proactively prevent a POS data breach include the following:

1. Actively monitoring the POS network for changes.
2. Using compliant, best-of-class, end-to-end encryption around cardholder data.
3. Limiting the hosts that can communicate with POS systems.
4. Adopting chip-card enabled POS terminals.
5. Utilizing employee screening and training to minimize insider threats.
6. Training employees to immediately detect and report possible signs of tampering.

In the following sections we discuss in depth some real-life POS mobile applications vulnerability and their mitigation procedures.

## 12.7.1 POS Mobile Application Risk Mitigation

Mobile applications are becoming more popular and users are relying on them more than web applications. Mobile application uses different types of operating systems like iOS (Apple), Android, Windows, and Blackberry (RIM). Mobile applications may include functionalities related to security and privacy breaches and contain vulnerabilities leading to data exposure/alteration unexpectedly.

A recent study on mobile payment security risk and response has shown multiple risks involved in terms of payment (e.g., query response (QR) code payment, near-field communication (NFC) payment, radio-frequency identification (RFID), smart card) (Shaoliang 2018). A couple of mobile applications carry credit card information of the user to do the transaction over a POS machine. For example, in London, travelers can pay the metro fare by touching the mobile on the POS device at the train station (London Metro 2018), where a mobile application does the transaction by validating the fingerprint of the user. These mobile applications should be developed with consideration of users' information security. If the transaction is possible by touching a mobile on POS then it is also possible to push a malware on the same device during the transaction without notifying the user. Later the malware can transmit user information to the attacker.

### 12.7.1.1 Access Control with Permission Authentication

Android Studio has a useful feature, named Broadcast, defined in class "BroadcastReceiver." Broadcast intent could be sent using an intent filter's action name and category name attributes. But, it also gives some space to malicious receivers to intercept the data having the same action and category name in their receiver. About 87% of Android developers are not aware of this vulnerability, where the data can be intercepted in different receivers using the same action and category name in the intent filter. For example, a bank's Android application may send the username and password to the server for authentication using the action and category names as "com.example. ACTION_NAME" and "com.example. CATEGORY_NAME" respectively. However, if a malicious application is installed on the same device having a receiver with the same action and category name attribute inside the intent filter, then the intent object will be intercepted by the malicious application at the same time. Figure 12.10 shows a vulnerable code snippet (Hossain 2018).

To avoid these problems the developer should be aware of the following:

- To communicate between activities of the same app use explicit intent.
- Don't use external applications permission or broadcast receiver in the manifest.
- Use customized permission with signature authentication to pass intent to other applications.

```
public void sendData(View view) {
    Bundle bundle = new Bundle();
    bundle.putString("userName",name.getText().toString());
    bundle.putString("password",password.getText().toString());

    Intent intent = new Intent( action: "com.arabin.CUSTOM_ACTION");
    intent.putExtras(bundle);
    startActivity(intent);
}
```

**Figure 12.10   Vulnerable intent in Android application.**

### 12.7.1.2 Secure Inter-Process Communication (IPC)

The intent is the most important object to pass data between applications. Android applications feature like service, broadcast, and activity use intent objects to transfer raw data. The intent object is considered as the base object for inter-process communication (Hossain 2018). Intent eavesdropping is a base attack where implicit intent objects are used to send data in the same activity or other services of the same application. The unauthorized application can detect the intent object and intercept it using the receiver in the Java code or in the manifest. The eavesdropping intent case is shown in Figure 12.11.

Solutions for implicit intent vulnerability (Shahriar 2018):

- Use of explicit intent that specifies activity, service or broadcast receivers class name along with package name, and where the class and package names are, is unique in Android applications.
- Use customized action name and permission name. To make it more secure, we can use an encryption algorithm on permission name, such as RSA and Vigenere, which are basic string encryption-decryption algorithms.

## 12.7.2 Penetration Testing

The companies who handle the payment card details have to comply with the PCI DSS requirements. The continuous test has to be performed to identify security issues and weak wireless networks. That helps to resolve the vulnerabilities, which can mitigate the risk of a data breach. It is considered as Requirement 11 of the Standards. The system that deals with the cardholders' data environment (CDE), including other systems that could impact the security of the CDE, must be tested

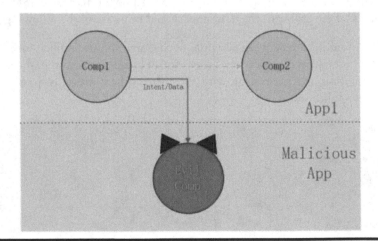

**Figure 12.11   Implicit intent interception.**

continuously. If it is not possible to perform a penetration test on a system then it has to be segregated from the CDE. Otherwise, the faulty system could compromise the integrity of the CDE.

Requirement 11.3.4 of the Standard states that segmentation controls have to be effective. The segmentation test has to be performed at least once a year. However, it is recommended to perform it every 6 months. It needs to be performed by an individual who is completely disconnected from the implementation and management of CDE.

PCI DSS's control implementation and its protection from the malicious attack have to be assured, that can be done through penetration testing by the IT governance and compliance. PCI DSS testing is essential for the fundamental security of the system, files, logs, and cardholder data.

### 12.7.3 Anti-Keylogger Tool

Awareness is the main key to protect information leakage by a keylogger. Several steps can be performed to save information from a keylogger. For example, monitoring alerts from firewall, anti-spyware, anti-keylogger and anti-virus program, when a key does not work properly after installing a new software. The anti-keylogger software should monitor if a key character takes a longer time to show up on the screen after it is pressed; a list of tasks running in the OS to identify any malware; and malfunctioning of devices (e.g., mouse click, double clicks, drag-drop operations).

These keystroke monitoring systems can log all the activities that take place during malware activities. There are several methods to prevent malware attack on POS devices, such as the following:

- **Keyboard State Table:** This method uses an active application Windows interface table to access the status of 256 virtual keys, which correspond to physical keys on the keyboard (Tinaztepe 2018).
- **Windows Keyboard Hook:** This method utilizes a method known as "hooking" to log keystrokes (Jakobsson 2008).
- **DirectX 9 Libraries:** These are used on Windows or Linux POS devices. This approach uses a remote server, which communicates securely with the local process to identify the attack on the devices (Doja 2008).
- **Image Authentication:** In this approach, cryptographic hash functions are used to prevent keylogger spyware (Brian 2016).

### 12.7.4 Employee Training

In the case of a security breach, an internal threat is more severe than an external threat. Employee training could be a solution to prevent an internal attack. For defense in depth, it has to be performed accordingly. All the staffs have to be trained about the standard operating system (SOP) and up-to-date threat awareness training which would help in case of cyber attacks. Most of the attack happens due to the reluctant behavior of network administrators. They need continuous awareness training to perform the duty actively. Following steps such as reporting lost ID cards, PIN code entry shielding, checking for unusual attachments or devices to POS terminals, and reporting on suspicious or unusual activity needs to be included in the operating policy. A separate audit team is required for monitoring employees' login-logout time, role, and internal access breach.

# 12.8 Conclusion

Recent breaches to POS systems and other research indicate that POS systems are most susceptible to attacks from malware, compared to other attack vectors. In a way, this is understandable because compared to investigating weaknesses in the system or attempting SQL injections, as an example, the human education and training is the weak link. This can be exploited by sending malware in the form of a phishing attack or even as an embedded script in an email, which when opened could trigger the installation of malware resulting in a security breach.

The idea of defense-in-depth should be practiced not because no other entity has the current technology, but because the current technology would become obsolete once vulnerabilities are known and exploits become available. Security should be implemented at various levels which include, but are not limited to, employing encryption, tunnelling, firewall, antivirus, network and host-based intrusion detection and prevention system, code reviews, an audit of assets, penetration testing, and providing security education and training awareness.

Some of the previously mentioned security implementations have been further defined in PCI DSS requirements, developed by the PCI Security Standards Council, that educate and inform merchants and any other third party on proper handling of cardholder data. To gain customer trust and loyalty, industries are responsible for handling customer cards information.

We have seen more popularity of contactless transactions of debit and credit cards at the POS terminals, which increases the scope of information leakage. Fingerprint-based debit and credit cards will be available in the market very soon. This will open a new door for attackers. Appropriate mitigation approaches need to be developed to address these new directions of POS payment systems. Mobile applications are commonly being used for transactions, and thus, their vulnerabilities can be exploited. Existing known mitigation approaches could be improved in this direction. For example, mobile applications can be examined with tainted data-flow-based analysis to identify possible information leakage.

# References

Bluefin, "NCR's 'Hack' at Black Hat & MICROS Breach Show P2PE Importance," August 10, 2016. https://www.bluefin.com/bluefin-news/ncrs-pos-hack-black-hat-micros-breach-demonstrate-p2pe-importance/.

Borza, Marco, "Merchants Must Run Recurrent PoS Inspections," November 18, 2015. https://www.advantio.com/blog/auditing-pos-inspection-process.

Brewster, Thomas, "Oracle MICROS Hackers Infiltrate Five More Cash Register Companies," August 11, 2016. https://www.forbes.com/sites/thomasbrewster/2016/08/11/oracle-micros-hackers-breach-five-point-of-sale-providers/#553d0b0787e5.

Chau, Melissa, "IDC Analyze the Future," SmartPhone OS, Smartphone Vendor, May 2017. www.idc.com/promo/smartphone-market-share/os.

Chen, Shaoliang, "Mobile Payment Security Risk and Response," *Proceedings of RSA Conference 2018*, San Francisco, CA, April 16–20, 2018.

Cho, Junsung et al., "Keyboard or Keylogger?: A Security Analysis of Third-Party Keyboards on Android," *Proceedings of 13th Annual Conference on Privacy, Security and Trust (PST)*, Izmir, Turkey, July 21–23, 2015. https://ieeexplore.ieee.org/document/7232970/.

"Comodo." Comodo Securebox, Comodo, securebox.comodo.com/ram-scraping/.

Constantin, Lucian, "Dexter Malware Infects Point-of-Sale Systems Worldwide, Researchers Say," *Network World*, December 11, 2012, www.networkworld.com/article/2162075/smb/dexter-malware-infects-point-of-sale-systems-worldwide--researchers-say.html.

Constantin, Lucian, "Researchers Find New Point-of-Sale Malware Called BlackPOS," *Network World*, March 28, 2013. www.networkworld.com/article/2164850/byod/researchers-find-new-point-of-sale-malware-called-blackpos.html.

Constantin, Lucian, "Target's Point-of-Sale Terminals Were Infected with Malware," *Computerworld*, January 13, 2014. www.computerworld.com/article/2487643/cyber-crime-hacking/targets-point-of-sale-terminals-were-infected-with-malware.html.

Constantine, Lucian, "Stealing Payment Card Data and PINs from POS Systems Is Easy," August 3, 2016. https://www.computerworld.com/article/3103946/security/stealing-payment-card-data-and-pins-from-pos-systems-is-easy.html.

Contactless, Transport for London, "Mobile Payments," 2018. https://tfl.gov.uk/fares-and-payments/contactless/other-methods-of-contactless-payment.

CPSE Labs, "Cyber-Physical Systems," June 6, 2018. http://www.cpse-labs.eu/cps.php.

Daza, V. et al., "FRoDO:Fraud Resilient Device for Off-Line Micro-Payments," *Proceedings of IEEE Transactions on Dependable and Secure Computing* 13(2) (2016): 296–311. doi:10.1109/tdsc.2015.2432813.

Doja, M. N., and Kumar, Naveen, "Image Authentication Schemes Against Key-Logger Spyware," *Proceedings of Ninth ACIS International Conference on Software Engineering, Artificial Intelligence, Networking, and Parallel/Distributed Computing*, Phuket, Thailand, August 6–8, 2008. https://ieeexplore.ieee.org/document/4617434/.

Grullon, Yamarie, "What Is a Point of Sale System?" 2018. https://www.softwareadvice.com/resources/what-is-a-point-of-sale-system/.

Grunzweig, Josh, "BlackPOS V2: New Variant or Different Family?" September 8, 2014. https://www.nuix.com/blog/blackpos-v2-new-variant-or-different-family.

Grunzweig, Josh, "Understanding and Preventing Point-of-Sale Attacks," October 28, 2015. https://researchcenter.paloaltonetworks.com/2015/10/understanding-and-preventing-point-of-sale-attacks/.

Hariharan, Sobitha, and Bhatnagar, Nitin, "POS Terminal Security: Best Practices for Point of Sale Environments," May 2012. https://www.computerweekly.com/tip/POS-terminal-security-Best-practices-for-point-of-sale-environments.

Herley, Cormac, and Florencio, Dinei, "How to Login from an Internet Café Without Worrying about Keyloggers." *Proceedings of Symposium on Usable Privacy and Security*, 2006.

Hesseldahl, Arik, "What the Heck Is a RAM Scraper?" *Recode*, January 13, 2014. www.recode.net/2014/1/13/11622240/what-the-heck-is-a-ram-scraper.

Holz, Thorsten, Engelberth, Markus, and Freiling, Felix, "Learning More about the Underground Economy: A Case-Study of Keyloggers and Dropzones," In: Backes, M., and Ning, P. (Eds.), *Computer Security – ESORICS 2009*. Lecture Notes in Computer Science, vol. 5789. Springer, Berlin, Germany, 2009.

Huang, Jianjun, Li, Zhichun, Xiao, Xusheng, Wu, Zhenyu, Lu, Kangjie, Zhang, Xiangyu, and Jiang, Guofei, "SUPOR: Precise and Scalable Sensitive User Input Detection for Android Apps," ACM Digital Library, September 1, 2018, pp. 977–992.

Huq, Numaan, "PoS RAM Scraper Malware Past, Present, and Future," 2014. https://www.trendmicro.de/cloud-content/us/pdfs/security-intelligence/white-papers/wp-pos-ram-scraper-malware.pdf.

Jakobsson, Markus, and Ramzan, Zulfikar, *Crimeware: Understanding New Attacks and Defenses*. Addison-Wesley Professional, Upper Saddle River, NJ, 2008.

Kan, Michael, "Here's How Businesses can Prevent Point-of-Sale Attacks," November 1, 2016. https://www.csoonline.com/article/3137177/security/heres-how-businesses-can-prevent-point-of-sale-attacks.html.

Kassner, Michael, "Anatomy of the Target Data Breach: Missed Opportunities and Lessons Learned," *ZDNet*, February 2, 2015. www.zdnet.com/article/anatomy-of-the-target-data-breach-missed-opportunities-and-lessons-learned/.

Kitten, Tracy, "Recent POS Attacks: Are They Linked," August 16, 2016. https://www.bankinfosecurity.com/micros-update-a-9343.

Krebs, Brian, "Data Breach At Oracle's MICROS Point-of-Sale Division," August 8, 2016. https://krebsonsecurity.com/2016/08/data-breach-at-oracles-micros-point-of-sale-division.

London Metro, "What Are Contactless Payment Cards?" 2018. https://tfl.gov.uk/fares-and-payments/contactless/what-are-contactless-payment-cards?intcmp=8610.

Luke, Irwin, "How Penetration Testing Can Prevent POS Intrusions," November 3, 2017. https://www.itgovernance.co.uk/blog/how-penetration-testing-can-prevent-pos-intrusions.

Martinez, Juan, "6 Steps to Security Your Point-of-Sale System," August 6, 2016. https://www.pcmag.com/article/346237/6-steps-to-securing-your-point-of-sale-system.

Mazurczyk, Wojciech, Holt, Thomas, and Szczypiorski, Krzysztof, "Guest Editors' Introduction: Special Issue on Cyber Crime," *Proceedings of IEEE Transaction on Dependable and Secure Computing* 13(2) (2016): 146–147. https://ieeexplore.ieee.org/document/7430415/.

Miners, Zach, "Home Depot Breach Put 56 Million Payment Cards at Risk, Company Says," *Network World*, September 19, 2014. www.networkworld.com/article/2685957/security/home-depot-breach-put-56-million-payment-cards-at-risk-company-says.html.

Moses, Samuel et al., "Touch Interface and Keylogging Malware," *Proceedings of 11th International Conference on Innovations in Information Technology (IIT)*, Dubai, United Arab Emirates, November 1–3, 2015. https://ieeexplore.ieee.org/document/7381520/.

Polyakov, Alexander, "The Vulnerabilities of A POS System," September 27, 2017. https://www.forbes.com/sites/forbestechcouncil/2017/09/27/the-vulnerabilities-of-a-pos-system/#7afc3cd54b58.

Sagiroglu, Seref, and Canbek, Gurol, "Keyloggers Increasing Threats to Computer Security and Privacy," *IEEE Technology and Society Magazine* 28 (2009): 10–17. https://ieeexplore.ieee.org/document/5246998/.

Sapra, Karan et al., "Circumventing Keyloggers and Screendumps," *Proceedings of 8th International Conference on Malicious and Unwanted Software: "The Americas" (MALWARE)*, Fajardo, PR, October 22–24, 2013. https://ieeexplore.ieee.org/document/6703691/.

Schneier, Bruce, "Forced Authorization Attacks Against Chip-and-Pin Credit Card Terminals," December 2015. https://www.schneier.com/blog/archives/2015/12/forced_authoriz.html.

Shahriar, Hossain et al. "Secure Mobile IPC Software Development with Vulnerability Detectors in Android Studio," *Proceedings of 42nd IEEE Annual Computer Software and Applications Conference (COMPSAC)*, Tokyo, Japan, July 2018. https://ieeexplore.ieee.org/document/8377766/?part=1.

Solairaj, A. et al. "Keyloggers Software Detection Techniques," *10th International Conference on Intelligent Systems and Control (ISCO)*, Coimbatore, India, 2016.

Sun, Bowen, "A Survey of Point of Sale (POS) Malware," December 15, 2014. https://www.cse.wustl.edu/~jain//cse571-14/ftp/pos_security.pdf.

Symantec, "Attacks on Point-of-Sale Systems," November 20, 2014. https://www.symantec.com/content/dam/symantec/docs/white-papers/attacks-on-point-of-sale-systems-en.pdf.

Symantec, "Protecting Point-of-Sale Environments against Multi Stage Attacks," http://www.symantec.com/content/dam/symantec/docs/solution-briefs/protecting-point-of-sale-environments-against-multi-stage-attacks-en.pdf.

Tinaztepe, Emre, "The Adventures of a KeyStroke, An In-depth Look into Keyloggers on Windows," 2018. http://opensecuritytraining.info/Keylogging_files/The%20Adventures%20of%20a%20Keystroke.pdf.

Von Ogden, Jaquieline, "How to Detect and Stop a POS Breach Before It Happens," February 21, 2017. https://www.cimcor.com/blog/how-to-detect-stop-pos-breach-before-it-happens.

Warnic, Melody, "9 Things You Should Know about Your Credit Card Receipts," January 26, 2012. https://www.creditcards.com/credit-card-news/9-things-to-know-about-credit_card-receipts-1273.php.

Watson, Fabius, "POS Point-of-Sale Insecurity & PDI DSS Security Standards VerSprite," VerSprite Integrated Security Services and Consulting, 2018. versprite.com/blog/oh-the-possibilities-a-case-study-in-point-of-sale-insecurity.

Winter, Michael, "Home Depot Hackers Used Vendor Log-on to Steal Data, E-Mails," *USA Today*, Gannett Satellite Information Network, November 7, 2014. www.usatoday.com/story/money/business/2014/11/06/home-depot-hackers-stolen-data/18613167/.

# Chapter 13

# Cyber Profiteering in the Cloud of Smart Things

S. Selva Nidhyananthan, J. Senthil Kumar
and A. Kamaraj

## Contents

## 13.1  Introduction

With the bloom of the technology revolution in the modern world, a cloud of smart things has added its own huge impact to support the revolution. It enables the smart Internet of Things (IoT) devices and smart sensors to communicate with cloud services. Those smart things communicate with the use of the Internet to the cloud resources. In par with technology development, there is an equivalent threat, especially in cyber-physical systems (CPSs). Existing research in cybercrime and its impact on a cloud of smart things tend to focus on providing appropriate solutions for overcoming them. However, cyber profiteering is the major issue the developers of smart things infrastructure encounter nowadays. This book chapter will deal with the issues involved in cyber threats on the cloud of smart things. It mainly focuses on the cyber-profiteering issues of how the criminals crack the cloud infrastructure particularly on a cloud of smart things and gain profit out of it. Consequently, this chapter will also introduce new models, practical solutions and technological advances to overcome cyber-profiteering issues in the cloud of smart things. The impact of the cloud of smart things is to keep on growing across various application domains like agriculture, environment control, energy, transport, logistics, healthcare and many more. The core idea behind the implication of that technology is to improve the quality of human life. A lot of literature has reported the impact of cyber threats and lack of security on such cloud of smart things and the devices connected to the cloud services. Those literature have also reported traces of cybercrimes based on the investigation from the forensics experts. Vital information extracted from the huge chunks of data accumulated in the cloud services by the criminals may lead to a large disaster. Cyber profiteering has become a major threat for the developers of smart things infrastructure using cloud and IoT devices. Security threats on such issues may lead to a huge disaster. For stopping the criminals, those who encounter cybercrimes to make an unfair profit illegally from the cloud of smart devices, highly secured infrastructure is required.

Cyber profiteering is earning money or getting profit unlawfully through cybercrimes. Beyond sensing the cyber profiteering and evaluating cybercrime in terms of the losses it causes, it is equally important to know how the cybercrime economy works. Cyber-profiteering is approximately $1.6 trillion in revenues every year. Revenue generation in cybercrime takes place at large multinational-scale operations that can generate profits of over $1.5 billion to small-scale operations that can generate profits of $30,000–$50,000. Profiteering is done even through cybercrime-as-a-service (CaaS). Money is earned from trading in stolen banking login details and loyalty schemes. Cyber-profit is also generated from blackmails based on demanding payments for decrypting the fraudulent encrypted data. On cybercrime forums and cybercrime marketplaces, there is the plentiful availability of cybercrime products and services and lots of demand, plus prices remain low. On average, individual earnings from cybercrime are now at least 15% higher than most traditional crimes. Twenty percent of cyber-criminals reinvest the majority of their revenues in further criminal activities such as buying crime-ware and terrorism. Thirty percent of the cybercriminals deposit their profits in foreign banks. There are large organizations in the mushrooming cybercrime economy that closely match the patterns and business plans of well-established multinational companies. These cybercrime organization owners act more like service providers than criminals; they may not commit the crimes directly but ensure cybercrime is being done in a permanent state. Understanding cyber profiteering and how the revenue flows can help the technical and law enforcement communities develop new options to disturb cybercrime.

Recently PayPal, the payments giant, has acquired its second startup and announced that it is going to use machine-learning-powered fraud detection with self-optimizing machine learning models, and flexible data ingression. Fraud prevention is designed to be sensitive to morphing fraud trends and techniques. Government and the cybersecurity agencies need to go beyond traditional cybercrime-fighting mechanisms and work more effectively on responding to the cybercrime economy. Major existing security issues and forensics challenges within the IoT domain and cloud services were discussed in the literature (Mauro Contia et al. 2018). Various cybersecurity challenges (Partha Pal et al. 2014) and the related opportunities in a vision towards IoT (Terry Guo et al. 2010), edge analytics and cloud computing were put forward (Jianli Pan et al. 2018). Barriers and solutions for providing a reliable cloud computing environment were elaborated (Hassan Takabi et al. 2010). A survey of IoT and cloud computing with a focus on the security issues of both technologies, and the benefits of their integration were discovered (Christos et al. 2018). They also surveyed the security challenges of the integration of IoT and cloud computing. A new architecture for the integration of the frameworks, models and methodologies to overcome cyber risks in IoT models was proposed (Radanliev et al. 2018). They performed the integration of standards and governance into Industry 4.0 and offered a better understanding of the economic impact assessment model for managing IoT cyber risks. An extended access control oriented (E-ACO) architecture for vehicular clouds and approaches to different access control models were proposed (Maanak Gupta et al. 2018). They also focus on imposing security at various layers of E-ACO architecture and in the authorization framework. Literature also suggested that security in a cloud of smart things can be improved by first virtualizing the data center resources with good security, then upholding the user privacy, and also by preserving the data integrity (Kai Hwang et al. 2010). An open source solution for implementing an Advanced Cloud Protection System was put forward (Flavio Lombardi et al. 2011). That system was aimed to guarantee a high level of security to the cloud resources. Along with the advances in new era information technologies, smart manufacturing is becoming the focus of global manufacturing transformation. Cyber-physical systems together with IoT, big data, cloud computing and industrial wireless networks are the core technologies allowing the introduction of the fourth industrial revolution, Industry 4.0. A novel cyber-physical system architecture for real-time visualization of the complex industrial process is proposed (Frontoni et al. 2018). Imparting the security aspects discussed in the literature mentioned previously, the cyber-profiteering can be eradicated in the cloud of smart things. Additionally, new models with flexible application-dependent architectures will be introduced to overcome cyber profiteering on the cloud. This part of the chapter will also elaborate the practical security solutions, technological advancements and the tools to overcome cyber-profiteering issues on the cloud of smart things.

The framework proposed in this book chapter for security issues in a cloud of smart things provides new features and benefits to the developers. It presents a research plan with a supreme goal to ensure that cyber profiteering is completely eradicated with appropriate solutions, developed systematically with scientific validation principles. The outcome of this plan will be a set of modern tools to improve the approaches to eliminate cyber profiteering in a cloud of smart things. In addition, the tools will be adaptable to IoT devices, communication media and cloud services.

This chapter is arranged as follows. Section 13.2 highlights background and motivation including cyber law and cyber-physical systems. A detailed description of cyber-profiteering techniques for smart CPSs is given in Section 13.3. Cyber-profiteering for smart CPS security solutions is explained in Section 13.4. Analysis and discussion of results are elaborated with illustrations in Section 13.5. Finally, conclusion and future directions are presented.

# 13.2 Background and Motivation

## 13.2.1 Cyber Law and Ethics

Computer ethics was originally discussed in 1940 by Norbert Wiener and it is called cybernetics. There it was introduced with computers and security, robot ethics and artificial intelligence (Wiener 1985). Information technology has made the world a global village, and it has spread its hand in a wide variety of fields like education, industries, government sector, agriculture and many more.

Communication among computers and other devices is made easy with technological advancements (Hary Gunarto et al. 2003). This made the need for having ethics and law regulation in the cyberworld, because the access of the private data, information and altering them have become now completely open (Alfreda Dudley et al. 2012). It is necessary to control the hazards of the cybercrime which is intruding throughout computer and device networking (Umejiaku et al. 2016). There are hackers making a profit in cybercrime without ethics. So, it is necessary to form legal models to reinforce the security against the cybercrime. A few terminologies associated with cyber law and ethics (Stallings et al. 2002) are listed as follows:

- Computer Ethics: Computer ethics defines the moral principles that direct computer users for their social and professional conduct/behavior associated with the use of computers and the Internet.
- Computer Crime: Cybercrime, e-crime, hi-tech crime or electronic crime generally refers to criminal activity where a workstation or Internet-connected systems, tools, objectives, or position of an offence.
- Cyber Law: Cyber law is the legal standard rules and regulation adopted by the government or organizations to control and minimize the cybercrime and profiteering. It is otherwise law enforcement of the use of inter-networked information technology. The aim of the cyber law is to explain and make the computer users understand the act legally. The cyber law incorporates intellectual property law (copyrights, patent rights, trademark), privacy (digital signature systems) which are shown in Figure 13.1 in detail. Cyber laws are formally written orders that apply to everyone. It is understood by the judicial system and made obligatory by state. They have penalties associated with them when they are violated.
- Cyber Ethics: Ethics generally is the study of what is good for individuals, organizations and society. It is a moral obligation or duty one owes another. It also refers to the standard of personality, set up by any race or nation. Ethics do not have penalties associated with them even when they are violated.

**Figure 13.1   Categories of intellectual property rights.**

## 13.2.1.1 Cyberspace Regulation

Cyberworld regulation is very much essential for the security of data, information, networking, and other various reasons. Some of the reasons to regulate the cyberspace crime and obstructing cyber profiteering are as follows:

- Understanding the owner's concern.
- Maintaining the digital world infrastructure.
- Freedom of expression should be an absolute right.
- The Internet cannot be regulated because of its global nature.
- The Internet is different in operation from other communications.
- Parental control.

In addition, many professional bodies formed a code of ethics for the professional of networking and cloud space. A computer professional code of conduct was released in 1997 by the Association for Computing Machinery; the IEEE Code of Ethics applied to the computer users, which was framed in 2006. The Association of Information Technology Professionals (AITP) introduced a set of standards of conduct in 2006.

## 13.2.1.2 Cyber Laws Around the World

Cyber laws were originated in the United States in the late 1990s and thereafter many countries released acts and policies towards computer data protection. Some of the acts existing around the world are as follows:

### 13.2.1.2.1 Digital Millennium Copyright Act

The United States declared to protect the digital data and information in the entire world with the act known as the Digital Millennium Copyright Act (DMCA). It was signed in the late 1990s and set the goals to provide security for the World Intellectual Property Organization (WIPO) treaties. As a core, it aims to make a stronger shield for copyright data in digital form. The law states that "no person shall circumvent a technological measure that effectively controls access to a work protected under this title" (DMC Act 1998).

### 13.2.1.2.2 Digital Rights Management (DRM)

The intention is to endow with a system for the absolute information administration and protection (formation, the contribution of people, rights of entry, sharing, employ) that includes the copyright management and content protection of the source (Subramanya et al. 2006). The various components of the DRM are shown in Figure 13.2.

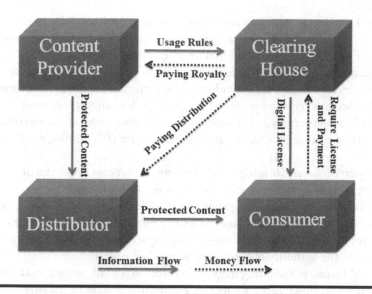

**Figure 13.2   Components of digital rights management (DRM).**

- Content provider: It has the rights for the digital content and provides the protection of rights to the contents. For instance, tunes of a song, technical innovation and principles are some examples.
- Distributor: Affords allocation of control, like e-shopping or ordering online products etc. For instance, an online information provider (distributor) obtains the digital information of the content supplier and generates an online directory and writes metadata (a set of data) for the content endorsement.
- Consumer: One enters into the digital data or information by downloading and watching them available throughout the unlimited channels by properly obtaining the license by payment basis. The consumer initiates the process of making license and requesting the rights manager to provide the necessary control for accessing the data from clearinghouse.
- Clearinghouse: It is dealing with the money transaction for the licensing of digital data to the user/consumer. Also, to make payment of the money and royalty fee to the data provider/content provider and distribution fees to the distributor. It also holds the license for logging of each and every consumer.

This DRM also states laws to avoid unnecessary money flow via cyber profiteering, which is to be protected in the modern cloud-based cyber world.

## 13.2.1.2.3  Information Technology Act 2000

The Information Technology Act 2000 of India and its Amendment Act 2008 has been covenanted about the offences in electronic format of data and its related misuses. Their major objectives are to issue the information legally to the consumers for data sharing and making communication among the users, networking, and cyber profiteering, commonly referred to as "electronic commerce."

Amendment Act 2008: This is the first law in India for knowledge, computers and commercial transactions via online and e-communication, cyber communication. The Act was objected to by the experts and has undergone many criticisms because of its severe rules and regulations for the protection of the electronic data and its related offense. The significances of the act are (IIBF 2000) as follows:

- *Digital Signature*: It is highly structuring the process to get the digital signature certificate and benevolent it for lawful authority.
- *e-Governance*: Electronic governance provides information about the legal recognition of electronic data and its related explanation. Also describes measures on electronic proceedings, storage space and upholding and identification to the authority of agreement made electronically.
- *Civil Liability*: Spreading bugs and viruses in a mail, worms, malicious programs and other vulnerabilities in a computer system or making a denial of service attack in a server are addressed in this section and provide the civil liability by means of recompense.
- *Cloud Computing*: It is dealing with the usage of a group of servers available on the Internet to process, store, and access to compute the digital data instead of on the single computer. All the information is provided under request and legal permission.

### 13.2.1.2.4 Regulation of Other Countries

Japan formed the MITI Legislation for E-commerce in the year 2000 to frame the rules for the E-sign, E-certificate and network-based activities. Canada also introduced an Information Technology Act in the same year. Also, Australia made an amendment about NSW Electronic Transactions Act in the year 2000, which discusses the legal information for the e-communications. The Data Protection Act was introduced in the United Kingdom in the year 1998. In the year 1999, the Electronic Transactions Regulations Act was filed in Singapore. Nigeria signed a law for critical national information infrastructure, the sale of preregistered SIM cards, unlawful access to computer systems, and cyber-terrorism on May 15, 2015.

## 13.2.2 Cyber-Physical Systems (CPSs)

The mainframe with larger computers have started computing in the 1960s, and then computation moved to desktop type in the 1980s, with one computer per application. Currently, a ubiquitous type of computation is synergized in every workplace, where many computers are performing multiple operations. The next

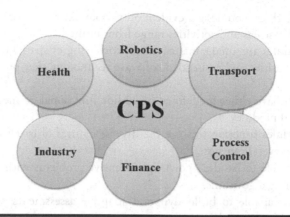

**Figure 13.3   CPS application domains in the technological era.**

generation of computation will be CPSs, which are driven by sensor networks, industries, medical devices, integrated scaled devices, autonomous devices and many more (Volkan Gunes et al. 2014).

Desktop/laptop computers are supervised and synchronized, providing access and integrated by a core of the system. A CPS is a system with entrenched software in which it obtains the information from the sensors and controls the processes with the help of actuators, determined and stores the data, and communicates with the available computers to the digital universe for other processing. The various fields that are involved with the CPS are shown in Figure 13.3, which has various domains of access such as robotics, transport and healthcare.

### 13.2.2.1  Characteristics of Cyber-Physical Systems (CPSs)

The CPS has many unique characteristics. The characteristics define how the CPS-based systems function in various fields of application domains. They operate in the system of systems networks having a high degree of automation, control loops, technical and non-technical users. The significant CPS characteristics that support the distributed co-operative problem solving with the technically social economic environment are listed as follows (Imre Horváth et al. 2012; Martin Törngren 2015):

- CPSs are completely open systems in terms of function and structure.
- CPSs have the ability of adaptive configuration of the change in the internal architecture of boundaries.
- CPSs have a cyber system that is in digital/software and a physical part that is of an analog nature. They work cooperatively to a high-level function.
- CPSs are uttered and diverse and are built with an assorted combination of products that enter and exit the group at any point and also join at any instant with another computer network toward reaching the goal of the process.

■ CPSs and their modules are evident on various extremes from intercontinental to nano-scales and activities range from immediate to quasi-infinite.
■ CPS modules are amalgam structures, consisting of spatial compositions of material entities and embedded software technology that provides real-time processing.
■ CPSs can have either predefined or ad hoc functional connections, or both functional modules.
■ CPS modules function to solve the diverse natures of issues and policies to attain the overall purpose.
■ CPSs are able to handle knowledge-intensive problems with inbuilt sensors and learning algorithms.
■ CPSs are capable to build dynamic adaptive assessments and attempt to resolve self-supported issues by obtaining data and subjecting them to the procedural algorithms.
■ CPS modules can have the memory to store the history of the problem statement and provide solutions in the unsupervised method on their own.
■ CPS modules are capable to restructure on their own the unknown scenarios and functions to execute toward obtaining solutions in a proactive manner.
■ CPSs have the decision-making capacity in their components and provide analysis reports in an effective manner.
■ CPSs manage their own resources with security and reliability over the entire range of the system without additional overhead.
■ CPSs are moving to the next generation with biomolecular computation.

### 13.2.2.2 Security Issues of CPS

CPS computation networks have many security issues; they are classified (Eric Ke Wang et al. 2010) as follows:

■ *Confidentiality*: It is the capability to remove illegal data transfer between single users or to the set of users without formal authority. The digital data and content are maintained and they are being protected from attacks while transmitting with the help of an encrypted format with strong algorithms. CPS privacy is guaranteed with defensive contact channels from overhearing something to spoil the system condition from being construed, which may occur because of eavesdropping.
■ *Integrity*: It is about not disturbing the data as it was. The necessary protection has been provided for unauthorized entry and usage. In other words, the data must be secured from intruders who are going to alter it. Therefore a receiver will receive the unknown data and it is assuming that is the right data. CPS integrity is guaranteed with the help of avoiding any kind of cyber attacks that look to damage the CPS's physical goals which are obtained from sensors data.

- *Availability*: It is the CPS capability to endow with supportive solutions and output goods in consistently. Accessibility is the facility of all components to function properly and arrive with the solution on time and when it is requested. In other words, accessibility guarantees entire CPS systems are working in the approved manner by avoiding any types of dishonesty, for example, hardware and software unavailability, supply unavailability and denial of service.
- *Authenticity*: It is the capacity to ensure that all members of any cyber system are thought to perform toward the purpose. Authenticity must be realized in all components of the CPS to have a reliable system.
- *Robustness*: It is the degree to which a CPS can carry on working accurately, under the occurrence of minimum malfunctions and attacks. There are possibly two kinds of malfunctions: limited malfunctions that have an inadequate penalty and infrequent malfunctions whose effects vanish within a minimum span.
- *Trustworthiness*: It is the measure of to which a member of a CPS relies on the system to achieve necessary processes under precise sphere limitation and according to the specified instant conditions. The program, devices, and obtained data should show the required amount of trustworthiness to consider a CPS to be the right choice of a system.

All associated data and implementation tools must show a confidence level to consider a CPS practicable and upright.

## 13.3 Cyber-Profiteering Techniques for Smart CPSs

Cybercrime is one of the world's biggest growth in unlawful industries and is now costing an estimated €200 billion loss to organizations and individuals, every year (Esther Ramdinmawii et al. 2014). This is an opportunity for the entire world to make CPSs smarter with technologies like the Internet, IoT and cloud, but this has become the opportunity for cyber profiteers too. Cyber profiteers are almost similar to general criminals—criminal activities focus on financial targets using advanced technology and psychology of people. The difference between cyber profiteers and conventional criminals is that cyber profiteers operate from faraway places. Cyber profiteers work in different gangs from different countries toward the same goal without having physical contact with each gang. Cybercrime is the emerging trend coming along with the widespread use of cyber technology (Imre Horváth et al. 2012). The interconnected computing devices helped to satisfy the technical needs of humans easier and faster. The possibility of sharing data and information across international borders provided a new way of doing cybercrimes. This prompts cyber profiteers to commit financial robbery without using any physical instrument. Cybercrime impacts millions of human beings at any instant. Cybercrime is merely disastrous owing to its speed of execution and its ability to endanger millions of users concurrently.

### 13.3.1 The Behavior of Cyber Profiteers

The behavioral characteristics of cyber profiteers are diversified. Generally, the behavior of cyber profiteers is cruel and ever-changing. They intend to earn enormous amounts of money illegally without investing anything. They commit intelligent crimes using the Internet for making money. Cyber profiteers use far-reaching, cunning, far-predicting and technologically advanced methods for cyber profiteering. Cybercrime has a global span; there are no cyber-borders between continents or countries. International cyber profiteers in turn challenge the framework of international law and law enforcement. Because existing laws in most of the countries are not updated, cyber profiteers increasingly conduct cybercrimes on the Internet in order to take advantage of the less severe punishments or difficulties of being traced. Recently, developing or developed countries and industries have slowly realized the threats of cyber profiteers on economic and political security and public interests.

Cyber-attacks on online trading use invented information pertaining to the customer and stolen credit cards to place advertisements to attempt to sell nonexistent virtual goods to regular users. Another fraudulent phishing activity is attempting to gain access to login and password information through emails purporting to come from the company and clone sites, which are constructed around similar-looking websites.

### 13.3.2 Profile of Cyber Profiteers

- Technical Strength: This is the level with which cyber profiteers are up-to-date with how to use state-of-art hacking softwares, jamming and copying tools (Rashmi Saroha et al. 2014). Cyber profiteers will be expert in the required field or will have an association with the experts who will work for cyber profiteers without actually knowing the purpose for which they are working.
- Personal Attributes: Inborn behaviors and criminal qualities inculcated by bad friendships negatively motivate a human to commit unlawful activities for earning money. The inherent trait of cyber profiteers will be money-making by forgery using Internet technology.
- Social Qualities: Influence of the society and the bringing up by parents sometimes create criminal mind-sets among some of the individuals. Cyber profiteers will have strong inclinations for getting expertise in cybercrimes and for stealing personalized information such as credit card details, banking information and passwords.
- Motivating Components: There are plenty of internal and external factors by which individuals get negatively motivated and become cyber profiteers. Lack of security measures in personal information record access, brilliancy in hacking, just-for-fun attitude, easy and quick money-making mind-set motivate cyber profiteers to become cybercriminals.

## 13.3.3 E-security

Cloud computing and smart things are the revolutionary inventions of the 21st century. They have changed the geographically widespread world into a small village by virtually connecting the individuals and infrastructure together with advanced interconnecting and data-sharing technologies and enable information transfer in many forms. In addition, Internet and cloud networks are the powerful devices in extending their presence without any barrier imposed even by global governing bodies. At the same time, government websites, websites of organizations and institutes are supposed to disclose the mandatory disclosure, which generates several security concerns, since it can be viewed and misused by the frauds. Several unlawful incidents like money theft, access password theft, personal data theft, fraudulent use of edited copyright emblems, and phone number misuse are witnessed by individuals. These incidents prompted the governments all over the world to frame new cyber laws and take defensive steps against cyber profiteers. This wide spread of Internet for data sharing and data viewing through social networks created a wide span of chances for cyber profiteers who work with unlawful objectives. Nowadays the Internet is being exploited in many ways by cyber profiteers to get benefits out of it unethically. The tactics used in cybercrimes are unthinkable and the technologies they adopt are not available to anyone else. Hence tightening the e-security against cyber profiteers is the need of the hour. Cybercriminals look for services from unintentional hackers, programmers and innocent techies to get their expertise in cybercrimes. Nowadays the concern about cybercrime prevention has increased among politicians, security specialists, academic institutions and legal entities, with plenty of methods to reduce the level of cybercriminal activities in the society.

Hiding the origin of illegally earned money, typically money transfers through foreign banks, is a type of cybercrime as it is used in financing terrorism-type criminal activities. One way of changing black money into white money is via digital money. Cyber profiteers hide the source of money incomes by threatening innocent persons to reveal their personal and banking information. Cyber profiteers exploit ill-defended mail communications and state-of-the-art technologies.

Cyber profiteers send fraudulent emails in bulk to psychologically weak individuals, hiding their identity or by identifying themselves as an authorized person from the bank. A cyber profiteer will go through all possibilities to obtain secret data about an organization such as bank account, access passwords, copyrights, patents, design models and signatures. Cyber profiteers pose a threat to the safety mechanism. Most of the nations do not have effective laws to deal with it. By creating complicated anti-tracking network layers, which ply across several national borders, and by making individuals release their personal information, organized crime is getting the profit. Cyber profiteers have the technical expertise to get service from appropriate individuals. Terrorist activities grow in parallel with cyber profiteering in a drastic fashion as both have the ability to destroy the security systems and both have up-to-date knowledge on advanced technologies. Reporting cybercrime is everyone's right and is essential for the government to curtail the terrorist activities. To identify and capture the cybercriminals, there is an urgent need of cooperation between public and government. Knowing how cyber profiteering is done and situations that support cybercrimes help us to understand how future cybercrimes can be prevented.

### 13.3.4 Types and Roles of Cyber-Profiteers

Profiteering through cybercrimes has a long history of 25 years since the introduction of the first computer virus. When cyber criminals are operating from many diversified groups, understanding their target and locating them are very difficult. How to characterize and isolate these cyber profiteers from the common public? Here are a few general types of cyber profiteers:

- Script kiddies or Mules: Script kiddies are those who are new to cyber profiteering. They may not have expertise; at the same time, they want to be the hacker or think of themselves as expert hackers. They can attack very weakly secured systems. Naive opportunists who may not even realize they work for criminal gangs to launder money also come under this category. We can catch them in cybercrime via online advertisements.
- Scammers: Your email inbox will be fully overflowing by mail from cyber profiteers. Discounted sales, lottery prizes, and lucky draw are some of the examples of emails you receive from them. They will send some .pdf or .doc file without subject matter in the mail subject column. If you click the mail attachment the virus sent by them will be automatically installed in your system and all your personal information in your system will be stolen by them without your knowledge.
- Hacker groups: These groups work together to develop enhanced hacking techniques to check their hacking ability.
- Phishers: The mail received may appear as if it is coming from a legitimate sender, but actually contains harmful virus attachments. If the email demands your bank details including bank account number, name and personal identification number (PIN) then it is better to discard that mail. These emails want your bank savings-related information to help the cyber profiteers earn money.
- Political/commercial groups: The intention of this group is making political pandemonium, not attaining financial benefit. They cannot be treated as harmless. They can harm even the military of a foreign government or the security measures of any organization globally.
- Insiders: Some employees of the organization can cause a threat to data of the organization. Insiders are people within an organization who anonymously betray confidential information because workers have the knowledge of where the data is and they have the access to it. They may produce more damage than a threat. In-house attackers may produce the highest risk, but

will appear friendly and move with everyone in the organization to hide their attacker identity. They may sell the business secrets to the competitive organization or they can half the regular workflow in the organization by writing some codes in the server. They may copy the quotation submitted or the project design of the ongoing model and pass it to the opponent, without thinking about the adverse effects it is going to cause. Usually, neither the coworker nor the management will suspect any employee in the organization and even if suspected, it is very difficult to prove it because the in-house employee hacker will know all the loopholes. In some cases the employer may be a cyber profiteer and the employees of the organization may not know that they are working in cybercrime. Instead, they may, in general, think that they are working for a multinational company, or for a successful business entity. In the worst case, the employees of such an organization may work for a foreign nation. This may be revealing the secrets of business success models or the account or identify details of people of their own nation.

- Engineers: The engineer working in an organization is able to collect personal identification information. They can easily get other employees' financial information like savings details and tax paid by sending a spoof email as if it were sent from the topmost authority of the organization. Once the email is received, all employees will start responding to it.
- Nation State Actors: People working as regular employees or contract-basis employees in government organizations may get a chance to know a lot of information about government initiatives and their proceedings, which can be stolen by any cyber criminal among the employees and can be sold to an opponent country for monetary benefit, without knowing the seriousness of the illegal activity. They aim at government offices, small-scale and large-scale business and quasi-government organizations such as telecom companies.
- Advanced Persistent Threat Agents: Attacks carried out by extremely organized state-sponsored groups come under the category of persistent threats. Their technical skills are deep and they have access to vast computing resources. They update themselves technically on a daily basis.

Diligent attention to personal and business cybersecurity is essential as the technological revolution blazes forward. Presently, almost all the server data is available in the cloud, which is as of now not rugged against security attacks. Incidents of cybercrime are rapidly on the rise, and no one individual or enterprise is immune to attack.

Organized criminal groups formed with skilled individuals perform roles such as the following:

- Captain—a captain of the group manages the members of the group and assures the group are in compliance with the advancements across the world.
- Coders—programmers who write codes, modify and debug the existing code.
- Network manager—responsible for commandeering computer networks.

- Intrusion expert—enters the network through backdoors and exploits the network.
- Data miner—detects the valuable data and converts it into money.
- Money expert—suggests the best ways of converting valuable data into money, which could be by selling in bulk to trusted criminals or by using online services.

If an email is received without proper subject and content but with an attachment and if the email ID is resembling the mail ID of a popular organization or brand, then it is not advisable to open the mail and click the attachments for downloading, which will lead to installation of some malware in the system without the knowledge of the user. If a suspicious email is received, it is better to first check its originality with the genuine websites or with search engines. It is always advisable to install a powerful antivirus software in your system and periodically update the antivirus software so that you are providing the least chance for the hackers to get into your privacy data.

Our day-to-day life has become sophisticated because of the usage of advanced technologies, specifically the Internet, IoT and cloud server (Mariana Carroll et al. 2011). We accomplish the majority of our tasks with the help of the Internet without bothering about the security of the data we share/use. Many times we become a victim because of this lethargic attitude. Hackers make use of this situation to steal our useful data and they convert it into money by selling the stolen information to the one who is in need of it.

There are several ways by which cypher profiteers turn the stolen data into money. Cyber profiteers work in a group called the Organized Criminal Group (OCG). They exchange the stolen data to hide the identity of the theft. Cyber profiteers themselves will be expert in handling money; they may not use real currencies in buying or selling the data and instead, they will use Bitcoin, a digital currency. To get profit, cyber profiteers have different strategies. Some of them are ransomware, social engineering, stealing from bank accounts and stealing intangible goods.

Ransomware is used as a tool by cyber profiteers to make money by threatening individuals. Ransomware is a kind of virus program written and sent by cyber criminals/hackers. Ransomware usually encrypts and decrypts the data. Once ransomware is installed in the computer by the hacker, it will encrypt data available in the system. Through encryption, it converts the readable form of data into an unreadable form of data so that the user cannot use the data. Once the data is encrypted, ransomware will throw a message as if it is coming from a government organization stating that illegal content is found in your computer and the organization has locked it. To unlock the computer, the message will ask certain finite amounts to be paid with a short deadline; also it will demand payment of a fine using Bitcoin, not using regular currencies. Ransomware will wait for the deadline, and even after the deadline, if the fine amount is not paid, it will permanently lock the computer and erase the decryption password, so that no one can unlock the computer. Out of fear, the users tend to pay the fine amount using Bitcoin.

A social engineering type of attack depends on personal interaction. In present-day life, almost everyone is interacting with someone else through social networking such as Hi-5, Facebook, WhatsApp, or Twitter with the help of an Internet connection. Cyber profiteers exploit this scenario as an opportunity to deceive people. They also try to steal data through phishing kind of attacks. Network banking facilities are a boon to bank customers. Without going to the bank, they can do any transaction they want. But at the same time, the customers may become a victim if the account details are known to a third party. Stealing intangible goods is an additional technique to gain profit with botnets. This criminality involves taking over a target's accounts, like those of people using gaming online, and marketing the intangible goods they have assimilated in the game. If the market demands certain requirements and if somebody is keen to pay for those, then those things can be taken away by the offenders. A few Trojans spread out by the criminals are focused towards theft of credentials thereby they access online games to steal the intangible goods that people have assimilated in their process of gaming.

## 13.3.5 *Preventing Cybercrimes*

Cybercrimes may contain the aloof theft of vital information through illegal trespassers into remote systems present in the framework. Cybercrime involves everything from transferring prohibited music to burglary of huge numbers of bank accounts online. Cybercrime also contains offenses that don't involve money, such as generating viruses on systems through a network or launching private information about commercial businesses on the Internet.

Authenticated social network users and users those require vital IoT device information are often prone to cybercrimes, offenders and unidentified hackers in cybernetic market. The duty falls on these users to guard themselves and their loved ones through harmless online actions. Following are the few instructions and simple techniques of how the to ensure protection from cyber profiteers who are committing cybercrimes.

■ Up-to-Date System: Cyber profiteers may use software faults made by employees or individuals to attack computer systems anonymously. We can configure systems to download software add-ons and updates routinely. By performing this update of systems, monitoring all activities performed online with the intent to spread viruses and other malicious tasks can be avoided. We can also upgrade to defend against alongside adware and spyware.

■ Personal Information Protection: Most of the services provided online today warrant personal information including phone number, name, email ID, address and account number. Financial transaction management websites should be highly secured with the aid of secure sockets in their links. This ensures that vital information is highly encrypted. A lock at the bottom of browsers ensures an additional level of protection to the personal data.

- Antivirus Software: Installation of antivirus packages averts malicious software programs from embedding on your computer. If such software senses a virus or a worm, it isolates them securely. Viruses can also affect systems without the awareness of users. To be harmless, the antivirus packages must be upgraded when new patches are available.
- The configuration of the system has to be secure enough, and it is significant that systems are configured to the level of security that is adequate and of content for the end user. Enhancing too many security features can have the opposing effect of annoying the end user and making the system sluggish.
- Strong unpredictable passwords with better usernames and identification numbers are highly mandatory for every financial transaction performed online. A robust password must be many characters in length and with a combination of special characters, letters and numbers. It is never a good practice to use a similar password for numerous sites, which may increase the risk of their discovery and possible misuse. Changing the password quite often is a good practice.
- Firewall Turned On: A firewall is designed to prevent unauthorized access to or from a private computer. It helps to defend the system from criminals who may try to gain admittance to bang it, erase vital data, or snip passwords and additional delicate information.
- Website Privacy Policies: In many websites, there is phraseology on the confidentiality policies that permit the website to retain material and data posted to the website indeterminately, even subsequently after the original data has been removed by the user, which may pose some severe issues.
- Financial Statements: Monitoring bank and credit card transactions periodically will decrease the influence of identity theft and fraud on bank credits by determining the problem quickly after the information has been stolen or when the usage of the stolen information is pursued by the offenders. Change the passwords often to be vigilant of cyber profiteers. Banks provide protection services for credit cards; they request their customers to be vigilant when there is uncommon activity happening on their accounts.
- Turn Off Your System: Ultrahigh-speed Internet access may tempt the users to keep it always on for action. Defaulters may use this loophole to invade and steal vital information.

## 13.4 Cyber Profiteering for Smart CPS Security Solutions

Smart things are able to sense and aggregate data, process the data, and interact with other smart things in the network. Since most of the smart things are highly heterogeneous regarding their abilities and roles in the IoT system, this also influences the privacy checks they can accomplish. When dealing with the complexity of hardware and requirements of larger resources for IoT devices, virtualizing IoT devices over the cloud services, deployment of the cloud of smart things is a good choice. Those smart things associate with a few software agents operating in the environment provided by cloud services. Most smart things connected to the Internet often lack physical protection (Hernandez-Ramos et al. 2015) Those smart things are usually a simple embedded system that is economical and mostly used for smart automation and industrial. The protocols deployed on the smart things may provide assurance for the integrity of data, confidentiality, securing communication channels using encryption, but still, noteworthy challenges prevail associated with IoT-based access control (Marta Beltrán et al. 2018).

For connecting smart things, an identity and access management methodology based on Internet specifications to specific requirements of a certain environment is discussed. This methodology lets IoT-related services be deployed in the cloud or locally to recognize, to validate and to allow access to smart things using Hyper Text Transfer Protocol (HTTP) and Constrained Application Protocol (CoAP). It also lets target users to be recognized, validated and authenticated via those IoT devices and smart things based on request and demand. Figure 13.4 shows the connection establishment between the smart things and the cloud services.

## 13.4.1 HTTP Authentication to the Cloud of Smart Things

Concerning smart things, the authenticated user such as an administrator or the owner is accountable for registering the smart devices with the identity

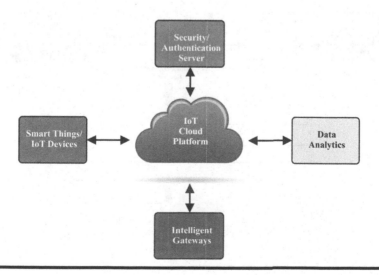

**Figure 13.4  Internet of Things and its interaction with cloud services.**

provider, as well as to the target users those who are allowed to access IoT-cloud facilities through that authenticated smart object. Following are the stages for registering and authenticating the smart things to the cloud services (Bogdan-Cosmin et al. 2018):

- Authorization request message is prepared by the IoT cloud service.
- The authorization server receives the request from the IoT cloud service.
- The authorization server requests the smart things to authenticate for service.
- Smart things authenticate to the authorization server.
- The authorization server sends the authorization code to the smart things and it will be forwarded to IoT cloud service.
- IoT cloud service grants this code to the identity provider which was received by the token endpoint.
- The token endpoint sends the ID tokens and access privileges to the IoT cloud service.
- Finally, the received tokens are validated by the IoT cloud service to acquire the identity of smart things.

Figure 13.5 shows the authentication stages of smart things to the IoT cloud service, provided that the smart things have the HTTP support for making authorization appeals to the server that provides authorization and for subsequent validation by the IoT cloud services to acquire the identity of smart things.

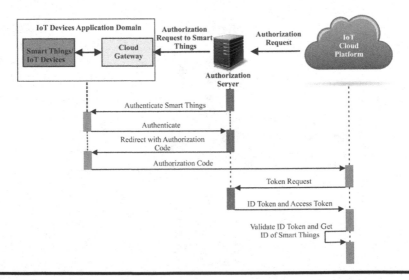

**Figure 13.5   Authentication stages of smart things to the IoT-cloud service.**

## 13.4.2 Threats to IoT Cloud of Smart Things

In spite of making successful authentication of the smart things to the cloud services, the smart things are also prone to certain threats such as the following:

- *ID spoofing* targets validation of ID token validation by managing the signatures of the token. If the opponent is able to create a malicious ID token, the service will validate this fake token and the opponent may be authenticated.
- *ID token confusion* happens when the IoT service takes an ID token created for added services while proper validation of the token audience is not available.
- *ID token replay* targets the expiration of tokens at the IoT service. If this is not properly designed and deployed, an old token will be stolen or seized by the opponent.
- *Redirect uniform resource identifier (URI) manipulation* is performed at the redirected identity provider for validation of the authentication request.
- *Downgrading* targets the reception of the redirected fraud Authorization Request by the Identity provider
- *Brute force* targets the smart IoT devices by means of password stealing, by which the invader gains access to smart devices physically.

To overcome those threats, the robust architectures with the appropriate security authorization scheme are required. Such controls and countermeasures could increase the security level at the cloud of smart things during implementation and deployment stages.

## 13.4.3 Security Authorization Scheme

In the deployment of a system with an IoT framework, communication between smart devices and services that are primarily cloud-based are quite common. To have a perfect digital identity, establishing a communication with high security is absolutely mandatory from the isometric view of the smart devices and the authenticated users of the devices (Ronald et al. 2010; Raza et al. 2014). Owner of the smart devices use mostly the existing infrastructure that cannot be directly controlled based on the flexibility the holder desires. For example, in the scenario of the smart house, it requires a security stack that remains appropriate for an assorted range of devices, which can be combined along with already existing IoT frameworks. The security scheme discusses a lightweight authorization stack for IoT applications with smart devices connected to the cloud and conveys commands to the user's device for authorization. This architecture is particularly device centric with respect to the users and addresses the security concerns in the framework of untrusted cloud services.

### 13.4.3.1 System Architecture

System architecture for the security authorization encompasses three key units: the application domain area where IoT devices are deployed, the cloud service platform and a handheld device management application. Cloud platform architecture takes care of processing the data acquired from IoT devices and the handheld device management applications are used to remotely control the IoT devices. This cloud platform–based architecture may be considered as an untrusted body, not essentially a malicious kind. This cloud platform can be considered to be a third-party service or managed by the specific device manufacturer. The data processing service provided by this system can comprise aggregation and machine learning services. Data transportation enables the communication between the IoT devices and the user or between the IoT devices located at different places.

Initially, the Fast Identity Online (FIDO) protocol was used for web sources, but it can also be used for smart IoT devices. Figure 13.6 shows the overall security authorization scheme structure using the FIDO protocol. This framework employs the authenticator with a certain cryptographic pair of keys for unlocking the security measures. FIDO provides good support from handheld device manufacturers for the authenticator modules.

### 13.4.3.2 Design and Implementation

Kaa is one of the IoT enablement technologies for cloud platforms applicable for any scale of enterprise IoT development. This cloud platform is deployed with a web access and interface for IoT management and a code generator component for supporting different programming languages. We can generate Kaa endpoint application program interface (API) for Android or iOS platforms or for any other embedded devices. Events and notifications are types of Kaa messages generated by endpoints to be distributed to other endpoints. After the process of FIDO registration is finished, the cloud platform's succeeding information between the IoT devices and the user can be exchanged. FIDO authentication protocol is used to protect all the information transferred.

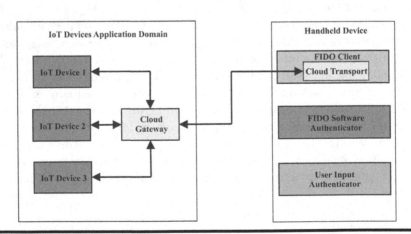

**Figure 13.6　The security authorization scheme overall structure.**

### 13.4.3.2.1 Authentication of the User

If a particular user required to connect to the smart IoT device, the user needs to send JavaScript Object Notation (JSON) based command to the IoT device through the cloud platform. The IoT device will respond back with the FIDO reply to verify the authentication of the command message. During the initialization of the IoT devices, a pair of elliptic curve cryptography (ECC) keys are produced and the public key can be recovered through a URL accessed using the CoAP protocol. During the device registration stage a similar procedure is performed by the controller side and generated public keys are exchanged. All exchanged information between the IoT devices and the controller is wrapped up in a cryptographic lightweight datagram with a series of encoded fields.

### 13.4.3.2.2 Cloud Confederated Authentication

Based on the extension of authentication for validating the end user, cloud federated authentication adds an additional layer of protection. This procedure will also be based on the FIDO authentication messages. The confederated setup is utilized while the requirement of the cloud platform is to process the data acquired from the IoT device that requires the strict authorization of the valid user. The IoT device processes the JSON formatted command received from the cloud platform and it sends a demand for FIDO authentication to the valid end user.

### 13.4.3.2.3 Security Scheme for Theft Resistance

Theft resistance is one of the most significant security scheme features for the IoT application. In this scheme, an authenticated user can identify if a particular IoT device that was issued with a command from the cloud environment was detached from the application environment. This resistance can be deployed with a similar security scheme using a Pico protocol for the theft resistance solution. Pico siblings security scheme for cloud of smart things applications shown in Figure 13.7 is one such variant, where the theft resistance can be ensured in IoT applications. In the Pico sibling scheme, the user credentials are stored in the IoT device. The Pico protocol then uses the suitable authorizations to notify the Kaa IoT cloud service, over the secured cloud gateway on which the smart device has demanded the action, which the Kaa IoT cloud service can then authorize. The Pico also offers nonstop validation by upholding the session with the presence of Pico, and closing it rapidly when the Pico is not present. Such kind of integration of cloud and Pico with the smart things applications ensures theft resistance security authorization schemes.

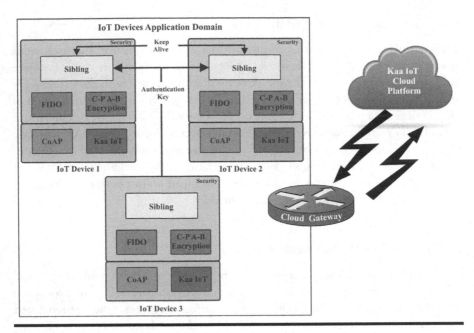

**Figure 13.7    Siblings security scheme for cloud of smart things applications.**

## 13.4.4 Trust and Local Reputation

IoT devices gain benefits from the social exposure of the software agents in the cloud services to cooperate and interact with the environment. In this framework, the selection of the IoT devices for collaborating is a delicate query, mainly in assorted IoT device environments (Yan et al. 2014). If an agent in the cloud does not hold appropriate evidence to perform a reliable communication, it can inquire required information from other reliable agents in the environment. Here the cooperation of the agents is essentially reinforced by a suitable model to ensure trust and reputation, which helps to select the potential pair of IoT devices for communication (Fortino et al. 2018). Based on trust relationships, further enrichment of this process can be done by isolating the agents in dissimilar groups. This provides each agent with the opportunity to choose the communications with the agents in the group for avoiding threats of different natures. By uniting the agents into groups in the cloud of things, it enables reliable operations with the support of an agent grouping algorithm. The trust metric measures the joint trust between the agents in the environment formed using the cloud of things.

### 13.4.4.1 Agent Grouping Algorithm

Every single member executes this algorithm and also while accepting new members in the group. Based on the average value of trust, in the first stage of the algorithm, each agent in the cloud of things environment joins the best group. The second stage of the algorithm ensures every member to act as group administrator for evaluating the joining possibility of the new member in the group. The assessment is done based on the mutually established trust between the agents in the group and the eligible fresh agents. Figure 13.8a shows the sequence of steps followed by the agents in each group for forming a trusted group. Figure 13.8b illustrates the execution steps of the administrator in each group for enrolling the agent members in the trust group.

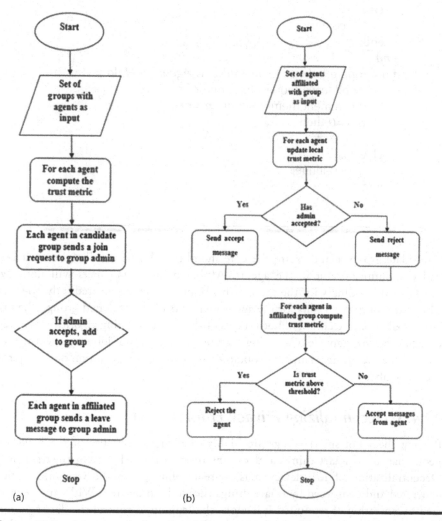

**Figure 13.8  Agent grouping algorithm (a) Execution by agents in the group (b) Execution by the administrators of the group.**

## *Algorithm: Agent Grouping Algorithm*

**Input:** Group **g**, Set **Z**, Set **S**, Set **H**;
**for** *Group **g** belongs to a set **Z** do*
      Compute Trust value **tg** by exploiting agents in **g**;
**end**
**for** *Group **g** belongs to a Candidate set **S** **do**
      Send join request to admin of **g**;
    **if** *g accepts request* **then**
          increment member count **m=m+1**;
    **else**
          continue;
    **end**
**end**
**for** *Group **g** belongs to an Affiliated Agent set **H** **do**
      Send leave request to admin of **g**
      decrement member count **m=m-1**;
    **if** *m==0* **then**
          break;
    **else**
          continue;
    **end**
**end**

Candidate group members are the core members who have full rights to access and communicate among the agents. Affiliated group members will not have full-fledged bonding with the group. This bonding can be ensured only after the administrator grants rights for the agent members of the affiliated group. The primary goal of our execution of both stages of the agent grouping algorithm is to test whether they are capable of forming a group with morals of joint trust within the groups that are greater than those obtained for diverse configuration of a cluster of agent members.

### 13.4.5 Decentralizing Privacy Enforcement

Privacy checks of smart things are highly challenging because of the heterogeneous nature of smart things and variant roles in the IoT system they perform. Decentralization of IoT devices has a great influence on the way information generated and consumed by smart things should be protected. Without a centralized organization of resources, it is more challenging to control the data acquired from the smart things (Ye et al. 2014). The smart things can be configured to specify their privacy preferences for ensuring decentralized enforcement of privacy

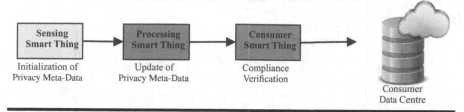

**Figure 13.9 Roles of smart things connected to the cloud for decentralized privacy enforcement.**

(Gokhan Sagirlar et al. 2018). A compliance check of the end user individual privacy preferences can be performed directly by smart things. Within the agent groups in the cloud services, this requires a high joint trust among their agent members; decentralized enforcement of privacy ensures reliability, local reputation and avoids high computational tasks.

The enforcement of privacy can be made at the smart things level to support the decentralized privacy of the cloud of smart things. First, for each data sensed by the IoT device or smart thing, privacy metadata need be enforced as shown in Figure 13.9. The setup consists of a group of sensing smart things (SSTs), processing smart things (PSTs) and consumer smart things (CSTs). In the proposed framework, the smart things in the scenario will be able to associate the privacy policy with the sensed data. The processing smart thing will generate additional local privacy preferences with specific roles for each set of smart things. It ensures that the PST be aware of new data inputs to the systems and their corresponding privacy preferences. Finally, the CST will be able to perform the compliance verification locally. In this framework, the sensed data from sensing smart things will continuously flow through other smart things that are supplemented with the privacy metadata. The smart things connected to the cloud perform authenticated operations on the data, such as aggregation, and update the privacy metadata. When sensed data reaches the CST, they check the authentication of privacy preferences of the data owner, implanted in privacy metadata, and the CST's privacy policy. After satisfied authentication using proper privacy preferences from owners, data are transferred to consumers' data cloud center.

## 13.5 Analysis and Discussion of Results

The latency in the network increases exponentially as a number of nodes in the IoT network increases. Figure 13.10 shows the observed delay in the open source Kaa cloud for broadcast and FIDO services. It is evident from the results that the impact caused by the protocol does not affect the delay drastically, but comparatively delay in FIDO is less than the broadcast service. Simulation is carried out by

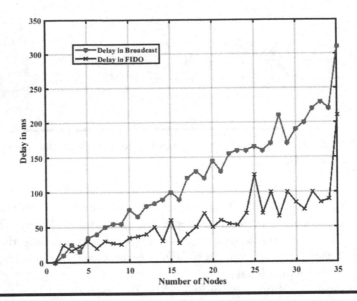

**Figure 13.10  Delay observation in Kaa cloud for broadcast and FIDO services.**

assuming that the network has a similar type of IoT devices. However, in a real-world scenario, the smart things may be of a heterogeneous type. Such cases are controlled in an asynchronous manner by utilizing one session of authentication per IoT device.

In the carried out experimental setup for enhancing the trust in the cloud of smart things, thousands of agents of different types are considered in the model. Each of those agents represents individual IoT devices in the network, and the trust among those thousand agents is assigned at random. By applying the parameters such that a number of feedbacks per step by setting the poison distribution as $\lambda = 50$, with low-performance normal distribution means as 0.9. The high-performance normal distribution mean is chosen as 0.2 and both with a standard deviation of 0.1. By varying the maximum number of new groups an agent in the group $M$ of agents in Figure 13.11, which shows the number of untrusted agents for each category with respect to the number of steps in the simulation, it can be observed from the figure that if the value of $M$ is greater than 5, nearly untrusted agents are kicked out of the groups, by replacing them with the trusted agents.

This experiments with the network of smart things such that the smart things execute stream-based queries. The complexity of the network is linearly proportional to the complexity of the varying query complexity. Investigations are performed with 10 queries for the smart things network with and without privacy preferences. Here the first query is assumed to be simplest and the tenth query to have high complexity. Other queries in between are assumed to have complexity in increasing order. It is obvious from Figure 13.12 with the

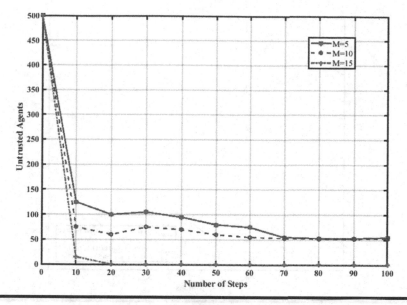

**Figure 13.11   Sum of untrusted agents for maximum number of new groups an agent (M) is able to analyze versus a number of simulation steps.**

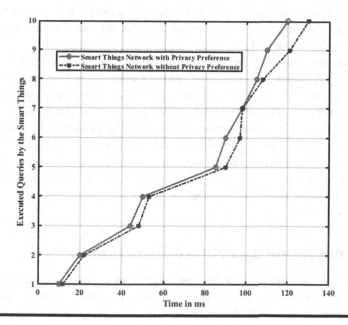

**Figure 13.12   Variable query complexity in the smart things network with time overhead.**

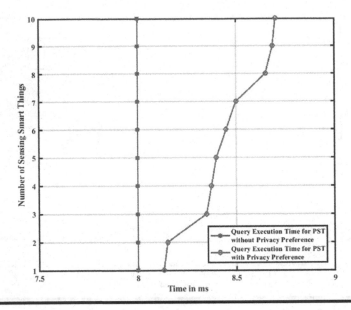

**Figure 13.13 Observation while varying number of SSTs with respect to PST for complex queries.**

proposed framework, while executing the queries when the privacy features are enabled, they consume marginally less time than a network without privacy settings. It was also observed that even for complex queries the time overhead of the system is less than 10%.

By experimenting with the varying numbers of SSTs in the network of smart things for a particular PST executing complex queries the observations made as shown in Figure 13.13. It is obviously such that the complex queries for all the SSTs take a constant time for the PST without privacy settings. The SSTs take variable time instants for the PST with privacy preferences. Even though in the overall case, the smart things with privacy preferences take less amount of time for executing queries, the PSTs take up a considerable amount of time for managing the privacy preferences. This ensures that decentralization of privacy enforcement makes the system less prone to attack by the cyber profiteers.

## 13.6 Conclusion and Future Directions

The discussed framework for enhancing security in a cloud of smart things provides new features and benefits to the developers. Existing research in cybercrime and its impact on a cloud of smart things tend to focus on providing appropriate solutions for overcoming them. However, cyber profiteering is the major issue the developers of smart things infrastructure encounter nowadays. Consequently, this chapter also presents frameworks for deployment of a secured cloud of smart things environment. Practical solutions and technological advances to overcome cyber-profiteering issues on the cloud of smart things are presented with a Kaa cloud service agent group algorithm. Robustness of the system was enhanced with the implications of decentralized privacy enforcement in the network of smart things. Deployment of the proposed technology for the cloud of smart things applications ensures to avoid cyber profiteers breaking into the system. This chapter discussed the issues involved in cyber threats on the cloud of smart things. It mainly focused on the cyber-profiteering issues of how the criminals crack the cloud infrastructure, particularly on a cloud of smart things, and gain profit out of it. Consequently, this chapter explained different threats to the IoT cloud of smart things and introduced new models such as security authorization scheme, decentralizing privacy enforcement, trust and local reputation and HTTP authentication to the cloud of smart things incorporating proposed system architecture and its design and implementation algorithm. Proposed practical solutions overcome cyber-profiteering issues on the cloud of smart things.

Future enhancements to help avoid cyber-profiteering issues can be performed by using artificial intelligence in the system. Recent advancements and strengths of the machine learning and deep learning approaches can add flavor to improved security in the cloud systems of smart things. The huge volume of data is required for analyzing the implications of deep learning algorithms for enhancing the security in the network. Its inherent features and the information learned by deep neural networks can completely eradicate the idea of breaking into the system by the cyber profiteers.

# References

Alfreda Dudley, James Braman and Giovanni Vincenti, "Investigating cyber law and cyber ethics: Issues, impacts, and practices" *Information Science Reference*, (2012), 255–263. doi:10.4018/978-1-61350-132-0.

Bogdan-Cosmin Chifor, Ion Bica, Victor-Valeriu Patriciu and Florin Pop, "A security authorization scheme for smart home internet of things devices," *Future Generation Computer Systems*, 6, (2018), 740–749.

Christos Stergiou, Kostas E. Psannis, Byung-GyuKim and BrijGupta, "Secure integration of IoT and Cloud Computing," *Future Generation Computer Systems*, 78, no.3, (2018), 964–975.

Digital Millennium Copyright Act of 1998, *U.S. Copyright Office Summary*, (1998).

Eric Ke Wang, Yunming Ye, Xiaofei Xu, Siu-Ming. Yiu, Lucas Chi Kwong Hui and Kam-Pui. Chow, "Security Issues and Challenges for Cyber Physical System," *IEEE/ACM International Conference on Cyber, Physical and Social Computing*, no.36, (2010), 733–738.

Esther Ramdinmawii, Seema Ghisingh and Usha Mary Sharma, "A study on the cyber-crime and cyber criminals: A global problem," *International Journal of Web Technology*, 3, (2014), 172–179.

Flavio Lombardia and Roberto Di Pietro, "Secure virtualization for cloud computing," *Journal of Network and Computer Applications*, 34, no.4, (2011), 1113–1122.

Fortino. G, F. Messina, D. Rosaci and G.M. L. Sarné, "Using trust and local reputation for group formation in the Cloud of Things," *Future Generation Computer Systems*, no.89, (2018), 804–815.

Frontoni E., Loncarski J., Pierdicca R., Bernardini M. and Sasso M. "Cyber physical systems for Industry 4.0: Towards real time virtual reality in smart manufacturing," *Lecture Notes in Computer Science, Springer*, 10851, (2018), 422–434.

Gokhan Sagirlar, Barbara Carminati and Elena Ferrari, "Decentralizing privacy enforcement for Internet of Things smart objects," *Computer Networks*, 143, (2018), 112–125.

Hary Gunarto, "Ethical issues in cyberspace and IT society," *Ritsumeikan Asia Pacific University*, 1, (2003), 1–8.

Hassan Takabi, James B.D. Joshi and Gail-Joon Ahn, "Security and privacy challenges in cloud computing environments," *IEEE Security & Privacy*, 8, no.6, (2010), 24–31.

Hernandez-Ramos, Jose L., Marcin Piotr Pawlowski, Antonio J. Jara, Antonio F. Skarmeta, and Latif Ladid, "Toward a lightweight authentication and authorization framework for smart objects," *IEEE Journal on Selected Areas in Communications*, 33, no. 4, (2015), 690–702.

Imre Horváth and Bart H. M. Gerritsen, "Cyber-physical systems: Concepts, technologies and implementation principles," *Proceedings of TMCE*, 1, (2012), 19–36.

Indian Institute of Banking & Finance, *Cyber Laws in India, IT Security of IIBF*, M/s TaxMann Publishers, (2000).

Jianli Pan and Zhicheng Yang, "Cyber-security challenges and opportunities in the new edge computing + IoT world," *Proceedings of the 2018 ACM International Workshop on Security in Software Defined Networks & Network Function Virtualization*, (2018), 29–32. doi:10.1145/3180465.3180470.

Kai Hwang and Deyi Li, "Trusted cloud computing with secure resources and data coloring," *IEEE Internet Computing*, 14, no.5, (2010), 14–22.

Maanak Gupta and Ravi Sandhu, "Authorization framework for secure cloud assisted connected cars and vehicular Internet of Things," *Proceedings of the 23nd ACM on Symposium on Access Control Models and Technologies* (2018), 193–204. doi:10.1145/3205977.3205994.

Mariana Carroll, Alta van der Merwe and Paula Kotzé, "Secure cloud computing: Benefits, risks and controls," *IEEE Conference on Information Security for South Africa*, doi:10.1109/ISSA.2011.6027519, (2011), 15–17.

Marta Beltrán, "Identifying, authenticating and authorizing smart objects and end users to cloud services in Internet of Things," *Computers and Security*, 77, (2018), 595–611.

Martin Törngren, "Cyber-physical systems: Characteristics, trends, opportunities and challenges," CPS overview, KTH; Stockholm—CPS summer school, June 22, 2015.

Mauro Contia, Ali Dehghantanhab, Katrin Frankec and Steve Watsond, "Internet of Things security and forensics: Challenges and opportunities," *Future Generation Computer Systems*, 78, no.2, (2018), 544–546.

Norbert Wiener, *Cybernetics: Control and Communication in the Animal and the Machine*, Cambridge, MA: MIT Press Cambridge, (1985).

Partha Pal, Rick Schantz, Kurt Rohloff and Joseph Loyall, "Cyber-physical systems security—challenges and research ideas," In *Proceedings of Workshop on Future Directions in Cyber-physical Systems Security*, Washington, DC: Academia Press, (2009), pp. 1–5.

Radanliev, Petar, Dave De Roure, Jason RC Nurse, Razvan Nicolescu, Michael Huth, Stacy Cannady, and Rafael Mantilla Montalvo, "Integration of cyber security frameworks, models and approaches for building design principles for the Internet-of-Things in Industry 4.0," In *Conference: Living in the Internet of Things: Cybersecurity of the IoT* (2018), pp. 28–29. doi:10.1049/cp.2018.0041

Rashmi Saroha, "Profiling a cyber criminal," *International Journal of Information and Computation Technology*, 4, no.3, (2014), 253–258.

Raza, Shahid, Simon Duquennoy, Joel Höglund, Utz Roedig, and Thiemo Voigt, "Secure communication for the Internet of Things—A comparison of link-layer security and IPsec for 6LoWPAN," *Security and Communication Networks*, 7, no.12, (2014), 2654–2668.

Ronald L. Krutz and Russell Dean Vines, *Cloud Security: A Comprehensive Guide to Secure Cloud Computing*, Hoboken, NJ: Wiley Publishing, 2010.

Stallings William, *Cryptography and Network Security: Principals and Practice*, fourth Edition, Upper Saddle River, NJ: Prentice-Hall, (2002).

Subramanya. S.R. and Byung K. Yi, "Digital rights management," *IEEE Potentials*, 25, no. 2, (2006), 31–34.

Terry Guo, Damon Khoo, Michael Coultis, Marbin Pazos-Revilla and Ambareen Siraj, "Poster abstract: IoT platform for engineering education and research (IoT PEER)— Applications in secure and smart manufacturing," *IEEE/ACM Third International Conference on Internet-of-Things Design and Implementation (IoTDI)*, 1, (2018) 17–20.

Umejiaku, Nneka Obiamaka, Anyaegbu and Mercy Ifeyinwa, "Legal framework for the enforcement of cyber law and cyber ethics in Nigeria," *International Journal of Computers & Technology*, 15, no.10, (2016), 7130–7139.

Volkan Gunes, Steffen Peter, Tony Givargis, and Frank Vahid, "A survey on concepts, applications, and challenges in cyber-physical systems," *KSII Transactions on Internet and Information Systems*, no.8, (2014), 4242–4268.

Yan, Zheng, Peng Zhang, and Athanasios V. Vasilakos, "A survey on trust management for Internet of Things," *Journal of Network and Computer Applications*, no.42, (2014), 120–134.

Ye, Ning, Yan Zhu, Ru-chuan Wang, Reza Malekian, and Lin Qiao-min, "An efficient authentication and access control scheme for perception layer of Internet of Things," *Applied Mathematics & Information Sciences*, no.8 (4), (2014), 1617–1623.

# Index

Note: Page numbers in italic and bold refer to figures and tables, respectively.

Printed in the United States
by Baker & Taylor Publisher Services